# Ecological Interactions and Biological Control

# Ecological Interactions and Biological Control

EDITED BY

David A. Andow,
David W. Ragsdale, and Robert F. Nyvall

Published in 1997 in the United States of America by Westview Press, 5500 Central Avenue, Boulder, Colorado 80301-2877, and in the United Kingdom by Westview Press, 12 Hid's Copse Road, Cumnor Hill, Oxford OX2 9JJ

Library of Congress Cataloging-in-Publication Data

Ecological interactions and biological control / edited by David A.
Andow, David W. Ragsdale, and Robert F. Nyvall.
p. cm.
Includes bibliographical references and index.
ISBN 0-8133-8758-2 (hc)
1. Pests—Biological control—Congresses. 2. Pests—Ecology—Congresses. I. Andow, David Alan. II. Ragsdale, David Willard, 1952– . III. Nyvall, Robert F.
SB975.E27 1997
632'.96—dc21 97-6907
CIP

Typeset by Letra Libre

The paper used in this publication meets the requirements of the American National Standard for Permanence of Paper for Printed Library Materials Z39.48-1984.

10 9 8 7 6 5 4 3 2

# Contents

# Tables and Figures

*Tables*

***Figures***

# Preface

Biological control of pests has been conducted for decades in entomology, plant pathology, and weed science. Despite the common goal, research has been conducted relatively independently in these fields. Has this independence impeded or facilitated the development of biological control? The practice of biological control has been largely an empirical process of finding the appropriate agent to control the identified target pest. Because the biology of arthropods, pathogens, and weeds is quite divergent, independence may have facilitated practical implementation of biological control. Yet the elucidation of underlying principles of biological control that span the fields may help guide future efforts, and efforts to integrate biological controls with current production practices may require interdependent development of multiple agents and targets. The contributions to this book address this theme by focusing on the practical problems of selecting, developing, implementing, and evaluating biological control agents.

This book deviates significantly from previous efforts by restricting the geographic scope of inquiry to work in cool temperate regions, with particular emphasis on the north temperate Nearctic region (Chapter 1). Biological control has been commonly observed to vary with weather, cropping system, and variation in population biology of the target pests. By emphasizing cool temperate regions, all contributors share experience with similar climates, similar cropping systems, and, in some cases, similar pest complexes. Ecological theoreticians may find the theoretical discussion foreign, but these common bases have facilitated our efforts to identify themes that interweave the disciplines and have grounded discussion in practical reality. This reality is concerned more with the problems of management and evaluation. Consequently, the ecology of management, the methodologies of evaluation, and the structure of feedback between evaluation and management inform our treatment of the theoretical basis of biological control.

We review parts of the history of biological control in cool temperate regions in Chapter 1. No compelling evidence indicates that classical biological control is less successful in temperate regions than in other biogeographic regions of the world, and even if such evidence existed, it would only imply that we would have to try harder for successes. The potential for biological control in cool temperate regions seems vast, but to facilitate our ability to learn from our efforts, we need to implement operational standards for the evaluation of success.

The next four chapters provide perspectives on the selection of potential biological control agents for further development. It is frequently suggested that natural enemies that control the pest in the native habitat are likely candidates for successful classical biological control. McClure (Chapter 2) provides an instructive counterexample. Several species of exotic hemlock-feeding homoptera are killing hemlock trees in the northeastern United States. These species are not pests in their native Japan, where they are controlled by natural enemies. None of these natural enemies are effective in the United States. McClure suggests that potential agents be sought not only in native habitats, but also in disturbed and artificial habitats in the native range. Sheldon (Chapter 3) reviews the search for agents to control Eurasian watermilfoil, an exotic weed in North America. Surprisingly, the most effective agent is a species native to North America, not a species from the native habitat of Eurasian watermilfoil. These examples suggest that in cool temperate regions the ecological domain of potential agents could be usefully expanded to include species outside of those found in native habitats.

Agent selection rarely incorporates consideration of management or development constraints, yet for inundative biological control agents used within existing pest management systems, these constraints may determine the ultimate utility of the agent. Andow (Chapter 4) evaluates the possible utility of *Trichogramma* wasps in several corn production systems using comparative analysis of costs and benefits of alternative pest control tactics. This analysis also enables estimation of the value of various potential technological improvements to *Trichogramma,* which can help guide future research efforts. Makowski (Chapter 5) identifies a number of factors that must be considered in agent selection, including market potential and registration costs. Although she suggests that registration regulations are the main factor stifling the development of effective biological control agents, most historical evidence from drug and pesticide development suggests that more generally, profit potential drives development decisions in the private sector. This more general perspective suggests that the rather modest registration costs associated with biological control agents could limit development of agents with small markets or benefits that cannot be privatized readily. The public sector could play a role in correcting these market failures.

The next ten chapters focus on topics related to the development and implementation of biological control. Some conceptual issues are addressed in Chapters 6–9. Milner et al. (Chapter 6) argue that effective biocontrol of plant pathogens will require greater understanding of microbial community ecology. They outline an intriguing hypothesis—the "camouflage" hypothesis—for effective biological control in the plant root rhizosphere. Agents that make the rhizosphere appear like the surrounding soil may camouflage the plant from root pathogens. Kinkel and Lindow (Chapter 7) provide a framework for using resource competition for plant disease suppression. Gilkeson (Chapter 8) presents the hypothesis that artificial rearing may select for characteristics that reduce the effectiveness of a biological control agent. In her view, not only may useful characteristics deteriorate in rearing environments, but traits that would be maladaptive in the field may actually be favored in rearing environments. Leonard (Chapter 9) describes criteria for long-term successful biological control weeds by pathogens. Chapters 10–12 provide overviews of potentially useful approaches to biological control, such as the use of rhizobacteria (Kennedy, Chapter 10) or smother plants (De Haan et al., Chapter 11) for biological control of weeds, and the use of microsporidia for long-term control of insects (Kurtti and Munderloh, Chapter 12). Chapters 13–16 provide interesting operational perspectives in biological control. Harris (Chapter 13) describes a sequential approach to monitoring impact and evaluating success of a weed biological control project, which could be applicable to other types of biological control efforts. Liu et al. (Chapter 14) describe some of their recent successes in biological control of potato scab. Bigler et al. (Chapter 15) discuss approaches for evaluating the quality of inundative biological control agents, using *Trichogramma* as a case in point. Walter and Lumsden (Chapter 16) provide an overview of the registration process for a microbial biological control agent.

The final three chapters are concerned with how biological control agents can be managed *in situ.* Lewis and Sheehan (Chapter 17) suggest that multitrophic interactions can be managed to enhance biological control by insect parasitoids, but their perspective may extend to other biological control systems. Windels (Chapter 18) illustrates the multitude of possible effects of soil amendments on plant diseases, and Lagnaoui and Radcliffe (Chapter 19) show how fungicides interfere with entomopathogenic fungi, disrupting natural control. The problem of *in situ* management of biological control agents needs greater research attention.

In summary, the development of biological control in entomology, plant pathology, and weed science does not appear to have been impeded by the lack of coordination among the disciplines. Although rather few disciplinary commonalities were revealed in our discussions, one emerged quite clearly. In most cases, biological controls are management systems that must be integrated into other, already existing management systems. These systems are comprised of biological, economic, and regulatory components that must be examined together for successful biological control. The contributions in this book make a first step toward that integration, focusing largely on the cool temperate regions. We believe that there is tremendous potential for biological control in these regions, but that greater attention could be paid to theoretical issues regarding the integration of management systems.

This book is the result of a conference entitled "Ecological Interactions and Biological Control" held at the University of Minnesota October 25–27, 1992. We would like to thank the organizing committee for their efforts and to extend special thanks and commendation to the members of the Legislative Commission on Minnesota Resources for their continued, long-term commitment to biological control in Minnesota. We also thank the anonymous peer reviewers of the chapters in this book for their efforts and helpful comments. These were instrumental in improving each of the chapters. Finally, we thank Tuan Le and Sheila Kelleher for their assistance in preparation of the manuscript for publication. Funding for the symposium and preparation of this book has been approved by the Minnesota Legislature Subdivision 6(a) as recommended by the Legislative Commission on Minnesota Resources from the Minnesota Environment and Natural Resources Trust Fund.

*David A. Andow,*
*David W. Ragsdale,*
*and Robert F. Nyvall*
St. Paul, Minnesota

# 1

# Biological Control in Cool Temperate Regions

*David A. Andow, David W. Ragsdale, and Robert F. Nyvall*

Since the first spectacular success more than a hundred years ago, biological control efforts have expanded, covering cropping systems worldwide. Biological control has grown to encompass several approaches: inoculation, inundation, and conservation. Inoculative methods involve introducing the biological control agent in relatively small numbers so that when it increases it will control the target pest. The classical inoculative method (abbreviated as the classical method) requires only one or a few inoculations to establish a self-reproducing population of biological control agents, while the seasonal inoculative method requires repeated seasonal introductions. Inundative methods involve introducing the agent in large enough numbers to suppress the target pest immediately. Conservation methods involve modifying the environment to retain and enhance the agent, thereby facilitating greater levels of control. Several books provide a summary of recent advances in the field (Wood and Way 1988; Baker and Dunn 1989; Hornby 1990; Charudattan 1991; Andrews 1992; Cook 1993; Yang and TeBeest 1993) and frame some of the contemporary conceptual issues (Mackauer et al. 1990), but because the field is so vast, it has not been reviewed uniformly since Clausen's 1978 effort, which summarized worldwide efforts using the classical inoculative method against weed and arthropod pests, and Cook and Baker's 1983 review of efforts related to plant pathogens. Here we break from previous treatments of biological control and, instead of attempting a worldwide coverage of the field, we focus on biological control in a restricted geographic area, the cool temperate region.

Worldwide, cool temperate regions occur polar to the Mediterranean and great desert climatic regions, which is generally north of 40°N latitude in North America and Asia and north of 45°N in Europe. These regions include the dry midlatitude steppes of Central Asia, the western Great Plains of North America, the humid mesothermal west coast marine climates of the Pacific Northwest and Western Europe, and the humid microthermal continental climates of North America and Eurasia (Trewartha 1957). These areas have at least one month with mean temperature less than 0°C, four months with mean temperature greater than 10°C, and more than 25 cm of annual precipitation. This climatic type is not common in the southern hemisphere. It occurs south of 43°S latitude in New Zealand, Tasmania, and the southern tip of South America, but in these southern regions the climate is moderated by the oceans. Cool temperate regions contain some of the most productive agricultural lands in the world. Wheat, barley, corn, rye, oats, soybeans, beet sugar, flax, alfalfa, and forest crops dominate the landscape. The crop systems are sometimes integrated with animal production, especially dairy cattle, beef cattle, and hogs. Many of the major potato production areas in the world are here, and cold-tolerant fruits and vegetables, such as apples and cole crops, are also abundant in suitable parts of the region.

Biological control efforts in this region were initiated early in this century against codling moth and expanded rapidly, targeting more than ninety species of arthropods by the 1970s with the classical method (Clausen 1978). Some early workers questioned the efficacy of biological control in temperate regions (for example, Taylor 1955). Their argument was rather imprecise, but the thrust of it was that because these regions have a variable climate, with much of the year unfavorable for reproduction, introduced agents would be subjected to greater hazards, making them less likely to establish and less likely to suppress the target populations. This argument has been largely dismissed (Greathead 1986), based on DeBach's (1964) analysis of successful or partially successful biological control attempts in various climatic regions. Many successes have been obtained in cool temperate regions, and successes have been more numerous where efforts have been more intensive; other factors, such as climate, have been unreliable guides to success (DeBach 1964; Munroe 1971).

More recently, a "disturbance" hypothesis has found some empirical support. DeBach's (1964) analysis focused on describing the number of successful biological control projects, but he did not consider the rate of success of biological control. Hall and Ehler (1979) and Hall et al. (1980) analyzed Clausen's (1978) summary of worldwide efforts in classical biological control of arthropods and plants and suggested that the rate of establishment of introduced arthropod natural enemies is 12.5 to 25 percent lower in annually disturbed habitats, such as annual and other short-season crops (0.28 establishment rate, n = 640), than in habitats with comparatively less disturbance, such as orchards and perennial crops (0.32 establishment rate, n = 913) or forest and rangeland (0.36 establishment rate, n = 535) (Hall and Ehler 1979). Moreover, the rate of success of introduced arthropod natural enemies is lower in annually disturbed habitats (0.43 success rate, n = 152) than in less disturbed habitats (0.72 success rate, n = 239 orchards and perennial crops; 0.47 success rate, n = 53 forest and rangeland) (Hall et al. 1980). Because the vast majority of agricultural lands in northern temperate regions are planted to annual crops, these results might imply that the rate of establishment and success of biological control agents will be lower in northern temperate regions than in other regions of the world. The analysis, however, does not control for variation in effort, and in some cases relatively little effort was made to establish the natural enemy. In other cases the natural enemy was misidentified, and a species that could not consume the intended host had been introduced, resulting in unmitigated failure. Clearly, a closer examination of establishment and success is essential before we can make reliable inferences from our previous experience.

In both North America and Europe, current national agricultural policies are strongly promoting nonchemical control strategies such as biological control (OTA 1995), stimulated in part by the declining number of effective pesticides available to control pests and by concerns about soil and water contamination. Consequently, there has been renewed interest in developing useful biological control systems for pest control, utilizing the entire range of biological control methodologies. This renewed interest has focused scientific attention on integrating biological controls with conventional pesticide-based pest management systems.

Several approaches have been used to integrate biological and pesticide-based management systems, but most of these efforts focus on operational issues (Hoy and Herzog 1985; Croft 1989). One methodological approach deals with how pesticides can be applied so that biological control agents can remain effective and is reminiscent of some early pathbreaking work in integrated pest management in Nova Scotian apple orchards (Pickett 1949; MacPhee and MacLellan 1972). The tactics include timing pesticide applications appropriately, using pesticides that have minimal impact on the agents, and using agents that are resistant to pesticides. Another approach considers how the occurrence of the agents can be used to modify pesticide action thresholds (Nakasuji et al. 1973; Ostlie and Pedigo 1987). Both of these operational concerns, however, mask an underlying fundamental problem: How is the success of a biological control agent evaluated?

## Evaluating Establishment and Success

Evaluation of establishment and success of a biological control effort has been idiosyncratic. In some cases, if the introduced natural enemy was successfully established and caused measurable mortality to the target pest, the effort was declared successful. In other cases, the natural enemy reduced the abundance of the target pest to such an extent that the pest no longer caused yield losses. Establishment has been claimed if the released natural enemy is found at the release site the next year. In other cases, only when the introduced natural enemy is geographically widespread has establishment been claimed. Thus, when we examine previous experiences with biological control, we find that the data are not entirely comparable.

In our view, a species is established when it has a self-perpetuating population in the habitat of introduction. Therefore, to be considered established, a population must persist for multiple generations; determination of this requires long-term observation. In practice, however, several factors, such as a high population growth rate and a large spatial range, mitigate in favor of persistence. Consequently, establishment has both a temporal and a spatial component (Table 1.1). Although most would agree that persistence over multiple generations is a minimal definition of establishment, less rigorous criteria may have been used. For example, Clausen (1978) states that a few recoveries of *Drino bohemica* Mesn. [Diptera: Tachinidae] were made on European pine sawfly in Canada, implying perhaps that it may have established on the sawfly. McGugan and Coppel (1962), who provide the data for Clausen's review, flatly state that recovery from European pine sawfly in the Maritime Provinces of Canada was incidental and that there is no evidence that *D. bohemica* established on this species. Clausen (1978) provides no indication of McGugan and Coppel's (1962) opinion. Given the number of authors who have used Clausen's (1978) compilation for synthetic reviews of the effectiveness of biological control agents, variation in meaning of establishment is a cause for concern.

Table 1.1. Proposed establishment criteria for the introduction of natural enemies to new habitats

| Spatial Component | Temporal Component | | | |
|---|---|---|---|---|
| Recovered at: | One generation evaluated, few individuals recovered | One generation evaluated, many individuals recovered | More than one generation evaluated, few individuals recovered | More than one generation evaluated, many individuals recovered |
| Site of release, or a few sites within ready dispersal range of an individual | Possibly will become established | Probably will become established | Possibly established or Established | Probably established or established |
| Sites within ready dispersal range of an individual, many sites | Possibly will become established | Possibly established | Possibly established or established | Established |
| Sites far from introduction site, few sites total | Possibly will become established | Possibly Established | Probably Established or established | Established |
| Sites far from introduction site, many sites total | Unlikely to be observed | Probably established | Established | Established |

*Note:* An introduced population could be designated to have many individuals if the population at any one site were greater than the number originally released. If few individuals are observed after more than one generation, the population would be established if it persists even at these low numbers for multiple generations. We suggest that it is not possible to determine if a population has established after only one generation or year.

Table 1.2. Possible criteria used for evaluating the success of a biological control agent

| Observed Outcome | Allowable Inference and Need For Information | Methods Needed To Collect Information |
|---|---|---|
| 1. Causes increased disease, defoliation, or mortality of target pest, or adversely affects other vital parameters | Success is possible. Minimal information to indicate the possibility or likelihood of success | Requires measurement of disease incidence, defoliation, or instantaneous mortality rates |
| 2. Causes partial supression of pest population or pest damage, or causes increased noncompensated mortality to target pest | Success: Partial control. Minimal information to allow integration of biological control into a pest management decision framework | Requires information on pest and natural enemy dynamics |
| 3. Causes occasional control of pest, or causes sufficient increased non compensated mortality to target pest | Success: Occasional control. Minimal information to allow integration of biological control into a pest management decision framework | Requires information on pest and natural enemy dynamics |
| 4. Causes constant control of pest | Success: Complete control. Minimal information for biological control to replace conventional control | Requires demonstration of change in pest status before and after biological control effort, and demonstrationthat the agent is sufficient to cause the change in pest status |

Evaluation of success is more complicated than evaluation of establishment, but it is essential to integrate biological control with pesticide-based controls. Several criteria, all of which have some merit and utility, have been used to evaluate the success of a biological control agent (Table 1.2). The first criterion is the easiest to demonstrate, requiring only one or at most two samples from the target population. Defoliation can be measured during one visit to a site; disease incidence from a pathogen can be determined from a single sample; and mortality from a parasitoid can be estimated from a single sample of hosts that is evaluated in the laboratory (see Van Dreische 1983). Measurement of mortality from a predator is somewhat more difficult to estimate. The problem with this criterion is that it evaluates only the potential for success. No matter what defoliation level, disease incidence, or mortality rate is observed, such measures do not demonstrate that the target pest population will be adversely affected, because subsequent density-dependent mortality could compensate for the observed additional infection, defoliation, or mortality. Success can be inferred only if the agent reduces populations of the target pest, thereby alleviating some portion of the damage caused by the pest.

Three kinds of success can be observed. Complete control results when the biological control agent constantly controls the pest. Of all the types of success, this is the most readily evaluated. If the success is not complete, it may vary temporally, resulting in occasional control, or it may not be very strong, resulting in partial control. These types of success are insufficient to control the pest by themselves and therefore must be integrated with other pest control measures. Both of these types of successes are much more difficult to demonstrate, and they require an analysis of pest and agent dynamics. This additional information requirement has hampered efforts to integrate biological control with conventional pest control systems.

### *Complete Control*

DeBach and Bartlett (1964) and Smith and DeBach (1942) provide minimal evidential criteria for demonstrating complete control. These are (1) average pest population is lower after establishment of the biological control agent than before establishment, (2) as the agent expands its geographic range, there follows a reduction in pest population, and (3) if the pest is protected from suppression by the agent, then pest survival increases. Criterion (2) is a special case of criterion (1) for those agents that expand geographic range on their own. Criteria (1) and (2) suggest but do not prove that the natural enemy is necessary to cause pest population decline, and criterion (3) demonstrates that the natural enemy is sufficient to cause pest population decline. The DeBach criteria provide sufficient but not necessary evidence that the natural enemy caused pest population decline. For complete control, however, these criteria are nearly irrefutable.

Winter moth, *Operophtera brumata* (L.) [Lepidoptera: Geometridae], invaded Nova Scotia from Europe and caused considerable damage to apple orchards and oak woodlands. Several species of natural enemies were introduced and *Cyzenis albicans* (Fallen) [Diptera: Tachinidae] and *Agypon flaveolatum* (Gravenhorst) [Hymenoptera: Ichneumonidae] established throughout the geographic range of the pest. Collapse of the pest population coincided with the increase in the natural enemies, especially *C. albicans* (Embree 1966). Using population models of the winter moth–*C. albicans* interaction and realistically estimated population parameters, Hassell (1980) demonstrated that *C. albicans* could reduce winter moth populations in Canada even though it was largely ineffective in Great Britain. Winter moth subsequently invaded British Columbia and Oregon and is one of the best-documented cases of complete control in northern temperate regions (Roland and Embree 1995).

An enigma is brown-tailed moth, *Nygmia phaerorrhoea* (Donovan) [Lepidoptera: Lymantriidae], which is native to Europe and was found in eastern Massachusetts in 1897, feeding on a number of deciduous trees (Clausen 1978). Nine of eleven species of natural enemies introduced against this species established in New England, and three of four introduced natural enemies established in the Canadian Maritime Provinces. In both areas, brown-tailed moth populations declined spectacularly and parasitism was quite high, confirming DeBach's first criterion. Unfortunately, there have been no attempts to determine the significance of the natural enemies, and it cannot yet be concluded that the introduced natural enemies caused the decline.

Classical uses of plant pathogen agents have been against introduced perennial weeds that grow in dense stands and infest large land areas, particularly when small residual populations of the host do not cause economic losses and other weed control practices are not economically or environmentally sound. The rust fungus *Puccinia chondrillina* was introduced from the Mediterranean region of Europe into Australia (Cullen 1978) and the western United States (Emge 1981; Supkoff 1988) to control rush skeletonweed (*Chondrilla juncea*) in wheat and ranges. Blackberry rust, *Phragmidium violaceum,* was introduced from Europe into Chile to control blackberries (*Rubus constrictus*) encroaching on ranges and pastures (Oehrens 1977).

Inundative applications of microorganisms for weed control are usually referred to as mycoherbicides or bioherbicides. These terms, however, have come to refer to any microorganism that controls or aids in controlling weeds, whether in the inoculative or inundative sense. The inundative method of bioherbicide application involves periodic application to weeds of fungal or bacterial pathogens similar to chemical herbicides. Application in these vast numbers overcomes or compensates for natural constraints to epidemic development. Although most bioherbicides are fungi or bacteria, a rose rosette disease is caused by a microorganism of unknown etiology that is transmitted by an indigenous mite (Epstein 1993). The microorganisms are usually indigenous microorganisms that cause endemic diseases of their respective hosts and do not build to a high population level early enough in the season to kill the host. There is frequently a low carryover and poor dissemination of inoculum. Most mycoherbicides use fungi that are facultative saprophytes that are only weakly competitive with obligate saprophytes. These fungi rely on saprophagy for survival when hosts are not growing.

Three fungi have been successfully developed as commercial mycoherbicides. These are *Colletotrichum gloeosporioides* f. sp. *aeschynomene* to control northern jointvetch (*Aeschynomene virginica*) in rice and soybeans (Boyette et al. 1973), trade name Collego; *Phytophthora palmivora* to control milkweed vine (*Morrenia odorata*) in citrus groves (Feichtenberger et al. 1983), trade name DeVine; and *Colletotrichum gloeosporioides* f. sp. *malvae* to control round-leaved mallow (*Malva pusilla*) and velvetleaf (*Abutilon theophrasti*), trade name BioMal in Canada (Makowski and Mortensen 1992; Mortensen 1988, Chapter 5 in this volume). At the present time, however, none are available for commercial sale.

The bacterium *Pseudomonas syringae* pv. *tagetis* successfully controls Canada thistle, *Cirsium arvense* (Johnson and Wyse 1991). It is also effective against common ragweed, *Ambrosia artemesiifolia;* dandelion, *Taraxicum officinale;* giant ragweed,

*Ambrosia trifida;* Jerusalem artichoke, *Helianthus tuberosus;* marigold, *Tagetes* spp.; and sunflower, *Helianthus annuus* (Johnson and Wyse 1992).

Like microbial biocontrol agents of weeds, introduced microbial biocontrol agents of plant diseases tend not to persist where they do not occur naturally or at population densities higher than occur naturally. The most successful biological control of plant pathogens with microorganisms involves suppressive soils, crop rotations, and organic amendments, which are based on management of natural communities of microorganisms (Baker and Cook 1974; Cook and Baker 1983; Papavizas and Lumsden 1980; Schneider 1982; Schroth and Hancock 1982; Chapter 17 in this volume). One of the best examples of this type of disease control is the suppression of take-all of wheat caused by *Gaeumannomyces graminis* seed treatments with the fluorescent pseudomonad *Pseudomonas fluorescens* (Weller and Cook 1983). Some populations of *G. graminis* were found to be sensitive to different compounds produced by strains of *P. fluorescens* (Mazzola 1994), which has led to using mixtures of pseudomonads in the seed treatments (Pierson and Weller 1994).

Some microbial agents are effective after a single application. Annosus root rot of pine is controlled by one application of *Phlebia gigantea* to the freshly cut surfaces of pine stumps (Risbeth 1975). Similarly, crown gall caused by *Agrobacterium tumefaciens* is controlled by a single application of *A. radiobacter* strain K-84 to the bare roots of susceptible orchard and ornamental plants (Kerr 1980). In both cases, however, the biocontrol agent does not move from the inoculated plant into the surrounding environment.

### *Occasional and Partial Control*

The weakness of the DeBach criteria is revealed when control is occasional or partial. In these cases, DeBach's criteria (1) and (2) are never clearly fulfilled. When the level of control varies from year to year or locality to locality, the variance in pest population when natural enemies are present and the occasional failure in control will weaken any conclusion about the net effect of the natural enemies on the pest population. Because control is occasional or partial, the additional control methods that inevitably will be used will confound interpretations, perhaps masking the effect of the biological control agent. In addition, natural enemy exclusion experiments may be insufficient to demonstrate the effects of the natural enemies, because their effects on the pest populations during such an experiment will be partial, occasional, or subtle, or will vary from year to year. Thus, demonstration of occasional or partial control requires a greater amount of research simply to fulfill DeBach's criteria.

From a practical perspective, occasional or partial biological controls must be integrated with other control tactics if they are to be used at all. Because growers tend to be averse to risk, they want to be sure that biological control will help control pests before they will base management decisions on it. This requires evidence that the biological control agent does cause a reduction in the target pest (necessary evidence), which is considerably more difficult than fulfilling the DeBach criteria.

Moreover, these occasional or partial controls must be predictable so that management decisions can be made with assurance. Our inability to predict the effects of such biological control agents is a major impediment to integrating biological control into integrated pest management systems.

As a case in point, considerable effort has been expended on biological control of European corn borer in the United States, but only three natural enemies, *Eriborus terebrans* Grav. [Hymenoptera: Ichneumonidae], *Macrocentrus grandii* [Hymenoptera: Braconidae], and *Nosema pyrausta* [Microspora: Nosematidae], are frequently encountered. Parasitism rates by any of these species can reach over 50 percent, yet parasitism by all three combined rarely exceeds 30 percent. Because corn borer populations and parasitism rates fluctuate considerably, there is no clear evidence that damage done by corn borers is less severe or less frequent as a result of this parasitism. This is a typical example of partial or occasional control, where none of DeBach's criteria can be met, yet it remains possible that the biological control agents are contributing to control of the target pest.

Evaluating the efficacy of such a biological control effort is difficult. The central difficulty lies in evaluating the dynamics of the target pest with and without the biological control agent on appropriately large spatial scales and for appropriately long periods of time. Because it is virtually impossible to remove and then replace the biological control agents in the appropriate spatial and temporal scales, other, more indirect methods must be used. One approach is to develop realistic models of the interaction between agent and target. If these models can be verified, they can be used to estimate the effect of the agent. An alternative approach is to develop an adaptive decision-making plan that uses current information about both agent and pest to guide pest management (see Nakasuji et al. 1973). The effect of the agent could be evaluated in terms of management decisions avoided or management options preserved. For example, if the agent is abundant enough that no other pest control tactic is needed, the agent can be valued as the cost of the alternative tactic that was avoided. In a similar way, a natural enemy could preserve future management options. By avoiding the need to apply an insecticide, the agent could preserve natural controls that suppress pests that would colonize later in the season.

Many inundative biological control agents exhibit variable or partial control. When they are evaluated like chemical pesticides, these agents are rarely used in actual production. For example, *Trichogramma nubilale* is variably effective at suppressing populations of European corn borer in the United States (Andow 1995), and there has been little interest in commercializing this species. Similarly, numerous microorganisms capable of suppressing plant diseases have been identified, but few have been commercialized, primarily due to their inconsistent performance (Weller 1988). Greater attention to quality control may help alleviate some of the variability in performance. Agents that are only partially effective could be combined with other partially effective agents, or they could be genetically improved to perform better. For example, applications of the microbial insecticide *Bacillus thuringiensis* are considerably less effective than available chemical pesticides for control of European corn borer in sweet corn, yet perhaps some combination of *Trichogramma* and *Bacillus thuringiensis* would be effective.

The great diversity of criteria for determining the establishment and success of a biological control agent implies that cursory analysis of establishment and success rates based on comprehensive reviews is unlikely to provide accurate conclusions. Retrospective analyses will need to distinguish among these ideas and provide a detailed analysis of previous biological control attempts. Here we provide an initial analysis of establishment of natural enemies during classical inoculative biological control in cool temperate regions.

## Establishment in Cool Temperate Regions

Failure of natural enemies to establish after introduction can be attributed to three major factors: lack of effort, physiological incompatibility, and ecological incompatibility. Because establishment is a probabilistic event, even a suitable, well-adapted species may not establish unless sufficient effort is expended. If sufficient efforts are made, a species may not establish if it is physiologically incapable of interacting with or using the target pest as food or a host. Because natural enemies and target pests can be misidentified or taxonomic revisions may split species, physiological incompatibility remains a contemporary concern. If the species are physiologically suitable, then ecological incompatibility may limit establishment. The species may be ecologically incompatible in many ways. Climate may be unsuitable; for cool temperate regions, inability to overwinter is a major barrier. If the species is able to complete an annual cycle, then dynamic incompatibilities may limit establishment. For establishment to occur, the net reproductive rate must be greater than one. Thus, if survival in association with the target pest is too low, fecundity is too low, or both, then establishment will not occur for dynamic reasons. It would be useful to have some understanding about the reasons for establishment failures in cool temperate regions to guide future work in biological control.

### *Arthropod Pests*

Clausen (1978) and Cameron (1989) provide reviews of biological control efforts in cool temperate regions. These reviews list ninety species of arthropod pests for which classical inoculative biological control has been attempted in cool temperate regions. We updated the information on these species (e.g., Drea and King 1981; Ferguson 1989; Hajek and Dahlsten 1981; Kennedy 1970; Klimaszewski 1984; Klimaszewski and Cervenka 1986; Pearson 1989) to the present. Various degrees of effort have been expended on these target species, from as few as forty individuals of *Lebia scapularis* Dejean [Coleoptera: Carabidae] released against elm leaf beetle, *Pyrrhalta luteola* (Muller) [Coleoptera: Chrysomelidae] to more than 4 million *Hunterellus hookeri* Howard [Hymenoptera: Encyrtidae] released against *Dermacentor variabilis* (Say) [Ixodidae]. Because the goal of biological control is to reduce pest damage, we evaluated each target host to determine if any released natural enemy established on the host. Subsequently, we analyzed the target species for which no introduction resulted in establishment to determine why failures occurred.

Of the ninety target species in cool temperate regions, sixty-two had at least one introduced natural enemy establish. All introductions failed to establish on twenty-three of these species, and the outcome remains unknown for the other five species (Table 1.3). These data suggest that biological control has the possibility of succeeding in about two-thirds of the potential target arthropod species in cool temperate regions. A detailed analysis of the reasons for successful establishment in these sixty-two species remains to be done.

The twenty-three target species that failed to have a natural enemy established are listed in Table 1.4. Of these twenty-three, eight species cannot be evaluated further. Four of these eight species (*Megoura viciae, Myzus persicae, Cnephasia longana,* and *Amphimallon majalis*) lack information on the numbers of individuals introduced, and four (*Dermacentor variabilis, Hyperodes bonariensis, Sitona cylindricollis,* and *Dendroctonus frontalis*) provide scant information on the biology of the pest–natural enemy interaction.

Table 1.3. Target arthropod species in cool temperate regions for which it is unknown if any introductions of biological control agents established successfully.

| Target species<br>  Introduced natural enemy, location of introduction, year(s) introduced |
|---|
| *Hyphantria cunea* (Drury) [Arctiidae] Fall webworm (Clausen 1978) |
|   *Merica ampelus* (Walker) [Tachinidae], Russia 1958--66 |
|   *Campoplex validus* (Cress.), Russia 1963--66 |
|   Eight species, Czechoslovakia 1960--63 |
| *Exoteleia pinifoliella* (Chamb.) [Gelichiidae] pine needle miner |
|   Parasitoids that were surplus from the lodgepole needle miner program: |
|   *Achrysocharoides* sp. [Eulophidae],Ontario 1950 (McGugan and Coppel 1962) |
|   *Apanteles californicus* [Braconidae], Ontario 1950 (McGugan and Coppel 1962) |
|   *Zagrammosoma* sp. [Eulophidae], Ontario 1950 (McGugan and Coppel 1962) |
| *Neodiprion fulviceps* (Cress.) [Diprionidae] |
|   *Dahlbominus fuscipennis* Zetterstedt [Eulophidae], South Dakota 1941 (Dowden 1962) |
| House/stable fly |
|   Mixture of *Muscidofurax* spp. introduced, primarily *M. zaraptor* Kogan and Legner, and *Pachycrepoideus vindemiae* (Rondani)[Pteromalidae] Mid-Canterbury, Dunedin, and Southland, New Zealand (Marwick et al. 1989) |
| *Austrosimulium* spp. [Simuliidae] black fly |
|   Mixture of perhaps four species of *Austroaeschna* [Aeshnidae] from Australia (probably *A. atrata* Martin, *A. inermis* Martin, *A. multipunctata* Martin, and *A. unnicornis unicornis* Martin), Cass, New Zealand, 1930, 1931 (Crosby 1989) |

Table 1.4. Host species for which no natural enemy species was established after attempts to introduce in cold temperate regions

| **Target Host**<br>Introduced natural enemy | Origin of enemy | Number released | Release locality and year | Reference |
|---|---|---|---|---|
| ***Dermacentor variabilis* (Say) [Ixodidae] American dog tick** | | | | |
| *Hunterellus hookeri* Howard [Encyrtidae] | France | many | Nashon Island, Mass. | (1) |
| | Texas | 90,000 f | Martha's Vineyard, Mass. | (2) |
| ***Dermacentor andersoni* Stiles [Ixodidae] Rocky Mountain wood tick** | | | | |
| *Hunterellus hookeri* Howard [Encyrtidae] | France | >4,000,000 | Montana, Colorado, Idaho, and Oregon | (3) |
| **Acrididae species Primarily *Camnula pellucida* (Scudder) and several species of *Melanoplus*.** | | | | |
| *Blaesoxipha caridei* (Brèthes) [Sarcophagidae] | Argentina and Uruguay | 48+ adults | Ontario | (4,5) |
| *Blaesoxipha australis* (Blanch.) [Sarcophagidae] | Argentina and Uruguay | few 1000 | Ontario, Nova Scotia, Alberta, Saskatchewan, and British Columbia | (4,5) |
| *Blaesoxipha neuquenensis* (Blanch.) [Sarcophagidae] | Argentina and Uruguay | few | Canada | (4,5) |
| *Mantis religiosa* L. [Mantidae] | Ontario | >11,000 | eastern and western Canada | (6) |
| ***Megoura viciae* Buckton [Aphididae] Vetch aphid** | | | | |
| *Aphidius megourae* Stary [Braconidae] | Moscow | unknown | several localities in Czechoslovakia | (7) |
| ***Myzus persicae* (Sulz.) [Aphididae] Green peach aphid** | | | | |
| *Aphidius matricariae* Hal. [Braconidae] | France and Middle East | unknown | several U.S. eastern and Pacific Coast states | (6) |
| ***Coleotechnites starki* (Freeman) [Gelechiidae] lodgepole needle miner** | | | | |
| *Eriplatys ardeicollis* (Wesm.) [Ichneumonidae] | Europe | 58 adults | Banff 1949 | (8) |
| *Pnigalio* sp. [Eulophidae] | Europe | 103 adults | Banff 1949 | (8) |
| | | 52 adults | Kootenay National Park 1949 | (8) |
| *Dicladocerus* sp. [Eulophidae] | Idaho on *C. mulleri* | 142 adults | Banff 1950 | (8) |
| | | 419 adults | Kootenay National Park 1951 | (8) |

| | | | | |
|---|---|---|---|---|
| ***Laspeyresia youngana* (Kearfott) [Olethreutidae] Spruce seed moth** | | | | |
| *Ascogaster quadridentata* Wesm. [Braconidae] | surplus from codling moth biocontrol project | 754 adults | Angus, Ontario | (6) |
| *Macrocentrus ancylivorus* Rohw. [Braconidae] | surplus from from oriental fruit moth biocontrol project | 7,768 | western Ontario | (6) |
| ***Choristoneura fumiferana* (Clemens) [Torticidae] Spruce budworm** | | | | |
| Sixteen species of parasitoids | BC and and Europe | ranging from 3 to >50,000 | eastern Canada and northeastern United States | (6) |
| ***Cnephasia longana* (Haworth) [Tortricidae] Omnivorous leaf tier** | | | | |
| At least 9 species of larval parasitoids | France | unknown | 4 localities in Oregon | (9) |
| ***Pyrrhalta luteola* (Muller) [Chrysomelidae] Elm leaf beetle** | | | | |
| *Tetrastichus galerucae* (Fonsc.) [Braconidae] | France | 1,800 | Massachusetts | |
| | | unknown | Virginia, and Washington, DC | |
| | | 36,000 | northeastern U.S. and California | (10) |
| *Lebia scapularis* Dejean [Carabidae] | Europe | 40 adults | Woburn, Mass. | (6) |
| *Erynniopsis antennata* Rondani [Tachinidae] | Europe | 1,212 adults | Massachusetts and Connecticut | (11) |
| ***Bruchus pisorum* (L.) [Bruchidae] Pea weevil** | | | | |
| *Triaspis thoracicus* (Curt.) [Braconidae] | France and | 66,000 | Idaho | (9) |
| | Austria | >3000 | Austrian peas at Forest Grove, Or, 1940 | (9) |
| | | 3,764 | British Columbia 1942 | (5) |
| | | 470 | Ontario 1942 | (5) |
| ***Epilachna varivestis* Mulsant [Coccinellidae] Mexican bean beetle** | | | | |
| *Aplomyiopsis epilachnae* (Ald.) [Tachinidae] | cent. Mexico | 82,000 | 85 localities in 19 U.S. states from Texas to Connecticut | (12) |
| *Lydinolydella metallica* Townsend [Tachinidae] | Argentina | 3,482 adults | New Jersey | |
| | and Brazil | small numbers | Maryland | (13) |
| *Pediobius foveolatus* (Cwfd.) [Braconidae] | India from *Henosepilachna sparsa* (Herbst) | 25,000 | New Jersey and south | (6) |
| ***Hyperodes bonariensis* Kuschel [Curculionidae] Argentine stem weevil** | | | | |
| *Anaphes atomarium* (Brèthes) [Mymaridae] | Argentina | ≤1,150 f | Lincoln, New Zealand | (14) |
| ***Sitona cylindricollis* (Fahraeus) [Curculionidae] Sweet clover weevil** | | | | |
| *Microctonus aethiops* (Nees) [Braconidae] and *Campogaster exigua* (Meig.) [Tachinidae] | Europe from other *Sitona* species | unknown | North Dakota and Manitoba | (6,15) |

Table 1.4. Continued.

| **Target Host**<br>Introduced natural enemy | Origin of enemy | Number released | Release Locality and Year | Reference |
|---|---|---|---|---|
| *Pygostolus falcatus* (Nees) [Braconidae] and *Perilitus rutilus* (Nees) [Braconidae] | Sweden | 2,800 first instar larvae | Brandon, Manitoba 1958 | (15) |
| ***Amphimallon majalis* (Razoumowsky) [Scarabaeidae] European chafer** | | | | |
| *Dexilla rustica* (F.) [Tachinidae] | France | unknown | probably near New York | (6,16) |
| *Dexilla vacua* (Fall.) [Tachinidae] | France | unknown | probably near New York | (6,16) |
| *Microphthalma europaea* Egger [Tachinidae] | France | unknown | probably near New York | (6,16) |
| *Tiphia morio* F. [Tiphiidae] | France | unknown | probably near New York | (6,16) |
| ***Anomala orientalis* Waterhouse [Scarabaeidae] Oriental beetle** | | | | |
| *Tiphia biseculata* A.&J. [Tiphiidae] | Japan | unknown | Long Island, New York | (6) |
| *Tiphia bicarinata* Cam. [Tiphiidae] | Korea | unknown | Long Island, New York | (6) |
| *Tiphia notopolita* Rob. [Tiphiidae] | Korea | unknown | Long Island, New York | (6) |
| ***Dendroctonus frontalis* Zimm. [Scolytidae] Southern pine beetle** | | | | |
| *Thanasimus formicarius* (L.) [Cleridae] | Germany | 2,200 adults | West Virginia | (17) |
| ***Dendroctonus obesus* (Mann.) [Scolytidae] Eastern spruce beetle** | | | | |
| *Rhizophagus* sp. [Rhizophagidae] | England from another species of Scolytidae | 800 adults | Quebec | (8) |
| *Rhopalicus tutela* (Wlkr.) [Pteromalidae] | England from another species of Scolytidae | 1,500 | Quebec | (8) |
| ***Neodiprion tsugae* Midd. [Diprionidae] Hemlock sawfly** | | | | |
| *Dahlbominus fuscipennis* (Zett.) [Eulophidae] | origin | 400,000 adults | near Vernon and Yale, B.C. | (8) |
| *Drino bohemica* Mesn. [Tachinidae] | origin | 515 adults | Queen Charlotte Islands | (8) |
| ***Neodiprion virginiana* Rohw. [Diprionidae]** | | | | |
| *Aptesis basizona* (Grav.) [Ichneumonidae] | Europe | 3,000 adults | Ontario | (8) |
| *Monodontomerus dentipes* (Dalm.) [Torymidae] | unspecified (originally Europe) | 3,805 | Canada | (8) |

| | | | | |
|---|---|---|---|---|
| ***Hemichroa crocea* (Geoffroy) [Tenthredinidae] Striped alder sawfly** | | | | |
| *Drino bohemica* Mesn. [Tachinidae] | surplus from European spruce sawfly biocontrol program | 500 | near Kingston, Ontario | (8) |
| ***Pristiphora geniculata* (Htg.) [Tenthredinidae] Mountain-ash sawfly** | | | | |
| *Drino bohemica* Mesn. [Tachinidae] | surplus from European spruce sawfly biocontrol program | 1,500 | near Wellington, Ontario | (8) |
| ***Psila rosae* (Fabricius) [Psiliidae] Carrot rust fly** | | | | |
| *Dacnusa gracilis* (Nees) [Braconidae] | England | 26,460 | British Columbia, Ontario, Prince Edward Island, and Quebec | (6, 18) |
| *Basalys tritoma* (Thoms.) [Proctotrupoidea] | England | 135,708 | British Columbia, Ontario, Prince Edward Island, and Quebec | (6, 18) |

(1) Larrouse et al. 1928
(2) Smith and Cole 1943
(3) Cooley 1928, Cooley and Kohls 1933
(4) Smith 1940, Lloyd 1951
(5) McLeod 1962
(6) Clausen 1978
(7) Stary 1964
(8) McGugan and Coppel 1962
(9) Richter 1966
(10) Berry 1938, Dowden 1962
(11) Flanders 1941
(12) Landis and Howard 1940
(13) Berry and Parker 1949, Clausen 1956
(14) Dymock 1989
(15) Loan 1965
(16) Parker 1959, Hertig 1960
(17) Hopkins 1893
(18) Maybee 1954

*Sources:* Data summarized from Clausen (1978), Cameron et al. (1989), and other cited references.

Although lack of effort may be the reason that establishment failed, this determination is difficult. When insufficient effort is expended, physiological and ecological incompatibilities are often unknown, so the true reason for failure remains indeterminant. For two of the target species (*Coleotechnites starki* and *Hemichroa crocea*), fewer than 1,000 individuals of each natural enemy were introduced. In these cases, insufficient effort may have been part of the reason that establishment did not occur.

Introductions on five species probably failed to establish because the introduced natural enemies were poorly adapted physiologically to the target host. Two target species (*Laspeyresia youngana* and *Pristiphora geniculata*) received natural enemies that were surplus production from a biological control effort targeting another species. One target species, *Bruchus pisorum,* received a natural enemy that was probably poorly adapted physiologically to it. The parasitoid, *Triaspis thracicus,* has very low parasitism rates on *B. pisorum* in its native Austria yet is quite abundant, implying the existence of a more important alternative host in the native habitat. Another target species (*Dendroctonus obesus*) received enemies collected from another species of scolytid. Because the suitability of the target host was not evaluated, it is quite possible that the natural enemies were poorly adapted physiologically to the target host. Two species of natural enemies were released against *Neodiprion virginiana* in Canada, but neither established. *Aptesis basizona* (= *Pleolophus basizonus*) appears to be poorly adapted to many sawfly hosts because it lays its egg loose in the sawfly cocoon, where it is frequently crushed by the moving pupa (Reeks 1953), and *Monodontomerus dentipes* has a greater affinity for *Diprion similis.* Both may be poorly adapted to *N. virginiana.*

Introductions on the remaining eight target species failed in part for ecological reasons. Physiological incompatibility can be ruled out because in all of these cases, at least one of the associated introduced natural enemies could complete a generation on the target host, frequently in great numbers. Climatic mismatch is implicated for the three natural enemies introduced against *Epilachna varivestis* and for *Tetrastichus galerucae* released against *Pyrrhalta luteola. Epilachna varivestis* is a subtropical to tropical species, and all of the natural enemies originated from these climatic regions. None of these natural enemies could successfully overwinter in cool temperate regions. For the other six species, it is unclear if lack of establishment was caused by poor climatic adaptation or by dynamical reasons. Several natural enemy species successfully overwintered on the target host (*Hunterellus hookeri* on *Dermacentor andersoni, Erynniopsis antennata* on *P. luteola,* and *Basalys tritoma* on *Psila rosae*) but failed to establish nonetheless, implying that some dynamical factors limited establishment.

A major biological control effort was mounted in eastern Canada and the northeastern United States against spruce budworm, *Choristoneura fumiferana.* Sixteen species of natural enemies from closely related species in western Canada and Europe were introduced but none established. This is one of the most spectacular failures in biological control, but the reasons for failure remain obscure. McGugan and

Coppel (1962) suggest that the environment of eastern Canada is unsuitable for these parasitoids, but Clausen (1978) suggests that interspecific competition with native species of natural enemies may have excluded the introduced species. An additional possibility is that the introduced agents were not well adapted to spruce budworm.

In summary, of the twenty-three target arthropod species with total establishment failures, two may have failed for lack of effort, five may have failed because of physiological incompatibility, eight failed because of ecological incompatibility, and the cause of eight failures could not be determined. These results highlight the need for uniform evaluation standards for biological control introductions. Minimally, physiological compatibility, release effort, and overwintering ability after release should be evaluated.

### *Weed Pests*

Julien (1992), Cameron (1989), and Kelleher and Hulme (1984) summarized releases of insects for biological control of weeds in cool temperate regions. We updated these reviews with more recent examples found in a variety of sources. As in attempts at biocontrol of arthropod pests, varying degrees of effort have been expended to establish biological control agents for a particular target weed. For some weeds, several insect species and thousands to millions of individuals have been distributed over a wide geographic area; examples are leafy spurge and purple loosestrife biocontrol agents. For other target weeds the effort has been minimal, with a few individuals of a single species released in one or two locations. Establishment rates of arthropods released to control a target weed are likely to be influenced by founder size, physiological condition of the insects upon release, climatic conditions at the site of release, and the overall effort employed to establish a natural enemy.

In cool temperate regions there have been thirty-five target weed species for which seventy-four candidate insects representing twenty-five families and five orders were released and evaluated as biological control agents (Tables 1.5 and 1.6). On average, 3.1 insect species were released for any one target weed with a range of one to twelve species (Table 1.7). For those target weeds where insects became established, regardless of whether or not the target weed was controlled, an average of 4.2 insect species were released. Contrast this to the six target weeds with no established insect where the average number of insect species released was just 1.2 (Table 1.7). Clearly, as the effort intensifies on a target weed, more species are released, with a significantly ($P = 0.03$) greater chance of having at least one natural enemy establish. As a target weed receives more attention, there is an increase in the number of individuals released at a given release site, with greater pre- and postrelease evaluation and site maintenance, all of which translates into higher establishment rates.

For twenty-nine (83 percent) of the thirty-five target weeds, at least one insect successfully established. Out of the seventy-four insect species used as biological control agents, forty-nine species (66 percent) became established in at least one of

Table 1.5. Insect biological control agent species released against a particular target weed.

| Target Host Family | Target Host | Number of Agent Species Released |
|---|---|---|
| Asteraceae | *Ambrosia artemisiifolia* | 4 |
| | *Ambrosia psilostachya* | 1 |
| | *Baccharis halimifolia* | 3 |
| | *Carduus acanthoides* | 3 |
| | *Carduus nutans* | 2 |
| | *Carduus pycnocephalus* | 2 |
| | *Carduus tenuiformis* | 2 |
| | *Carduus thoermeri* | 2 |
| | *Centaurea diffusa* | 8 |
| | *Centaurea maculosa* | 7 |
| | *Centaurea solstitialis* | 5 |
| | *Centaurea virgata spp. squarrosa* | 3 |
| | *Chondrilla juncea* | 1 |
| | *Cirsium arvense* | 5 |
| | *Cirsium palustre* | 1 |
| | *Cirsium vulgare* | 3 |
| | *Onopordum acanthium* | 1 |
| | *Senecio jacobaece* | 4 |
| | *Silybum marianum* | 1 |
| | *Sonchus arvensis* | 3 |
| Caryophyllaceae | *Silene vulgaris* | 1 |
| Chenopodiaceae | *Halogeton glomeratus* | 1 |
| | *Salsola australis* | 2 |
| Clusiaceae | *Hypericum androsaemum* | 1 |
| | *Hypericum perforatum* | 7 |
| Convolvulaceae | *Convolulus arvensis* | 1 |
| Euphorbiaceae | *Euphorbia cyparissias* | 9 |
| | *Euphorbia x pseudovirgata* | 12 |
| Fabaceae | *Cytisus scoparius* | 2 |
| | *Ulex europaeus* | 2 |
| Lamiaceae | *Slvia aethiopis* | 1 |
| Lythraceae | *Lythrum saicaria* | 4 |
| Scrophulariaceae | *Linaria dalmatica* | 2 |
| | *Linaria vulgaris* | 1 |
| Zygophyllaceae | *Tribulus terrestris* | 2 |

the release sites. These data are similar to establishment rates reported by Crawley (1989); he reported a 65 percent establishment rate for pre-1980 releases. Some degree of control was achieved for seventeen (59 percent) of these target hosts. Complete control was obtained for five target weeds (14 percent) (Table 1.8).

Focusing on those few weeds where complete control was obtained, there are some interesting observations. Clearly, with only five weeds out of thirty-five under complete control, inferences drawn are based on limited data. For target hosts where complete control was obtained, a single insect species generally was responsible for

Table 1.6. Insects used in weed biological control programs in cool temperate regions

| Order (number of species) | Family (number of species) | Species |
|---|---|---|
| Coleoptera (34) | Anthribidae (1) | *Trigonorhinus tomentosus* (Say) |
| | Apionidae (2) | *Apion fuscirostrre* Fabricius |
| | | *Apion ulicis* (Forster) |
| | Buprestidae (2) | *Agrilus hyperici* (Creutzer) |
| | | *Sphenoptera jugoslavica* Obenberger |
| | Cerambycidae (1) | *Oberea erythrocephala* (Schrank) |
| | Chrysomelidae (16) | *Altica carduorum* Guerin-Meneville |
| | | ***Aphthona cyparissiae*** (Koch) |
| | | ***Aphthona czwalinai*** (Weise) |
| | | *Aphthona flava* Guilleeau |
| | | ***Aphthona nigriscutis*** Foudras |
| | | *Cassida azurea* Fabricius |
| | | ***Chrysolina hyperici*** (Forster) [a] |
| | | ***Chrysolina quadrigemina*** (Suffrian) |
| | | *Chrysolina varians* (Schaller) |
| | | *Gallerucella calmariensis* (L.) |
| | | *Gallerucella pusilla* (Duft.) |
| | | *Lema cyanella* (l.) |
| | | *Longitarsus flavicornis* (Stephens) |
| | | ***Longitarsus jacobaeae*** (Waterhouse) |
| | | *Trirhabda baccharidis* (Weber) |
| | | *Zygogramma suturalis* F. |
| | Curculionidae (12) | *Bangasternus orientalis* (Capimont) |
| | | *Ceutorhynchus litura* (F.) |
| | | *Cyphocleonus achates* (Fahraeus) |
| | | *Eustenopus villosus* (boheman) |
| | | *Gymnetron antirrhini* (Paykull) |
| | | *Hylobius transversovittatus* Goeze |
| | | *Microlarinus lareynii* (Jacquelin du Val) |
| | | *Microlarinus lypriformis* (Wollaston) |
| | | *Nanophyes marmoratus* |
| | | *Phrydiuchus tau* Warner |
| | | ***Rhinocyllus conicus*** (Frolich) |
| | | ***Trichosirocalus horridus*** (Panzer) |
| Diptera (21) | Agromyzidae (1) | *Liriomyza sonchi* Hendel |
| | Anthomyiidae (3) | *Botanophila seneciella* (Meade) |
| | | *Pegomya curticornis* (Sterin) |
| | | *Pegomya euphorbiae* (Kieffer) |
| Diptera (21) | Cecidomyiidae (6) | *Cystiphora schmidti* (Rubsaamen) |
| | | *Cystiphora sonchi* (Bremi) |
| | | *Rhopalomyia california* Felt |
| | | *Spurgia capitigena* (Bremi) |
| | | *Spurgia esulae* Gagne |
| | | *Zeuxidiplosis giardi* (Kieffer) |
| | Syrphidiae (1) | *Cheilosia corydon* (Harris) |

*Note:* Species involved in a successful control program are in bold, large font, and insects that failed to establish on a target host are underlined. All other species established but failed to give complete control or are of no consequence in the management of the target weed.

Table 1.6. continued.

| Order (number of species) | Family (number of species) | Species |
|---|---|---|
| | Tephritidae (10) | *Chaetorellia australis* Hering |
| | | *Euaresta bella* (Loew) |
| | | *Tephritis dila cerata* (Loew) |
| | | *Urophora affinis* (Frauenfeld) |
| | | *Urophora cardui* (L.) |
| | | *Urophora jaculata* Rondani |
| | | *Urophora quadrifasciata* Meigen |
| | | *Urophora sirunaseva* (Hering) |
| | | *Urophora solsstitialis* (L.) |
| | | <u>*Urophora stylata*</u> (Fabricius) |
| Hemiptera (1) | Aphididae (1) | *Aphis chloris* Koch |
| Lepidoptera (17) | Aegeriidae (2) | *Chamaesphecia empiformis* Esper |
| | | *Chamaesphecia tenthrediniformis* (Denis & Schiff.) |
| | Arctiidae (1) | *Tyria jacobaeae* (L.) [b] |
| | Cochylidae (1) | *Agapeta zoegana* L. |
| | Coleophoridae (2) | *Coleophora klimeschiella* Toll |
| | | <u>*Coleophora parthenica*</u> Meyrick |
| | Gelechiidae (1) | *Metzneria paucipunctella* Zeller |
| | Geometridae (1) | *Aplocera plagiata* (L.) |
| | Lyonetiidae (1) | *Leucoptera spartifoliella* Hubner |
| | Noctuidae (3) | *Calophasia lunula* (Hufnagel) |
| | | *Tarachidia candefacta* Hubner |
| | | <u>*Tyta luctuosa*</u> (Denis & Schiffermuller) |
| | Pterolonchidae (1) | *Pterolonche inspersa* Staudinger |
| | Pterophoridae (1) | *Oidaematophorus balanotes* (Meyrick) |
| | Sphingidae (1) | *Hyles euphorbiae* (L.) |
| | Tortriciadae (2) | *Lobesia euphorbiana* (Freyer) |
| | | *Pelochrista medullana* (Staudinger) |
| Thysanoptera (1) | Thripidae (1) | *Sericothrips staphylinus* Haliday |

[a] *Chrysolina hyperici* contributed to successful control of *Hypericum perforatum*, but it did not establish on a related species, *H. androsaemum*.

[b] This species is only partially effective alone but in concert with *Longitarsus jacobaeae*, tansy ragwort is effectively controlled. However, the flea beetle alone is capable of controlling tansy ragwort if plant competition is present (McEvoy, *et al.* 1989)

control of the weed in a given locale (soil type and climate). In addition, all agents that control these five weeds are univoltine Coleoptera (two Curculionidae and five Chrysomelidae). A possible exception to this general trend may be *Senecio jacobaeae* L., tansy ragwort: McEvoy et al. (1989) conclude that the best control is obtained when both the cinnabar moth, *Tyria jacobaeae* (L.), and the flea beetle, *Longitarsus jacobaeae* (Waterhouse), were established. However, their data also show that the cinnabar moth alone does not provide control under any situation, whereas the flea beetle alone will give complete control when there is plant competition.

Moreover, there does not appear to be any advantage in selecting an agent depending upon the tissue fed upon as an adult or larvae. For example, *Rhinocyllus conicus*

Table 1.7. Success of establishment and degree of target weed control by insect biological control agents, and the number of agents released per target host.

| Establishment of Insect Biocontrol Agent & Control of Target Weed | Mean No. Insect Species Released | Range |
|---|---|---|
| Succesful Control | 5.4 | 2-12 |
| Partial Success | 3.8 | 1-9 |
| No Control | 2.5 | 1-5 |
| No Insect Established | 1.2 | 1-2 |

(Frolich) controls musk thistle, *Carduus nutans* L., by destroying a large proportion of the seed heads so that musk thistle stands were reduced and other competitive plants displaced the target host (Harris 1980). In contrast, Klamath weed, *Hypericum perforatum,* is controlled by adult and larval leaf beetles defoliating the plant during a critical period in the year; apparently the plant desiccates and does not recover when rains begin in late fall (Goeden 1978; Wilson 1960; Hoy 1963; Harris 1980). In general, *Chrysolina quadrigemina* (Suffrian) is the principal species controlling Klamath weed, although *C. hyperici* (Forster) is often found in sites with higher moisture. For leafy spurge and tansy ragwort, root-feeding larval Chrysomelidae are the life stage responsible for control of the target weed. As more weeds are subjected to biological control efforts, it will be interesting to note if this trend of Coleoptera being the most important order holds true. However, the number of cases where weeds have been completely controlled is too few to make any reasonable conclusion as to which group of insects are likely to be successful biological control agents. We are not suggesting that only Coleoptera be imported for future biological control efforts in the north temperate zone, but if there is a choice, and limited resources, the focus might first be on Coleoptera, specifically the Curculionidae and Chrysomelidae.

The average number of agent species released against the five weeds where complete control has been obtained was 5.4 with a range of 2 to 12 (Table 1.7). For Klamath weed (*Hypericum perforatum* L.) and leafy spurge (*Euphorbia* x *pseudovirgata* (Schur) So), control at any one site is attributed to one species of beetle but multiple species are needed to control the weed throughout the weed's range where soil type, rainfall, and other climatic factors favor one insect species over the other. Thus, control over the entire geographic range of the target weed is often dependent upon more than one insect species. This finding implies that to effectively control any one species, a thorough understanding of the weed throughout its range and its associated herbivore fauna is needed. Work is still ongoing with leafy spurge biocontrol, and it may be optimistic to declare leafy spurge completely controlled throughout the weed's range, but for many areas in the northern Great Plains control is substantial.

Table 1.8. Introduction success and success of established agents to control the target weed

| Order | Family | Total Releases | Establishment Results: Under Evaluation | Establishment Results: Not Established | Establishment Results: Established | Success Results: Complete Control | Success Results: Partial Control | Success Results: No Control | Success Results: Unknown |
|---|---|---|---|---|---|---|---|---|---|
| Coleoptera | Anthribidae | 2 | 1 | 1 | 0 | 0 | 0 | 0 | 0 |
| | Apionidae | 5 | 0 | 0 | 5 | 0 | 5 | 0 | 0 |
| | Buprestidae | 12 | 0 | 4 | 8 | 0 | 4 | 0 | 4 |
| | Cerambycidae | 8 | 1 | 4 | 3 | 0 | 0 | 2 | 1 |
| | Chrysomelidae | 122 | 9 | 38 | 75 | 31 | 14 | 8 | 22 |
| | Curculionidae | 99 | 10 | 20 | 66 | 6 | 24 | 15 | 24 |
| Diptera | Agromyzidae | 2 | 0 | 1 | 1 | 0 | 0 | 0 | 1 |
| | Anthomyiidae | 12 | 1 | 7 | 4 | 0 | 1 | 2 | 1 |
| | Cecidomyiidae | 26 | 0 | 10 | 16 | 0 | 5 | 8 | 3 |
| | Syrphidiae | 1 | 1 | 0 | 0 | 0 | 0 | 0 | 0 |
| | Tephritidae | 74 | 11 | 23 | 40 | 0 | 7 | 19 | 14 |
| Hemiptera | Aphididae | 2 | 1 | 1 | 0 | 0 | 0 | 0 | 0 |
| Lepidoptera | Aegeriidae | 5 | 1 | 4 | 0 | 0 | 0 | 0 | 0 |
| | Arctiidae | 21 | 2 | 4 | 15 | 0 | 4 | 11 | 0 |
| | Cochylidae | 10 | 3 | 3 | 4 | 0 | 0 | 2 | 2 |
| | Colleophoridae | 19 | 0 | 18 | 1 | 0 | 0 | 1 | 0 |
| | Gelechiidae | 6 | 0 | 0 | 6 | 0 | 4 | 2 | 0 |
| | Geometridae | 5 | 2 | 2 | 1 | 0 | 0 | 1 | 0 |
| | Lyonetiidae | 2 | 0 | 0 | 2 | 0 | 0 | 2 | 0 |
| | Noctuidae | 20 | 2 | 12 | 6 | 0 | 0 | 1 | 5 |
| | Pterolonchidae | 7 | 2 | 5 | 0 | 0 | 0 | 0 | 0 |
| | Pterophoridae | 1 | 0 | 0 | 1 | 0 | 0 | 0 | 1 |
| | Sphingidae | 15 | 1 | 10 | 4 | 0 | 1 | 1 | 2 |
| | Tortricidae | 9 | 4 | 5 | 0 | 0 | 0 | 0 | 0 |
| Thysanoptera | Thripidae | 1 | 1 | 0 | 0 | 0 | 0 | 0 | 0 |
| **TOTAL** | | 486 | 53 | 172 | 261 | 37 | 69 | 75 | 80 |

*Sources:* Data compiled using all release sites (486).

Much of the current focus on leafy spurge biocontrol is redistributing various flea beetles in the genus *Aphthona* to provide control of leafy spurge in a variety of habitats, soil types, and climates.

In contrast with the successful biological control programs, no insect became established for six target weeds. For five of these six target weeds where no insects became established, only a single importation was conducted with insects distributed at most to two sites. The only exception was a slightly more intense attempt to establish a single insect against puncture vine, *Tribulus terrestris* L., where the curculionid, *Microlarinus lareynii* (Jacquelin du Val), was released in multiple states with a concerted effort to select for a cold-adapted strain for release in the north temperate zone. All attempts to establish *M. lareynii* in the north temperate zone, which has a severe winter, failed.

## Conclusions

There is no compelling evidence to suggest that the success of classical biological control is less in cool temperate regions. Indeed, even if there were such evidence, it would imply only that we would have to try harder for successful biological control. A lower probability of success does not imply that biological control should not be attempted.

Our retrospective analysis of classical biological control of arthropods and weeds revealed considerable variation in the meaning of successful establishment and successful control. We suggest that biological control practitioners analyze this variation and more rigorously evaluate success. To facilitate this discussion, we propose conceptual frameworks to compare and contrast these meanings and operationalize them with potential observations. We found that our frameworks could be applied to the historic literature on biological control in cool temperate regions, but the data are too few to allow any reasonable generalizations.

The potential for research on biological control in cool temperate regions seems vast. There are many unresolved cases of classical biological control, such as brown-tailed moth, that could yield interesting insights. There are several ongoing efforts to introduce natural enemies that could be enhanced. Inundative methods are just now being developed for control of arthropod pests, weeds, and pathogens, and conservation of natural enemies has been limited by our knowledge of the ecology of the natural enemies. Integration of biological and chemical control in integrated pest management systems has remained elusive. Research to develop adaptive decision-making plans could accelerate integration and alleviate much unnecessary research to develop precise estimates of the effects of partial and occasional biological controls.

## References

Andow, D. A., G. C. Klacan, D. Bach, and T. C. Leahy. 1995. Limitations of *Trichogramma nubilale* (Hymenoptera: Trichogrammatidae) as an inundative biological control of *Ostrinia nubilalis* (Lepidoptera: Crambidae). *Environmental Entomology* 24:1352–57.

Andrews, J. H. 1992. Biological control in the phyllosphere. *Annual Review of Phytopathology* 30:603–35.

Baker, K. F., and R. J. Cook. 1974. *Biological control of plant pathogens.* San Francisco: Freeman.

Baker, R. R., and P. E. Dunn. 1989. *New directions in biological control.* New York: Liss.

Berry, P. A. 1938. Laboratory studies on *Tetrastichus xanthomelaenae* Rond. and *Tetrastichus* sp., two hymenopterous egg parasites of the elm leaf beetle. *Journal of Agricultural Research* 57:859–63.

Berry, P. A., and H. L. Parker. 1949. Investigations on a South American *Epilachna* sp. and the importation of its parasite *Lydinolydella metallica* Tns. into the United States. *Proceedings of the Washington Entomological Society* 52:251–58.

Boyette, C. D., G. C. Templeton, and R. J. Smith Jr. 1973. Biological control of northern jointvetch in rice with an endemic fungal disease. *Weed Science* 21:303–7.

Cameron, P. J., R. L. Hill, J. Bain, and W. P. Thomas. 1989. *A review of biological control of invertebrate pests and weeds in New Zealand 1874 to 1987.* Oxon, U.K.: CAB International.

Charudattan, R. 1991. The mycoherbicide approach with plant pathogens. In *Microbial control of weeds,* ed. D. O. TeBeest, 24–57. New York: Chapman and Hall.

Clausen, C. P. 1956. Biological control of insect pests in the continental United States. *USDA Technical Bulletin* 1159.

______. 1978. Introduced parasites and predators of arthropod pests and weeds: A world review. *USDA Agriculture Handbook* 480.

Cook, R. J. 1993. Making greater use of introduced microorganisms for biological control of plant pathogens. *Annual Review of Phytopathology* 31:53–80.

Cook, R. J., and K. R. Baker. 1983. *The nature and practice of biological control of plant pathogens.* St. Paul, Minn.: American Phytopathological Society.

Cooley, R. A. 1928. Tick parasites. *Montana State Board of Entomology 7th Biennial Report* 10–16.

Cooley, R. A., and G. M. Kohls. 1933. A summary of tick parasites. *Proceedings 5th Pacific Science Congress* 5:3375–81.

Crawley, M. J. 1989. Plant life-history and the success of weed biological control projects. In *Proceedings of the VII International Symposium on Biological Control Weeds,* ed. E. S. Delfosse, 17–26. Rome: Istituto Sperimentale per la Patologia Vegetale, Ministero dell' Agricoltura e delle Forestale.

Croft, B. A. 1989. *Arthropod biological control agents and pesticides.* New York: Wiley.

Crosby, T. K. 1989. *Austrosimulium asutralense* (Schiner), New Zealand black fly and *Austrosimulium ungulatum* Tonnoir, west coast black fly (Diptera: Simuliidae). In *A review of biological control of invertebrate pests and weeds in New Zealand 1874 to 1987,* ed. P. J. Cameron, R. L. Hill, J. Bain, and W. P. Thomas, 375–79. Oxon, U.K.: CAB International.

Cullen, J. M. 1978. Evaluation of the success of the program for the biological control of *Chondrilla juncea* L. *Proceedings of the Ecological Society of Australia* 10:121–34.

DeBach, P. 1964. Successes, trends, and future possibilities. In *Biological control of insect pests and weeds,* ed. P. DeBach, 673–713. New York: Reinhold.

DeBach, P., and B. R. Bartlett. 1964. Methods of colonization, recovery and evaluation. In *Biological control of insect pests and weeds,* ed. P. DeBach, 402–26. New York: Reinhold.

Dowden, P. B. 1962. Parasites and predators of forest insects liberated in the United States through 1960. *USDA Agricultural Handbook* 226.

Drea, J. J. Jr., and E. G. King. 1981. The coordination of scientific and commercial aspects of biological control of filth breeding flies. In *Status of biological control of filth flies,* ed. J. J. Drea Jr. and E. G. King, 1–4. New Orleans: Science and Education Administration, U.S. Department of Agriculture.

Dymock, J. J. 1989. *Listronotus bonariensis* (Kuschel), Argentine stem weevil (Coleoptera: Curculionidae). *Commonwealth Institute of Biological Control, Technical Communication* 10:23–26.

Embree, D. G. 1966. The role of introduced parasites in the control of the winter moth in Nova Scotia. *Canadian Entomologist* 98:1159–68.

Emge, R. G., J. S. Melching, and C. H. Kingsolver. 1981. Epidemiology of *Puccinia chondrillina,* a rust pathogen for the biological control of rush skeleton weed in the United States. *Phytopathology* 71:839–43.

Epstein, A. H., J. H. Hill, and J. J. Obrycki. 1993. Rose rosette disease. Iowa State University Pm-1532.

Feichtenberger, E., G. A. Zentmyer, and J. A. Menge. 1983. Identity of *Phytophthora* isolated from milkweed vine. *Phytopathology* 73:50–55.

Ferguson, A. M. 1989. *Phthorimaea operculella* (Zeller), potato tuber moth (Lepidoptera: Gelechiidae). In *A review of biological control of invertebrate pests and weeds in New Zealand 1874 to 1987,* ed. P. J. Cameron, R. L. Hill, J. Bain, and W. P. Thomas, 119–28. Oxon, U.K.: CAB International.

Flanders, S. E. 1941. Observations on the biology of the elm leaf beetle parasite, *Erynnia nitida* (R. Desv.). *Journal of Economic Entomology* 33:947–48.

Goeden, R. D. 1978. Biological control of weeds. *USDA Handbook* 480: 357–414.

Greathead, D. J. 1986. Parasitoids in classical biological control. In *Insect parasitoids,* ed. J. Waage and D. Greathead, 290–318. London: Academic Press.

Hajek, A. E., and D. L. Dahlsten. 1981. First California record for *Dendrosoter protuberans* (Nees) (Hymenoptera: Braconidae) parasite of the bark beetle, *Scolytus multistriatus,* a vector of Dutch elm disease. *Pan-Pacific Entomologist* 57:504–5.

Hall, R. W., and L. E. Ehler. 1979. Rate of establishment of natural enemies in classical biological control. *Entomological Society of America Bulletin* 25:280–82.

Hall, R. W., L. E. Ehler, and B. Bisabri-Ershadi. 1980. Rate of success in classical biological control of arthropods. *Entomological Society of America Bulletin* 26:111–14.

Harris, P. 1980. Stress as a strategy in the biological control of weeds. *BARC Symposium* 5: 333–40.

Hassell, M. P. 1980. Foraging strategies, population models and biological control: A case study. *Journal of Animal Ecology* 49:603–28.

Hertig, B. 1960. Biologie der westpalärktischen Raupenfliegen Diptera, Tachinidae. *Monographien zur angewante Entomologie* 16:1–188.

Hopkins, A. D. 1893. Destructive scolytids and their natural enemy. *Insect Life* 6:123–29.

Hornby, D. 1990. *Biological control of plant pathogens.* Oxon, U.K.: CAB International.

Hoy, J. M. 1963. Present and future prospects for biological control of weeds. *New Zealand Science Review* 22:17–19.

Hoy, M. A., and D. C. Herzog. 1985. *Biological control in agricultural IPM systems.* Orlando, Fla.: Academic Press.

Johnson, D. R., and D. L. Wyse. 1991. Use of *Pseudomonas syringae* pv. *tagetis* for control of Canada thistle. *Proceedings of the North Central Weed Science Society* 46:14–15.

______. 1992. Biological control of weeds with *Pseudomonas syringae* pv. *tagetis. Proceedings of the North Central Weed Science Society* 47:13.

Julien, M. H. 1992. *Biological control of weeds: A world catalogue of agents and their target weeds.* Oxon, U.K.: CAB International.

Kelleher, J. S., and M. A. Hulme. 1984. *Biological control programmes against insects and weeds in Canada 1969–1980.* Farnham Royal, England: Commonwealth Agricultural Bureaux.

Kennedy, B. H. 1970. *Dendrosotor protuberans* (Hymenoptera: Braconidae) an introduced larval parasite of *Scolytus multistriatus. Annals of the Entomological Society of America* 63:351–58.

Kerr, A. 1980. Biological control of crown gall through production of agrocin 84. *Plant Discovery* 64:25–30.

Klimaszewski, J. 1984. A revision of the genus *Aleochara* Gravenhorst of America north of Mexico (Coleoptera: Staphylinidae, Aleocharinae). *Memoirs of the Entomological Society of Canada* 129.

Klimaszewski, J., and V. J. Cervenka. 1986. A revision of the genus *Aleochara* (Coleoptera: Staphylinidae) of America north of Mexico. Supplement 3. New distribution data. *Entomology News* 97:119–20.

Landis, B. J., and N. F. Howard. 1940. *Paradexodes epilachnae,* a tachinid parasite of the Mexican bean beetle. *USDA Technical Bulletin* 721.

Larrouse, F., A. G. King, and S. B. Wolbach. 1928. The overwintering in Massachusetts of *Ixodiphagus caucurtei. Science* 67:351–53.

Lloyd, D. C. 1951. A survey for grasshopper parasites in temperate South America. *Canadian Entomologist* 83:213–30.

Loan, C. C. 1965. Status of *Pygostolus falcatus* as a parasite of *Sitona* spp. following releases in Manitoba and Ontario. *Journal of Economic Entomology* 58:798–99.

Mackauer, M., L. E. Ehler, and J. Roland. 1990. *Critical issues in biological control.* New York: Intercept.

MacPhee, A. W., and C. R. MacLellan. 1972. Ecology of apple orchard fauna and development of integrated pest control in Nova Scotia. *Tall Timbers Conference on Ecological Animal Control by Habitat Management* 3:197–208.

Makowski, R.M.D., and K. Mortensen. 1992. The first mycoherbicide in Canada: *Colletotrichum gloeosporioides* f. sp. *malvae* for round-leaved mallow control. *Proceedings of the First International Weed Control Congress* 2:298–300.

Marwick, N. P., R. L. Hill, and D. J. Allan. 1989. Muscidae, muscid flies (Diptera). In *A review of biological control of invertebrate pests and weeds in New Zealand 1874 to 1987,* ed. P. J. Cameron, R. L. Hill, J. Bain, and W. P. Thomas, 387–94. Oxon, U.K.: CAB International.

Mazzola, M., D. K. Fujimoto, and R. J. Cook. 1994. Differential sensitivity of *Gaeumannomyces graminis* populations to antibiotics produced by biocontrol fluorescent pseudomonads. *Phytopathology* 84:1091 (abstr.).

McEvoy, P. B., C. S. Cox, R. R. James, and N. T. Rudd. 1989. Ecological mechanisms underlying successful biological weed control: Field experiments with ragwort *Senecio jacobaea.* In *Proceedings of the VII International Symposium on Biological Control of Weeds,* ed. E. S. Delfosse, 55–66. Rome: Istituto Sperimentale per la Patologia Vegetale, Ministero dell' Agricoltura e delle Forestale.

McGugan, B. M., and H. C. Coppel. 1962. A review of the biological control attempts against insects and weeds in Canada. Part II. Biological control of forest insects, 1910–1958. *Tech-*

*nical Communication* 2:135–216. Farnham Royal, Great Britain: Commonwealth Agriculture Bureaux.

McLeod, J. H. 1962. A review of the biological control attempts against insects and weeds in Canada. Part I. Biological control of pests of crops, fruit trees, ornamentals, and weeds in Canada up to 1959. *Technical Communication* 2:1–33. Farnham Royal, Great Britain: Commonwealth Agriculture Bureaux.

Mortensen, K. 1988. The potential of an endemic fungus, *Colletotrichum gloeosporioides,* for biological control of round-leaved mallow (*Malva pusilla*) and velvetleaf (*Abutilon theophrasti*). *Weed Science* 36:473–78.

Munroe, E. G. 1971. Part IV. Biological control in Canada, 1959–1968: Synopsis. Status and potential of biological control in Canada. *Commonwealth Institute of Biological Control, Technical Communication* 4: 213–55.

Nakasuji, F., H. Yamanaka, and K. Kiritani. 1973. Control of the tobacco cutworm, *Spodoptera litura* F., with polyphagous predators and ultra-low concentration of chlorophenamidine. *Japanese Journal of Applied Entomology and Zoology* 17:171–80.

Oehrens, E. 1977. Biological control of the blackberry through the introduction of rust, *Phragmidium violaceum,* in Chile. *FAO Plant Protection Bulletin* 25:26–28.

Ostlie, K. R., and L. P. Pedigo. 1987. Incorporating pest survivorship into economic thresholds. *Bulletin of the Entomological Society of America* 33:98–102.

OTA. 1995. *Biologically based technologies for pest control.* U.S. Congress, Office of Technology Assessment. Washington, D.C.: U.S. Government Printing Office.

Papavizas, G. C., and R. D. Lumsden. 1980. Biological control of soilborne fungal propagules. *Annual Review of Phytopathology* 18:389–413.

Parker, H. L. 1959. Studies of some Scarabaeidae and their parasites. *Boletin Laboratorio Entomologia Agricoltura "Filippo Silvestri" Portici* 17:29–50.

Pearson, W. D. 1989. *Coleophora frischella* L., whitetipped clover casebearer, *C. spissicornis* Haworth, banded clover casebearer (Lepidoptera: Coleophoridae). In *A review of biological control of invertebrate pests and weeds in New Zealand 1874 to 1987,* ed. P. J. Cameron, R. L. Hill, J. Bain, and W. P. Thomas. 73–86. Oxon, U.K.: CAB International.

Pickett, A. D. 1949. A critique on insect chemical control methods. *Canadian Entomologist* 81:67–76.

Pierson, E. A., and D. M. Weller. 1994. Use of mixtures of fluorescent pseudomonads to suppress take-all and improve the growth of wheat. *Phytopathology* 84:940–47.

Reeks, W. A. 1953. The establishment of introduced parasites of the European spruce sawfly *Diprion hercyniae* (Htg.) (Hymenoptera: Diprionidae) in the Maritime Provinces. *Canadian Journal of Agricultural Science* 33:405–29.

Risbeth, J. 1975. Stump inoculation: A biological control of *Fomes annosus.* In *Biology and control of soil-borne plant pathogens,* ed. G. W. Bruehl, 158–62. St. Paul, Minn.: American Phytopathological Society.

Ritcher, P. O. 1966. Biological control of insects and weeds in Oregon. *Oregon Agricultural Experiment Station Technical Bulletin* 90.

Roland, J., and D. G. Embree. 1995. Biological control of the winter moth. *Annual Review of Entomology* 40:475–92.

Schneider, R. W. 1982. *Suppressive soils and plant disease.* St. Paul, Minn.: The American Phyopathological Society.

Schroth, M. N., and J. G. Hancock. 1982. Disease-suppressive soil and root-colonizing bacteria. *Science* 216:1376–81.

Smith, C. N., and M. M. Cole. 1943. Studies of parasites of the American dog tick. *Journal of Economic Entomology* 36:569–72.

Smith, C. W. 1940. An exchange of grasshopper parasites between Argentina and Canada with notes on parasitism of native grasshoppers. *Entomological Society of Ontario 70th Annual Report* 1939:57–62.

Smith, H. S., and P. DeBach. 1942. The measurement of the effect of entomophagous insects on population densities of their hosts. *Journal of Economic Entomology* 35:845–49.

Stary, P. 1964. Biological control of *Megoura viciae* Bckt. in Czechoslovakia. *Acta Entomologica Bohemoslovaca* 61:301–22.

Supkoff, D. M., D. B. Joley, and J. J. Marois. 1988. Effect of introduced biological control organisms on the density of *Chondrilla juncea* in California. *Journal of Applied Ecology* 25:1089–95.

Taylor, T.H.C. 1955. Biological control of insect pests. *Annals of Applied Biology* 42:190–96.

Trewartha, G. T. 1957. Elements of physical geography. New York: McGraw-Hill.

Van Dreische, R. G. 1983. Meaning of "percent parasitism" in studies of insect parasitoids. *Environmental Entomology* 12:1611–22.

Weller, D. M. 1988. Biological control of soilborne pathogens in the rhizosphere with bacteria. *Annual Review of Phytopathology* 26:379–407.

Weller, D. M., and R. J. Cook. 1983. Suppression of take-all of wheat by seed treatments with fluorescent pseudomonads. *Phytopathology* 73:463–69.

Wilson, F. 1960. A review of the biological control of insects and weeds in Australia and Australian New Guinea. *Commonwealth Institute of Biological Control, Technical Communication* 1.

Wood, R.K.S., and M. J. Way. 1988. Biological control of pests, pathogens and weeds: Development and prospects. *Philosophical Transactions of the Royal Society of London* B318:109–376.

Yang, X. B., and D. O. TeBeest. 1993. Epidemiological mechanisms of mycoherbicide effectiveness. *Phytopathology* 83:891–93.

PART ONE

# Agent Selection

# 2

# Biological Control in Native and Introduced Habitats: Lessons Learned from the Sap-Feeding Guilds on Hemlock and Pine

*Mark S. McClure*

## Introduction

Many of our most important and destructive pests are herbivorous insects that were introduced from abroad. Much time and many resources have been expended during this century to control our many introduced herbivorous pests by importing natural enemies from abroad (DeBach 1974; Clausen 1978; Laing and Hamai 1976), yet examples of successful biological control have been relatively few (Caltagirone 1981). Many biological control efforts have failed by using a "shotgun" approach of releasing any and all natural enemies with little knowledge of their ecological requirements and with no system in place for evaluating their performance following introduction (Turnock et al. 1976; Pschorn-Walcher 1977).

Surprisingly few studies have examined the impact of natural enemies on the population dynamics of introduced pests in their homelands. Those that have been done have often focused on natural habitats where pests frequently persist at low, innocuous densities (Caltagirone 1981). This situation can be quite different from that of introduced habitats where pests commonly persist at damaging outbreak population levels (McClure 1983a). As we shall see, factors that affect the population dynamics of pests in native and introduced habitats differ greatly, and failing to understand and address these differences can impose significant constraints on our efforts to control introduced pest populations by introducing their native natural enemies.

There are many possible constraints on successful biological control of introduced pests. The common paucity of taxonomic and bionomic information on pests and their natural enemies in their homeland is among the most obvious and significant of these constraints (Delucchi et al. 1976). In native natural habitats, herbivorous insects often persist at low, innocuous densities due to a combination of various factors including host defenses and natural enemies. Consequently, such insects are rarely studied in any great detail and may be entirely unknown. This makes the task of accurately identifying introduced pests and their natural enemies and determining their homeland all the more difficult. The lack of bionomic information on pest and enemy life stages, phenological compatibility, and environmental tolerances in the native habitat greatly reduces the chances of successful establishment and control by native natural enemies in introduced habitats (DeBach et al. 1976; Zwölfer et al. 1976).

Another significant constraint on the biological control of introduced pests is the rapidity with which their populations often build and severely injure the new host plant (McClure 1983a). As we shall see, populations of introduced pests often quickly attain outbreak levels on a defenseless new host plant in the absence of their native natural enemies. High population densities of these introduced pests usually persist only for a short time, until the deteriorating host itself adversely affects herbivore performance. Therefore, natural enemies that are effective control agents in native natural habitats where their hosts occur at low, innocuous densities may have limited ability and utility as biological control agents for pest populations that outbreak in introduced habitats.

In their native habitats, herbivorous insects often reach injurious outbreak population densities only in cultivated or disturbed habitats or on susceptible exotic host species (Turnock et al. 1976). Reduced host resistance in response to environmental stress, changes in host nutritional chemistry, and disruption of the natural enemy complex, along with numerous other factors, may explain such outbreaks (Barbosa and Schultz 1987). As we shall see, explorations abroad for native natural enemies and evaluations of their potential for biological control of introduced pests should be conducted not only in natural habitats, but also in cultivated habitats and, if possible, on susceptible exotic hosts where factors that affect herbivore population dynamics are apt to resemble more closely those that characterize introduced habitats. Comparative studies of two such populations using life table analyses (Bellows et al. 1992) and experimental methods (Luck et al. 1988) can identify factors that can be manipulated to control damaging populations of introduced pests and can provide a sound basis for evaluating the performance of natural enemies introduced for biological control. Thus, we may avoid many of the constraints and pitfalls of previous biological control failures.

Two species of adelgids and three species of scales that are native to Japan constitute two destructive sap-feeding guilds of hemlock and pine in eastern North America and have been the main subjects of my ecological investigations in Asia and North America during the past twenty years. Using these guilds as examples, I will illustrate how comparative studies of native and introduced pests in natural and cultivated habitats can generate useful theoretical predictions of biological control, but why some of these predictions fail in practice. I will discuss the various constraints on successful biological control in introduced habitats and will offer alternate strategies and approaches that may enhance the usefulness of studies of pests in their native habitats for biological control of introduced pests.

## Exotic Sap-Feeding Guilds of Hemlock and Pine

Two armored scales, *Fiorinia externa* Ferris and *Nuculaspis tsugae* (Marlatt) (Homoptera: Diaspididae) (Fig. 2.1), and an adelgid, *Adelges tsugae* Annand (Homoptera: Adelgidae) (Fig. 2.2), attack eastern hemlock, *Tsuga canadensis* Carriere, in the northeastern United States. Another scale, *Matsucoccus matsumurae* (Kuwana) (Homoptera: Margarodidae) (Fig. 2.3), and the adelgid *Pineus boerneri* Annand (Homoptera: Adelgidae) (Fig. 2.4), attack red pine, *Pinus resinosa* Aiton.

In Japan, their homeland, *F. externa* and *N. tsugae* are common inhabitants of two native hemlocks, *Tsuga diversifolia* Masters and *Tsuga sieboldii* Carriere. However, these scales seldom attain injurious densities in Japan. Both scales were introduced accidentally in the vicinity of New York City in the early 1900s and have since become serious pests of *T. canadensis* in several northeastern states. Both scales feed together on the undersides of hemlock needles by sucking cell fluids from the mesophyll. In the United States their densities often increase rapidly to levels that cause needles to discolor and drop prematurely and branches to die. Many ornamental and forest trees have been killed within ten years (McClure 1980).

*Adelges tsugae* is also native to Japan, where it is a harmless inhabitant of *T. diversifolia* and *T. sieboldii.* This adelgid was first noticed in North America in 1921 on *Tsuga heterophylla* Sargent in Vancouver, British Columbia. It now occurs throughout much of that province and the northwestern United States, where its damage has been rare, probably due to host resistance (McClure 1992). In the eastern United States, *A. tsugae* was first reported about forty years ago on *T. canadensis* in Virginia. Since that time it has spread primarily northeastward and now occurs as far north as New England (McClure 1987a). Unlike the needle-feeding scales, *A. tsugae* sucks cell fluids from the young branches and probably also injects a toxic saliva that causes rapid desiccation and drop of needles, dieback of main limbs, and death of the tree, usually within four years (McClure 1991).

Yet another Japanese species, *M. matsumurae,* now considered to be the same species as *M. resinosae* Bean and Godwin (McClure 1983b), was first discovered in Connecticut in 1946 in a dying plantation of *P. resinosa.* Historical evidence suggests that this scale was introduced into the United States on exotic pines planted at the New York World's Fair in 1939 (McClure 1977a). This scale sucks sap from the phloem parenchyma of the three-year-old wood, which causes discoloration and dieback of branches and tree death within two to five years. In China, where it was probably also introduced (McClure 1983c), *M. matsumurae* has caused extensive injury to *Pinus densiflora* Siebold, *P. thunbergiana* Franco, and *P. tabulaeformis* Carriere. In Japan injury from this scale is rare and occurs only on cultivated trees (McClure 1985a).

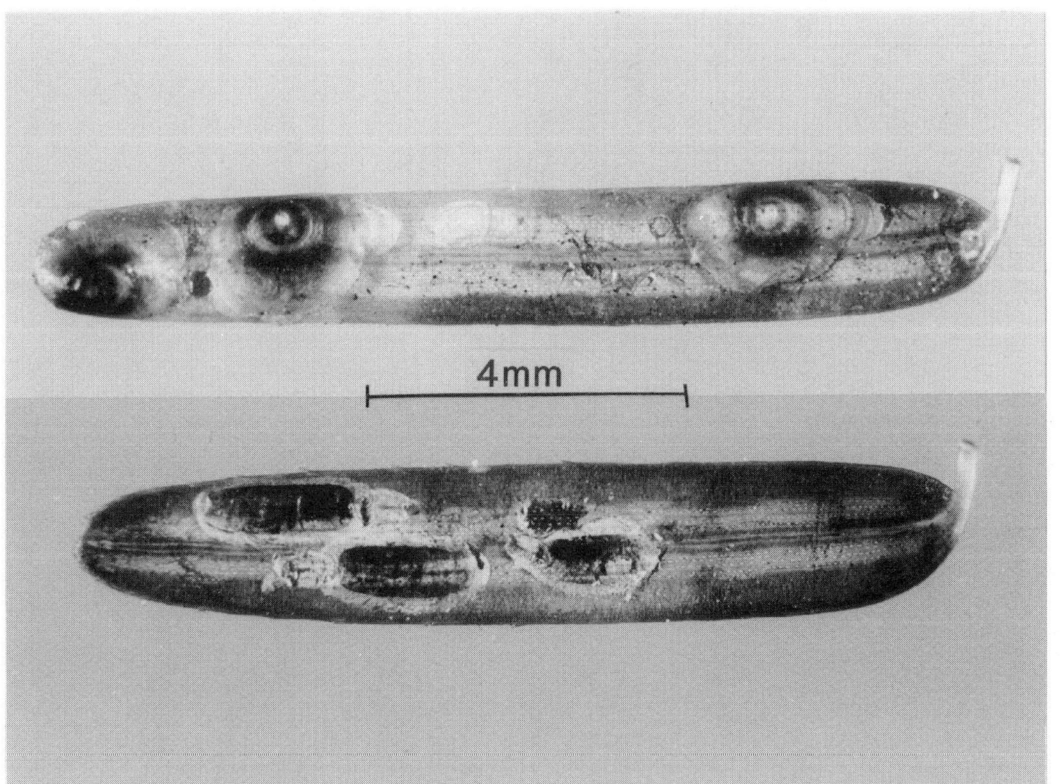


Figure 2.1. Introduced diaspidid scales on needles of *Tsuga canadensis:* nymphs and adults of *Nuculaspis tsugae* (upper) and *Fiorinia externa* (lower)

Figure 2.2. Introduced adelgid on branches of *Tsuga canadensis:* adults with woolly ovisacs of *Adeleges tsugae*

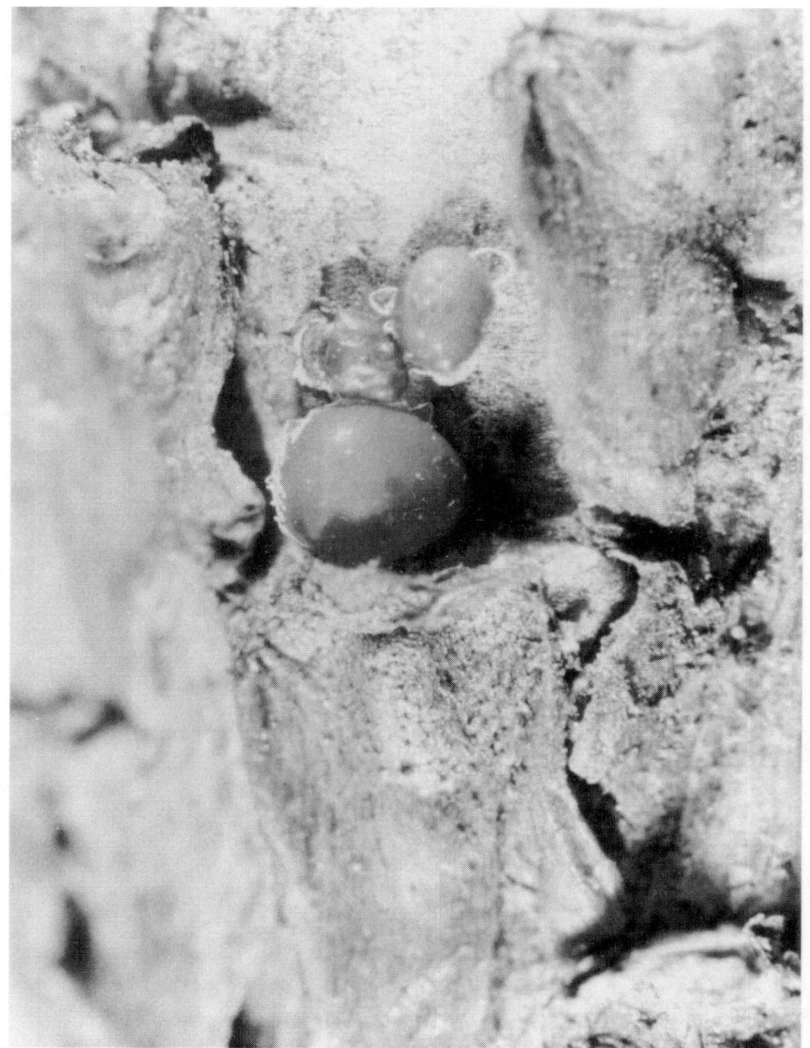

Figure 2.3. Introduced margarodid scale on branch of *Pinus resinosa:* cysts (second instar nymphs) of *Matsucoccus matsumurae*

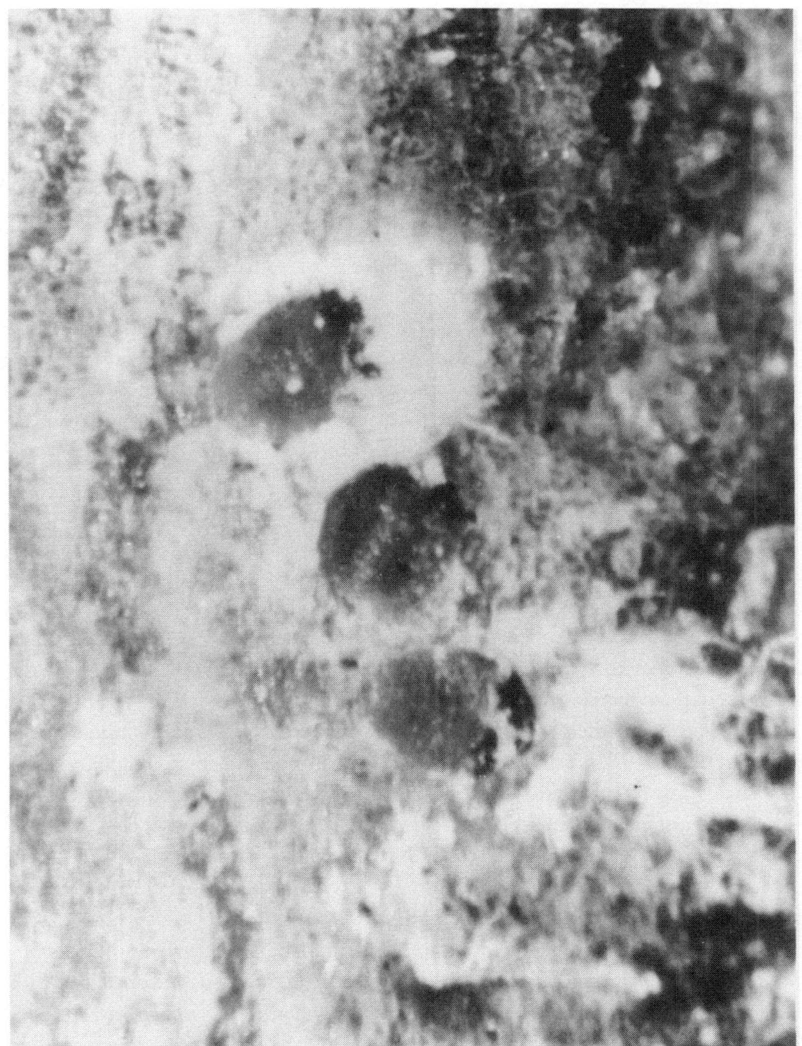

Figure 2.4. Introduced adelgid on branch of *Pinus resinosa:* adults of *Pineus boerneri*

The adelgid *P. boerneri* was probably introduced from Japan on the same pines that carried *M. matsumurae* to North America (McClure 1982). This adelgid also feeds primarily on the three-year-old wood, where it sucks sap from the phloem parenchyma and causes the same type of damage as *M. matsumurae.* Unlike the scale, however, *P. boerneri* also feeds within the needle sheaths. The adelgid has spread much more quickly than the scale because *P. boerneri* completes more generations each year and because its wind-dispersed nymphs occur at a time when prevailing winds are strongest (McClure 1989a).

## Factors That Affect Introduced Populations of Guild Species

None of the predators and parasites that attack the introduced adelgid and scale pests in eastern North America are effective biological control agents. In addition, there is no evidence that forest or ornamental hemlocks and pines are resistant to attack. Therefore, introduced populations of these insects multiply quickly and cause rapid deterioration of the host. The success of these introduced species is determined by their relative abilities to exploit deteriorating host resources. Consequently, density-dependent negative feedback and interspecific competition are common features of their population dynamics.

### *Density-Dependent Feedback*

Density-dependent feedback has a major impact on the population dynamics of exotic guild species on hemlock and pine. A pattern typical of all these insects in introduced habitats is reflected in the population trends of the pine guild species over a seven-year period in nine forests in Connecticut (McClure 1990). Three of the pine stands were pure infestations of *P. boerneri* throughout the period; the other six stands, which had been infested only with *P. boerneri* in 1979, were invaded by *M. matsumurae* between 1981 and 1983 (Fig. 2.5). Combined numbers of adelgids and scales rose steadily at first, but then fell sharply one to two years after at least 50 percent of the crown of each tree had been visibly damaged by these insects. The sharp decline in adelgid and scale abundance was probably due to the lowered nutritional status of the host for nymphs, indicated by discoloration and desiccation of branches and resinosis (McClure 1983b, 1989b).

The relationship between the density and performance of *P. boerneri* and *M. matsumurae* during the period from population increase to decline in other plantations of *P. resinosa* in New England also suggests a pest-induced reduction in host quality (McClure 1983b, 1989b). Initially when densities of adelgids and scales were low and injury to trees was minor, survival and developmental rate of nymphs and fecundity of adults were not related to density. However, as density increased and pines became significantly injured, each of these fitness parameters was negatively correlated with density. Even though adelgid and scale densities decreased sharply after the fourth year, performance continued to decline, indicating that the deterioration of *P. resinosa* as a host was progressive and irreversible.

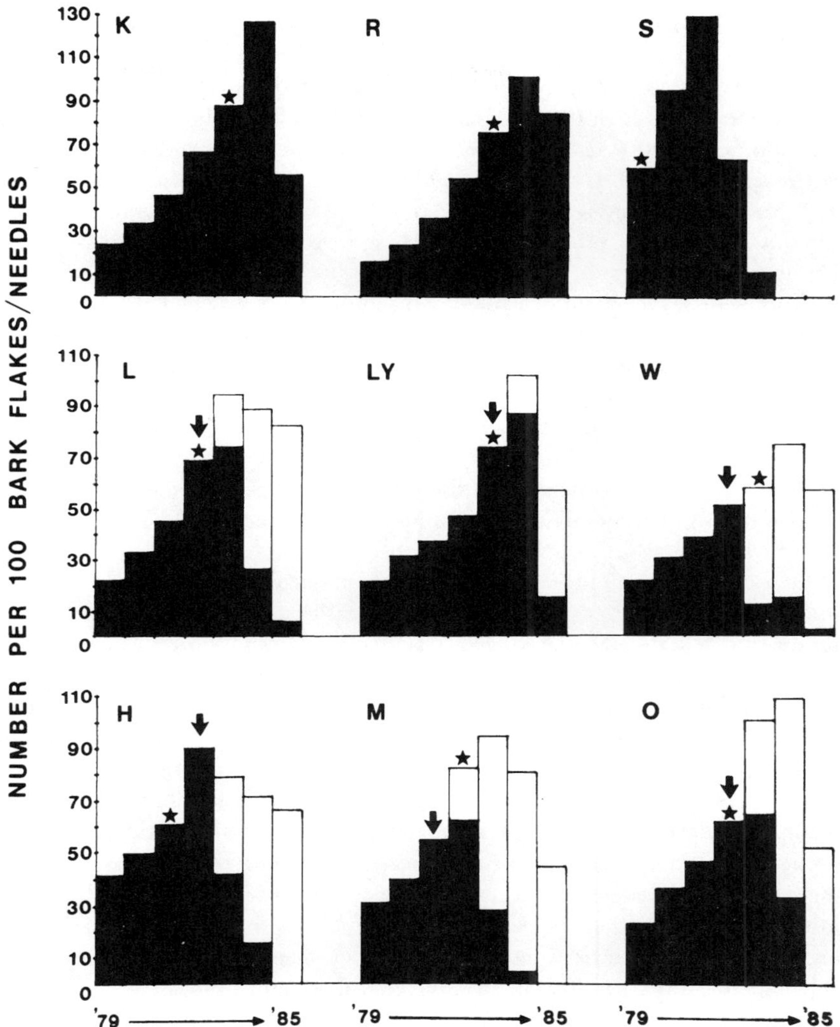


Figure 2.5. Mean number of adelgids and scales on 100 needles and beneath 100 bark flakes of *Pinus resinosa* in nine stands in Connecticut from 1979 through 1985. *Notes:* Within each bar are the proportions of the total insects that were *P. boerneri* (shaded area) and *M. matsumurae* (unshaded area). Stars indicate the first year during which the crowns of all trees were at least 50 percent visibly damaged. Arrows denote the year during which *M. matsumurae* first invaded the stand. Letters designate stands in the towns of Kent (K), Rocky Hill (R), Stonington (S), Litchfield (L), Lyme (LY), Waterford (W), Haddam (H), Middletown (M), and Old Lyme (O). *Source:* From McClure (1990) *Environmental Entomology* 19: 675.

There is substantial evidence that density-dependent feedback governs the dynamics of introduced pest populations on hemlock as well. On heavily infested hemlocks, mortality of *F. externa* nymphs was four times greater, several more days were required to complete nymphal development, and up to 30 percent fewer eggs were produced per adult than on sparsely infested trees (McClure 1979). Further indication that density adversely affects hemlock scale reproductive rates was seen when scale populations resurged rapidly following pesticide spraying (McClure 1977b). Scales that survived on sprayed trees had significantly higher fecundity than did those on controls, probably because of improved host quality following reduced herbivore pressure.

Studies in several hemlock forests in Connecticut revealed that the performance of *A. tsugae* was also adversely affected by density-dependent feedback (McClure 1991). The presence of this adelgid, even in low densities, inhibited production of new growth, which in turn caused high nymphal mortality in subsequent generations. Density also had a profound impact on the performance of the current generation of *A. tsugae* (Fig. 2.6). Mortality of nymphs was strongly correlated with nymph density on forty forest hemlocks and ranged from about 10 percent on sparsely infested trees to nearly 90 percent on the most heavily infested trees (Fig. 2.6).

Of even greater significance to the population dynamics of this adelgid in introduced habitats, however, was the impact of density and host deterioration on the production of sexuparae, the winged stage that migrates to spruce. The life cycle of *A. tsugae* in both Asia and North America includes a wingless generation that remains on hemlock and a winged generation that must feed on and complete its development on *Picea* spp. (McClure 1989c). All of the native and common exotic *Picea* spp. that occur in the northeastern United States are unsuitable hosts for the sexuparae of *A. tsugae* (McClure 1987a). Therefore, the portion of the adelgid population that develops into winged sexuparae each summer dies. As adelgid densities on hemlock increased, an increasingly greater proportion of the nymphs developing in summer became winged sexuparae (Fig. 2.6). For example, on the sparsely infested trees, only about 14 percent of the population became sexuparae, whereas on the heavily infested trees, nearly 90 percent of the population became sexuparae and dispersed from hemlock in an unsuccessful attempt to locate a suitable *Picea* host (McClure 1991). Clearly, this density-dependent feedback mechanism has a significant impact on the dynamics of introduced populations of *A. tsugae* in North America.

### *Interspecific Competition*

All the introduced guild species on hemlock and pine have overlapping geographic distributions in the northeastern United States, and two or more species often infest the same trees. However, my extensive studies of both guilds revealed that species compete intensely and that coexistence is transient. Interspecific competition leading

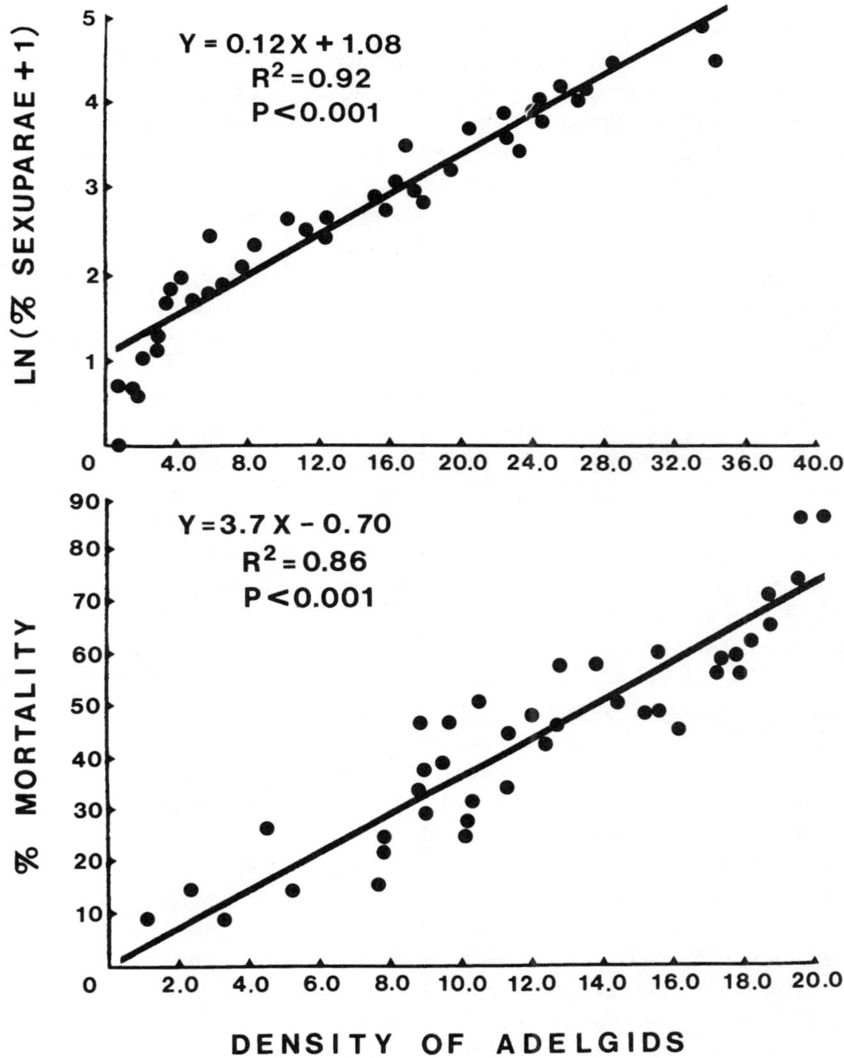


Figure 2.6. Relationship between the number of *A. tsugae* nymphs per $20mm^2$ of branch and percent mortality of nymphs (bottom) and the percent of progeny produced by the sisters that become sexuparae (top) on forty hemlocks in Connecticut forest. *Source:* From McClure (1991) *Environmental Entomology* 20: 261.

to competitive exclusion is common and obviously plays a major role in the population dynamics of these introduced pests.

Experiments revealed that the performance of *P. boerneri* feeding both on branches and on needles of *P. resinosa* was significantly reduced by the presence of *M. matsumurae,* which fed only on branches (McClure 1990). The presence of *P. boerneri* had no significant reciprocal effect on the fitness of *M. matsmurae.* Interspecific competition was a significant limiting factor for *P. boerneri* even though it fed on needles and *M. resinosae* did not. Population trends of these insects during a seven-year period in six cohabited pine stands substantiated the competitive superiority of *M. matsumurae* (Fig. 2.5). In each of six stands that were infested first by *P. boerneri* and subsequently by *M. matsumurae,* adelgid abundance on needles and bark declined sharply following scale invasion. In three of these stands *P. boerneri* was excluded by its competitor within three years.

Studies in the greenhouse and in hemlock forests in Connecticut established that *F. externa* and *N. tsugae* compete for food and space and that *F. externa* is the superior competitor (McClure 1980, 1983a). Comparison of mortality data from solitary and coexisting populations of these scales revealed that *F. externa* had a greater adverse effect on the survival of its competitor than *N. tsugae* had on itself, while *N. tsugae* had a less significant effect on *F. externa* survival than *F. externa* had on itself. This superior competitive ability of *F. externa* resulted from the nutritional advantage gained by colonization two to four weeks earlier than its competitor, when foliar nitrogen and water were more concentrated. Early feeding by *F. externa* not only reduced the amount of foliar nitrogen by the time that *N. tsugae* colonized the needles, but also forced *N. tsugae* to colonize less nutritious older growth where scale success was significantly reduced (McClure 1980).

Because of its competitive superiority, *F. externa* quickly excluded *N. tsugae* from mixed infestations in a field plot and in twenty hemlock forests in Connecticut (McClure 1980). In twelve forests in which *F. externa* outnumbered its competitor in the initial census, *N. tsugae* was eliminated after only three years (Fig 2.7). Even in eight forests in which densities of *N. tsugae* were initially sevenfold higher than those of *F. externa, N. tsugae* was eliminated after only four years (Fig. 2.7). Similar results were obtained in the field plot experiment, wherein the relative abundance of *N. tsugae* on cohabited trees was reduced from 66 percent of the total scales present to less than 1 percent in four years. The host-finding behavior of a parasitoid common to both scales also hastened the decline of *N. tsugae* populations (McClure 1980).

During the past ten years, *A. tsugae* has invaded many hemlock stands previously inhabited by one or both hemlock scales in southern New England. The rapid desiccation and premature drop of hemlock foliage in response to attack by this adelgid have had a major impact on the needle-feeding scales. The immobile stages of these scales, which are firmly attached to the needles for eleven months during the year, are immediately lost to the population when the needles drop prematurely. The more subtle species interactions that occur prior to needle drop involve changes in

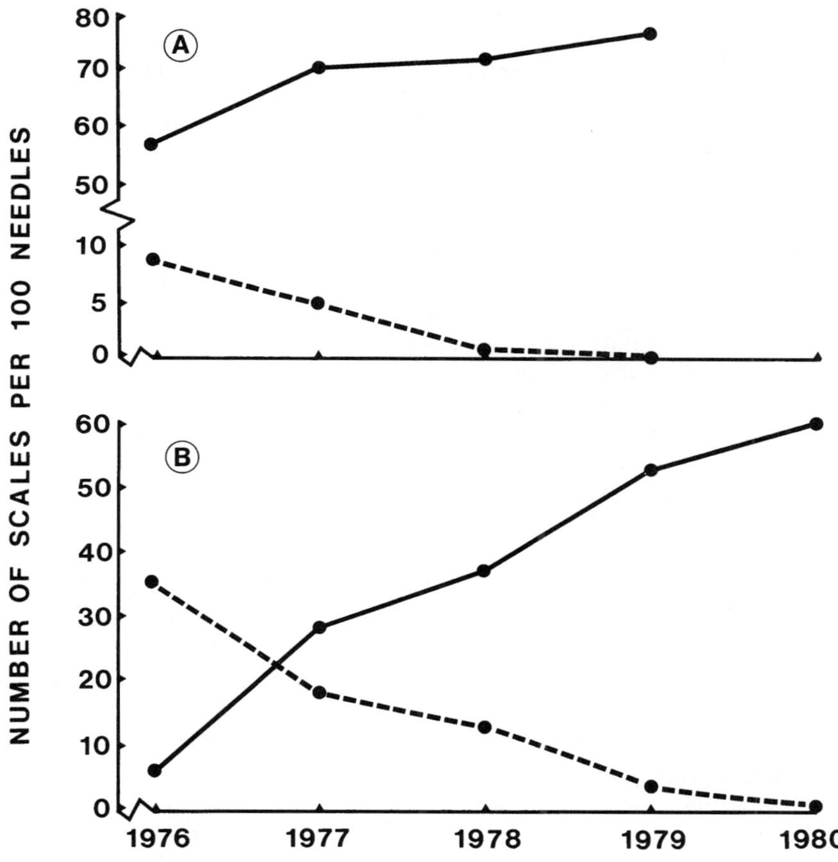


Figure 2.7. Population densities of *F. externa* (solid lines) and *N. tsugae* (dashed lines) in twelve cohabitated hemlock forests in Connecticut in which *F. externa* was initially predominant (A) and in eight forests in which *N. tsugae* was initially predominant (B). *Source:* From McClure 1980.

host nutritional quality and differential responses by species to these changes. As was the case on pine, the level of herbivory and host deterioration needed to bring about a significant reduction in the abundance of the introduced hemlock pests was severe and often lethal. Consequently, the intense competitive interactions among the introduced guild species have been of no benefit to hemlock and pine as these insects have expanded their distributions northward.

### Factors That Affect Native Populations of Guild Species

Studies on the population dynamics of the guild species on hemlock and pine in Asia revealed that in native habitats these species are usually maintained at innocuous densities by the combined influence of host resistance and natural enemies. Experiments were conducted in Japan in pure and hybrid stands of *Pinus* species of the subsection Sylvestres (the group to which all known host species of *M. matsumurae* belong) to determine their relative susceptibility to this scale. Survivorship of scale nymphs and fecundity of adults were signficantly lower on the two native Japanese pines, *P. densiflora* and *P. thunbergiana,* and their interspecific hybrid than on several pure and hybrid pines of related exotic species (McClure 1985a). Maternal parents of Japanese species did not confer resistance upon F1 progeny that resulted from hybrid crosses with susceptible exotic species. Overall trends in scale survivorship in ten pure and hybrid stands indicated a chemical basis for host resistance. However, interspecific variation in bark texture and significant differences in survivorship of nymphs in exposed and protected sites on bark indicated a physical basis for host resistance as well (McClure 1985a).

Life table data gathered from thirty ornamental and forest pine stands in Japan revealed that predators also have an important impact on the dynamics of native scale populations on pine (McClure 1986a). The coccinellid beetle *Harmonia axyridis* Pallas, native to Japan, comprised 84 percent of the total number (n = 3,071) of predators captured. It killed 97 percent of the scales in one heavily infested stand of an exotic pine species, *P. massoniana* Lamb., in less than four weeks. When realistic densities of beetles were caged on infested pines, 81 percent of the small, inconspicuous life stages of the scale and 98 percent of the larger, more conspicuous life stages were consumed. This resulted in a sixty-sevenfold decrease in scale population growth compared with control cages containing no beetles (McClure 1986a). On the basis of these studies, one might predict that this beetle would be an effective biological control agent of *M. matsumurae* in North America. However, as we shall see, the texture of *P. resinosa* bark and the lack of overwintering sites and alternate hosts impose significant constraints to biological control success.

Other studies in Japan revealed that host resistance and natural enemies were important regulatory factors for native populations of adelgids and scales on hemlock as well. Samples taken from thirteen forest and ornamental stands of *T. diversifolia* and *T. sieboldii* throughout Honshu revealed that densities of *F. externa* and *N. tsugae* were always innocuous and low relative to those of introduced populations on *T. canadensis* in North America (McClure 1986b). However, density, survivorship, and fecundity of both scale species were significantly higher on ornamental hemlocks planted outside their natural range in Japan than on naturally occurring montane trees, and significantly higher on *Tsuga* species exotic to Japan than on native Japanese hemlocks (McClure 1985b, 1986b). I infer from these and other results (McClure 1983d) that trees growing outside their natural habitats are less resistant to insect herbivores, presumably due to stress from less adequate growing conditions and to fewer natural enemies.

Results of a field plot study and life table data gathered from thirteen sites in Japan revealed that native hemlock scale populations on ornamental trees have the potential to attain densities as high as those of introduced populations in North America (McClure 1986b). This seldom occurs, however, because of a hymenopteran parasitoid, *Aspidiotiphagus citrinus* Craw, which regularly killed more than 90 percent of both scale species (McClure 1986b). From these studies we might predict that *A. citrinus* would be an effective biological control agent of introduced hemlock scales in North America. However, as we shall see, asynchrony of parasitoid and host life cycles in North America seriously undermines the success of biological control.

Native populations of *A. tsugae* on hemlock were sampled at seventy-six forest and ornamental sites in Japan. The adelgid occurred on *T. diversifolia* and *T. sieboldii* at all thirty-seven forest sites, but always at low innocuous densities, much like the hemlock scales. It also occurred at twenty-four of thirty-four sites where Japanese hemlocks had been planted and in all five cultivated plantings of two North American species, *T. canadensis* and *T. heterophylla.* Injury was observed only at one site on *T. canadensis* where adelgid densities were high (97 ovisacs per 500 cm of branch) and natural enemies were absent. Density, survival of nymphs, and fecundity of *A. tsugae* on Japanese hemlocks were all significantly higher on ornamental trees than on forest trees (Table 2.1), possibly the result of lowered host resistance from less than optimal growing conditions.

At six unusual ornamental sites where growing conditions were obviously quite poor (soil severely compacted, roots injured, irrigation insufficient or excessive, branches heavily sheared), *A. tsugae* abundance on Japanese hemlocks was as high as that observed on *T. canadensis* in North America (greater than 300 ovisacs per 500 cm of branch). Reduced performance of adelgids at these sites (lower survival and fecundity) (Table 2.1) probably indicate density-dependent negative feedback such as that previously discussed for introduced populations of the other guild species (McClure 1991). However, even at these sites where adelgid densities were at outbreak levels, no injury to the trees was observed. This indicates that Japanese hemlocks were tolerant to attack by *A. tsugae.* The intolerance of North American hemlock species to attack by *A. tsugae* (McClure 1992) may seriously constrain our biological control efforts, because exceptionally high mortality from natural enemies may be needed to maintain introduced pest populations at innocuous levels.

In addition to host resistance, natural enemies also played an important role in keeping native populations of *A. tsugae* at low densities in forest habitats. Four species of insect predators including a coccinellid beetle (*Pseudoscymnus* new species), a chrysopid (near *Mallada prasina* [Burm.]), a cecidomyiid fly (*Lestodiplosis* sp.), and an unidentified syrphid collectively killed more than 98 percent of the adelgid eggs on Japanese hemlocks at fifteen forest sites (Table 2.1). At four typical ornamental sites where adelgid density and performance were significantly higher than in natural areas, the insect predators killed 69 percent of available eggs and

Table 2.1. Percent survival of nymphs, number of ovisacs per 500 cm of branch, number of eggs per ovisac, and percent of adelgid eggs killed by insect predators for native populations of *Adelges tsugae* on *Tsuga diversifolia* and *T. sieboldii* in Japan.[a]

| Habitat | *n* | Percent Survival of nymphs | Density | Fecundity | *n* | Percent Predation of eggs |
|---|---|---|---|---|---|---|
| Natural | 39 | 3.6 ± 0.5a | 0.2 ± 0.1a | 207.6 ± 24.7a | 15 | 98.9 ± 12.6a |
| Cultivated | | | | | | |
| Typical | 20 | 87.6 ±11.2b | 21.5 ± 7.3b | 296.4 ± 31.3b | 4 | 69.6 ± 8.1b |
| Atypical | 6 | 46.8 ± 7.4c | 94.8 ± 16.5c | 138.9 ± 16.5c | 6 | 8.6 ± 1.7c |

[a] Numbers (means ± one SD) in each column followed by different letters differ significantly ($p<.005$) by ANOVA.

were less effective control agents (Table 2.1). At six unusual ornamental sites where adelgid populations on Japanese hemlock were at outbreak levels, these insect predators attacked only 8 percent of the available eggs (Table 2.1).

At nine other ornamental sites where *A. tsugae* was abundant, an oribatid mite, *Diapterobates humeralis* (Hermann) was also abundant and destroyed more than 98 percent of all adelgid ovisacs (McClure 1995). Included in this group were two sites where high populations of *A. tsugae* had been controlled on *T. canadensis,* a highly susceptible North American species, before any injury to the trees had occurred. Field and laboratory experiments revealed that the mite was not eating the adelgid, but rather was consuming the woolly material enveloping the eggs (Fig. 2.2). However, as the mite ate, eggs were dislodged from hemlock trees to the forest floor, where they fell victim to desiccation and generalist predators, including ants and spiders. This mite, which typically outnumbered egg masses two to one, dislodged more than 95 percent of the eggs from each ovisac within one week. In this capacity as a "pseudopredator," *D. humeralis* had a major impact on outbreak populations of *A. tsugae* in native cultivated habitats in Japan (McClure 1995).

On the sole basis of my studies at fifteen forest sites where predation was high and adelgids were few, we probably would have predicted that the four insect predators would be effective control agents in North America. However, data from ten ornamental sites in Japan revealed that these same predators were much less effective at high adelgid densities. Results from nine other ornamental sites where high population densities of *A. tsugae* had been ravaged by *D. humeralis* support the prediction that this mite would be a more effective biological control agent than the insect predators in North America, where adelgid populations are nearly always at outbreak levels. Studies are now under way in North America to test these predictions and to evaluate the effectiveness of these natural enemies as biological control agents.

## Biological Control of Introduced Pests in Theory and Practice

Clearly, natural enemies have an important impact on the dynamics of native populations of the guild species on hemlock and pine in Japan. On the basis of my studies there, we could reasonably expect that enemies such as *H. axyridis* on *M. matsumurae, A. citrinus* on *F. externa* and *N. tsugae,* and *D. humeralis* on *A. tsugae* would be effective biological control agents of these introduced pests in North America. Unfortunately, this has not been the case thus far. The reasons for the failure of potentially good biological control agents of exotic pests on hemlock and pine, as well as in other systems, reflect both the complexity of plant-herbivore-enemy interrelationships and the inherent differences between the dynamics of pest populations in native and introduced habitats.

For example, the failure of *H. axyridis* to establish itself as an effective predator of *M. matsumurae* in Connecticut was due to its limited ability to exploit all life stages of the scale and to locate suitable overwintering sites (McClure 1987b). Cage experiments in Connecticut revealed that *H. axyridis* significantly reduced the abundance of *M. resinosae* when the conspicuous life stages were present (Fig. 2.8). However, this beetle was ineffective during other times of the year when scales were predominantly first instar nymphs concealed in the cracks and crevices of the bark (Fig. 2.8). Unable to find scale hosts, beetles would either starve to death, become cannibalistic, or disperse from the stand and presumably perish in the absence of suitable alternative hosts (McClure 1987b). In Japan, *H. axyridis* is an effective predator throughout the year because the relatively untextured bark of Japanese pine species does not offer the refugia for first instar nymphs that the textured bark of *P. resinosa* does (McClure 1985a, 1986a, 1987b). In this case an unlikely feature of the introduced habitat, namely host bark texture, imposed a significant constraint on successful biological control.

Results of two overwintering experiments in Connecticut indicated that the ability of *H. axyridis* to survive winter conditions in its new environment also undermined biological control efforts (McClure 1987b). Less than 10 percent of the adult beetles (n = 762) placed in cages in the field survived from November through March, a period during which weather conditions were normal for Connecticut. In addition, none of the hundreds of marked adults that were released in autumn in a pine stand in Connecticut were observed the following spring.

Equally frustrating as our unsuccessful biological control efforts against *M. matsumurae* has been the failure of the widely established parasitoid *A. citrinus* to control introduced populations of hemlock scales in North America. Throughout Honshu, Japan, the parasitoid and both scales have two generations each year, and the occurrence of adult parasitoids and vulnerable stages of both scale species were highly synchronous (Fig. 2.9). This resulted in high parasitism rates and stable population regulation in Japan (McClure 1986b). In the northeastern United States, however, where the length of the growing season is somewhat shorter than in Honshu, *F. externa* is able to complete only a single generation each year (McClure 1978). Unfortunately,

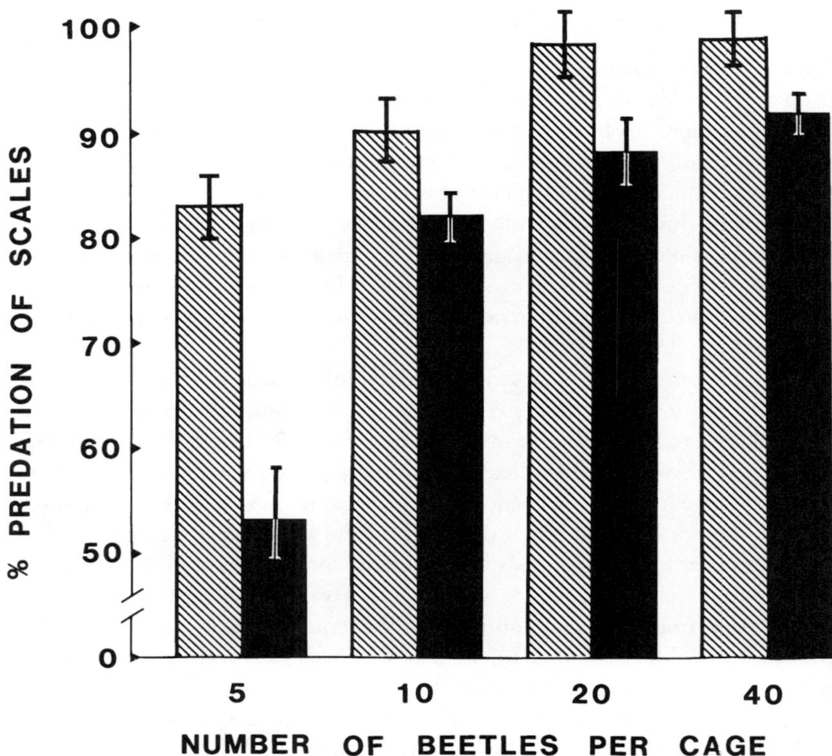


Figure 2.8. Mean (± SE) percent mortality of *M. matsumurae* in cages containing different densities of *H. axyridis* for the periods during which scales were predominantly small, inconspicuous first instar nymphs (solid bars) or larger, more conspicuous ovisacs, cysts, and adults (cross-hatched bars). *Source:* From McClure (1987b) *Environmental Entomology* 16: 226.

the parasitoid completes two generations each year in North America, and, as a result, the occurrence of adult parasitoids and vulnerable stages of hemlock scale were highly asynchronous (Fig. 2.9). This resulted in inconsistent parasitism rates and unregulated populations of introduced hemlock scales (McClure 1986b). In this case slight climatic differences between the native and introduced habitats and their impact on the development of hemlock scales and their primary parasitoid constrained the efforts of a potentially valuable biological control agent.

### Improving Biological Control Strategies

Perhaps the ultimate challenge confronting those of us in biological control is trying to manage populations of introduced pests. We have seen that in a hospitable climate,

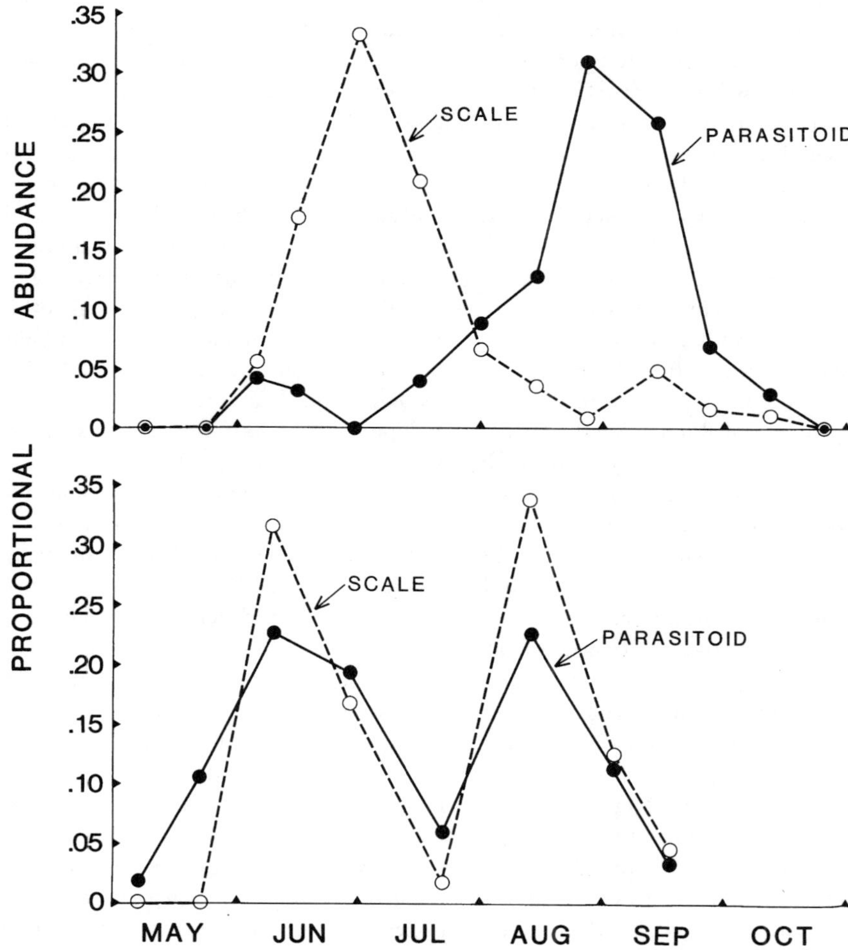


Figure 2.9. Seasonal abundance of adults of the parasitoid *Aspidiotiphagus citrinus* and susceptible host stages of *F. externa* on hemlock in the United States (above) and in Japan (below). *Source:* Redrawn from McClure (1986b) *Ecology* 67: 1411–1421.

on a defenseless host plant, and separated from their cohort of native natural enemies, introduced herbivorous insects often increase rapidly to high, injurious densities that become limited only through density-dependent negative feedback, competition, and host annihilation.

Explorations abroad for natural enemies of introduced pests have often concentrated in natural habitats where populations of native pests usually persist at low innocuous densities under the combined effects of various factors including host defenses and natural enemies. Accordingly, it may be unrealistic to expect natural enemies inhabiting such natural areas to be effective biological control agents in introduced habitats where pest populations are frequently at outbreak levels on their susceptible new host plants. Therefore, explorations abroad for native natural enemies and evaluations of their potential for biological control of introduced pests should be conducted not only in forests, but also in ornamental sites and on susceptible exotic hosts where factors that affect herbivore population dynamics more closely resemble those that characterize introduced habitats. The best biological control agents of introduced pests may not be those that help maintain pest populations at nonoutbreak levels in natural habitats, but rather those that are most responsive to pest outbreaks in cultivated and disturbed habitats. A good example is the case of *A. tsugae,* where four insect predators were effective on forest hemlocks where adelgids were few, but ineffective on heavily infested ornamental hemlocks (Table 2.1). Instead, the mite *D. humeralis,* which was absent from many of the forest sites in Japan, was an abundant and effective "pseudopredator" at most ornamental sites, especially on trees where adelgids reached outbreak levels (Table 2.1).

We have seen how population dynamics studies of native pests in natural and cultivated habitats can identify important enemies and potentially useful biological control agents. However, we have also seen how the effectiveness of these potentially useful enemies as biological control agents in introduced habitats can be undermined by unforeseen circumstances. Examples include the lack of alternate hosts and adequate overwintering sites; host bark texture providing refugia for herbivores from predators (as in the case of pine scale); and life cycle modification resulting in asynchrony between a parasitoid and its host (as in the case of hemlock scale).

I suggest that the success of biological control strategies against introduced pests would be enhanced by first thoroughly understanding the biotic and abiotic features of the introduced habitat and how they affect introduced pest populations, and then concentrating efforts abroad in climates, in habitats, and under circumstances that most closely resemble those of the introduced habitat. In the case of herbivorous insects, this often involves a situation where native pest populations have been stimulated either on susceptible exotic host species or on native host species by factors that lower resistance. Explorations abroad for natural enemies of introduced pests should include, in addition to the homeland, other areas of the world where the same or closely related pest species occur and where climate and other habitat features are similar to the introduced habitat.

As we have seen, studies that compare the population dynamics of pests in natural and cultivated native habitats and in introduced habitats can provide valuable insight into factors that regulate herbivore populations and can identify natural enemies and other factors that may be manipulated to control damaging populations of introduced pests. We must abandon the all too common and unsuccessful "shotgun" approach of blindly releasing any and all available natural enemies with little or no knowledge of their ecological requirements or follow-up appraisal of their effectiveness. Instead we should concentrate our efforts on developing a theory for selecting the most effective biological control agents as well as a system for evaluating the performance of natural enemies after they are introduced. Such an approach should greatly enhance our ongoing efforts for biological control of introduced pests.

## References

Barbosa, P., and J. C. Schultz. 1987. *Insect outbreaks.* San Diego, Calif.: Academic Press.

Bellows, T. S. Jr., R. G. Van Driesche, and J. S. Elkinton. 1992. Life table construction and analysis in the evaluation of natural enemies. *Annual Review of Entomology* 37:587–614.

Caltagirone, L. E. 1981. Landmark examples in classical biological control. *Annual Review of Entomology* 26:213–32.

Clausen, C. P. 1978. *Introduced parasites and predators of arthropod pests and weeds: A world review.* U.S. Department of Agriculture Handbook 480.

DeBach, P. 1974. *Biological control by natural enemies.* Cambridge: Cambridge University Press.

DeBach, P., C. B. Huffaker, and A. W. MacPhee. 1976. Evaluation of the impact of natural enemies. In *Theory and practice of biological control,* ed. C. B. Huffaker and P. S. Messenger, 225–85. New York: Academic Press.

Delucchi, V., D. Rosen, and E. I. Schlinger. 1976. Relationship of systematics to biological control. In *Theory and practice of biological control,* ed. C. B. Huffaker and P. S. Messenger, 81–91. New York: Academic Press.

Laing, J. E., and J. Hamai. 1976. Biological control of insect pests and weeds by imported parasites, predators, and pathogens. In *Theory and practice of biological control,* ed. C. B. Huffaker and P. S. Messenger, 685–743. New York: Academic Press.

Luck, R. F., B. M. Shepard, and P. E. Kenmore. 1988. Experimental methods for evaluating arthropod natural enemies. *Annual Review of Entomology* 33:367–91.

McClure, M. S. 1977a. Population dynamics of the red pine scale, *Matsucoccus resinosae* (Homoptera: Margarodidae): The influence of resinosis. *Environmental Entomology* 6:789–95.

______. 1977b. Resurgence of the scale, *Fiorinia externa* (Homoptera: Diaspididae), on hemlock following insecticide application. *Environmental Entomology* 6:480–84.

______. 1978. Seasonal development of *Fiorinia externa, Tsugaspidiotus tsugae* (Homoptera: Diaspididae), and their parasite, *Aspidiotiphagus citrinus* (Hymenoptera: Aphelinidae): Importance of parasite-host synchronism to the population dynamics of two scale pests of hemlock. *Environmental Entomology* 7:863–70.

______. 1979. Self-regulation in populations of the elongate hemlock scale, *Fiorinia externa* (Homoptera: Diaspididae). *Oecologia* 39:25–36.

______. 1980. Competition between exotic species: Scale insects on hemlock. *Ecology* 61:1391–401.

______. 1982. Distribution and damage of two *Pineus* species (Homoptera: Adelgidae) on red pine in New England. *Annals of the Entomological Society of America* 75:150–57.

______. 1983a. Competition between herbivores and increased resource heterogeneity. In *Variable plants and herbivores in natural and managed systems,* ed. R. F. Denno and M. S. McClure, 125–53. New York: Academic Press.

______. 1983b. Population dynamics of a pernicious parasite: Density-dependent vitality of red pine scale. *Ecology* 64:710–18.

———. 1983c. Temperature and host availability affect the distribution of *Matsucoccus matsumurae* (Kuwana) (Homoptera: Margarodidae) in Asia and North America. *Annals of the Entomological Society of America* 76:761–65.

———. 1983d. Reproduction and adaptation of exotic hemlock scales (Homoptera: Diaspididae) on their new and native hosts. *Environmental Entomology* 12:811–15.

———. 1985a. Susceptibility of pure and hybrid stands of *Pinus* to attack by *Matsucoccus matsumurae* in Japan (Homoptera: Coccoidea: Margarodidae). *Environmental Entomology* 14:535–38.

———. 1985b. Patterns of abundance, survivorship, and fecundity of *Nuculaspis tsugae* (Homoptera: Diaspididae) on *Tsuga* species in Japan in relation to elevation. *Environmental Entomology* 14:413–15.

———. 1986a. Role of predators in regulation of endemic populations of *Matsucoccus matsumurae* (Homoptera: Margarodidae) in Japan. *Environmental Entomology* 15:976–83.

———. 1986b. Population dynamics of Japanese hemlock scales: A comparison of endemic and exotic communities. *Ecology* 67:1411–21.

———. 1987a. Biology and control of hemlock woolly adelgid. *Bulletin Connecticut Agriculture Experiment Station* 851.

———. 1987b. Potential of the Asian predator, *Harmonia axyridis* Pallas (Coleoptera: Coccinellidae) to control *Matsucoccus resinosae* Bean and Godwin (Homoptera: Margarodidae) in the United States. *Environmental Entomology* 16:224–30.

———. 1989a. Importance of weather to the distribution and abundance of introduced forest insects. *Agricultural Forest Meteorology* 47:291–302.

———. 1989b. Biology, population trends, and damage of *Pineus boerneri* and *P. coloradensis* (Homoptera: Adelgidae) on red pine. *Environmental Entomology* 18:1066–73.

———. 1989c. Evidence of a polymorphic life cycle in the hemlock woolly adelgid, *Adelges tsugae* Annand (Homoptera: Adelgidae). *Annals of the Entomological Society of America* 82:50–54.

———. 1990. Cohabitation and host species effects on the population growth of *Matsucoccus resinosae* (Homoptera: Margarodidae) and *Pineus boerneri* (Homoptera: Adelgidae) on red pine. *Environmental Entomology* 19:672–76.

———. 1991. Density-dependent feedback and population cycles in *Adelges tsugae* (Homoptera: Adelgidae) on *Tsuga canadensis. Environmental Entomology* 20:258–64.

———. 1992. Hemlock woolly adelgid. *American Nurseryman* 175:82–89.

———. 1995. *Diapterobates humeralis* (Hermann) (Oribatei: Ceratozetidae): An effective control agent of hemlock woolly adelgid (Homoptera: Adelgidae) in Japan. *Environmental Entomology* 24:1207–15.

Pschorn-Walcher, H. 1977. Biological control of forest insects. *Annual Review of Entomology* 22:1–22.

Turnock, W. J., K. L. Taylor, D. Schroeder, and D. L. Dahlsten. 1976. Biological control of pests of coniferous forests. In *Theory and practice of biological control,* ed. C. B. Huffaker and P. S. Messenger, 289–311. New York: Academic Press.

Zwölfer, H., M. A. Ghani, and V. P. Rao. 1976. Foreign exploration and importation of natural enemies. In *Theory and practice of biological control,* ed. C. B. Huffaker and P. S. Messenger, 189–207. New York: Academic Press.

# 3

# Ecological Approaches for Biological Control of the Aquatic Weed Eurasian Watermilfoil: Resource and Interference Competition, Exotic and Endemic Herbivores and Pathogens

*Sallie P. Sheldon*

## Scope of the Eurasian Watermilfoil Problem

The accidental or intentional introduction of exotic plants into aquatic systems, as into terrestrial systems, has resulted in the rapid expansion of populations of exotic species and a reduction in abundance and species richness of native aquatic macrophytes (e.g., Aiken et al. 1979; Carpenter 1980; Arlington and Mitchell 1986; Room 1990). Such changes in the aquatic macrophyte community affect both littoral communities and human uses of the waters. Communities of native aquatic macrophytes significantly influence lakewide processes such as rates of sedimentation and succession, thermal structure, and dissolved oxygen concentration (Carpenter and Lodge 1986). Mixed native macrophyte beds support a diverse community of invertebrates (Campbell and Clark 1983; Cyr and Downing 1988; Miller et al. 1992; Morrow et al. 1992). Invertebrates use aquatic macrophytes as structure: some eat the epiphytes that grow on leaf surfaces (Bronmark 1989), and some feed on the plants directly (Lodge and Lorman 1987; Lodge 1991; Newman 1991). Finally, aquatic macrophyte communities are important to fish both for supporting invertebrate prey and for refuge (Crowder and Cooper 1982; Stein and Savino 1982; Nichols and Shaw 1986).

In contrast, the extensive, nearly monospecific stands of introduced macrophytes often support a low diversity of aquatic plants (e.g., Nichols and Shaw 1986; Madsen et al. 1991) and fewer macroinvertebrates (Keast 1984). From an economic standpoint, dense stands of nuisance macrophytes restrict recreational uses of the water such as fishing, boating, and swimming and also influence commercial navigation, flood control, power generation from hydroelectric dams, water supplies, and regional development (Henderson 1992). In addition, rafts of aquatic plants can support large populations of noxious insects such as mosquitoes (Cooney 1990).

The most widespread nuisance aquatic weed in North America is Eurasian watermilfoil, *Myriophyllum spicatum* L. (Haloragaceae) (Cofrancesco 1993). Eurasian watermilfoil is a rooted aquatic vascular plant with a long stem and many finely divided leaves. The plant is entirely below the water surface except the inflorescences, which are emergent. The original range of *M. spicatum* was Europe, Asia, and Northern Africa (Couch and Nelson 1986). Eurasian watermilfoil has spread throughout North American water bodies since its accidental introduction (Aiken et al. 1979; Couch and Nelson 1986). While there are reports of Eurasian watermilfoil in North America in the 1880s (Reed 1977), the earliest confirmed record was in 1942 in the District of Columbia (Couch and Nelson 1986). By the end of the 1940s Eurasian watermilfoil was found in California and Arizona; by the end of the 1980s it had spread to thirty-three states and three Canadian provinces (Couch and Nelson 1986). As of 1993, Eurasian watermilfoil is now found in forty states and three provinces.

The source of the original population of Eurasian watermilfoil is unclear, but at least in some cases it was introduced by people placing aquarium plants in bodies of water (Couch and Nelson 1986). Until recently, Eurasian watermilfoil was intentionally grown for the aquarium trade (Couch and Nelson 1986). It is now illegal to possess or transport Eurasian watermilfoil in the United States.

Eurasian watermilfoil reproduces primarily by vegetative means, both stoleniforously and by fragmentation (Aiken et al. 1979; Madsen et al. 1988); its seeds are also viable (Madsen and Boylen 1989). Smith (1971) estimated that a single plant can produce 250 million ramets by repeated fragmentation. Once Eurasian watermilfoil is established in an area, it can be easily transported by boats (on propellers, center boards, and trailers). Life history characteristics of Eurasian watermilfoil are typical of invasive species: rapid propagation, the ability to obtain nutrients both from the sediments and from water, and high photosynthetic efficiency (see Nichols and Shaw 1986 for a review).

Eurasian watermilfoil can withstand a broad range of abiotic conditions. It can grow in oligotrophic to eutrophic conditions, in 0.5 m to 8 m water depth, in sand, in organic substrate, under pH ranging from 5.4 to 10, in brackish water, in northern temperate lakes and in rivers, and in subtropical water bodies (Aiken et al. 1979; Nichols and Shaw 1986; Smith and Barko 1990). Eurasian watermilfoil's ability to grow over such a range of conditions and to reproduce rapidly and extensively has lead to the broadening distribution of this introduced species. In some areas in the continental United States and at least the southern Canadian provinces where Eurasian watermilfoil is not currently found, it seems to be only a matter of time until lakes are infected (P. Chambers, Canadian Ministry of the Environment, and A. Cofrancesco, U.S. Army Corps of Engineers, personal communication). For example, Eurasian watermilfoil populations were established in Wisconsin between 1962 and 1969, and a number of Wisconsin lakes were heavily infested throughout the late 1960s and 1970s (Carpenter 1980). Eurasian watermilfoil wasn't found in Minnesota until 1987, then suddenly, within three years, it spread from a few lakes to more than fifty (H. Krosch, Minnesota Department of Natural Resources, personal communication).

After *M. spicatum* is introduced into a lake, it often becomes the dominant plant (Aiken et al. 1979; Carpenter 1980; Painter and McCabe 1988; Madsen et al. 1991). For example, it was first found in Currituck Sound in North Carolina in 1965; 40 hectares were affected. In 1966 the Eurasian watermilfoil mat covered 3,200 hectares, and by 1974, 32,000 hectares were infested (Spencer and Lekic 1974). Established Eurasian watermilfoil beds have a variety of effects on littoral communities and local economies. Biotically, Eurasian watermilfoil beds are different from mixed native species beds. Instead of a complex, species-rich assemblage of macrophytes of variable height, growth form, and leaf shape resulting in many strata and microhabitats, which in turn support a highly diverse invertebrate community (Cyr and Downing 1988; Lodge et al. 1988), Eurasian watermilfoil beds grow as thick, impenetrable walls of uniform height and leaf form. Plants grow quickly in the spring, and by midsummer Eurasian watermilfoil density in established beds can exceed 300 stems/$m^2$. Once at the water surface, stems spread laterally, forming a canopy. Eurasian watermilfoil beds support a lower abundance and diversity of invertebrates compared to native aquatic plant beds (Dvorak and Best 1982; Keast 1984). While fish species usually use mixed native aquatic macrophyte communities as refuge, they do not use dense (> 250 stem/$m^2$) macrophytes (Savino and Stein 1982; Engel 1987). Fish density is lower in Eurasian watermilfoil beds than in native macrophyte communities (Keast 1984).

In addition to its impact on aquatic plants, invertebrates, and fish, Eurasian watermilfoil affects human use of the waters. Extensive mats of Eurasian watermilfoil preclude use of the water for navigation, recreational boating, swimming, and, to some extent, fishing. Rafts of Eurasian watermilfoil can create habitat for mosquitoes (Smith et al. 1967; Legner and Fischer 1980). There are also economic effects on enterprises related to tourism, for example, decreases in revenues from boat rentals, bait and tackle sales, food establishments, and lodging. Finally, lakeshore property values decrease as a lake becomes infested by Eurasian watermilfoil (Smith et al. 1967).

## Current Controls for Eurasian Watermilfoil

Possibly the best control method for Eurasian watermilfoil control is minimizing introduction to bodies of water not yet infected. Public education campaigns and warnings at all boat launches appear to have decreased the frequency of new infestations each year in Vermont, USA. When pioneer populations are found, hand pulling by scuba divers has eradicated some populations (H. Crosson, Vermont State Department of Environmental Conservation, Waterbury, Vermont, and L. Anderson, USDA, University of California, Davis, personal communication).

Once a Eurasian watermilfoil population is established, a variety of physical, mechanical, and chemical methods are used to control the plant. The most common physical controls are water drawdowns and placement of bottom barriers. Both are somewhat effective (drawdowns Webb 1990; bottom barriers Cooke 1986), but they are not specific to Eurasian watermilfoil, and bottom barriers are expensive, ranging from $5,000 to $12,000 per hectare.

Plant harvesting (using an underwater "lawn mower" that cuts and removes plants up to 2.5 m from the water surface; they are transferred to a barge and then to trucks for disposal) is the most commonly used mechanical control. Weed harvesting can be successful in the short run (Nichols and Cottam 1972; Nichols 1973), but the process increases the number of plant fragments. Recent harvesting increases the number of fragments caught on boats and trailers (personal observation), and the fragments may be blown to previously uncolonized areas of the lake (Newroth 1974).

Herbicides are used to control Eurasian watermilfoil in some states, but aquatic herbicides are illegal in other states. Of the herbicides available, Triclopyr appears to be the most specific to Eurasian watermilfoil (Getsinger et al. 1993).

Physical, mechanical, and chemical control methods can provide short-term reductions in the extent of aquatic weeds, but they must be used repeatedly. Perhaps most importantly, most of these control methods are not specific to a single weed species, and they do not eradicate the target weeds nor provide long-term weed control.

## Diverse Approaches to the Biological Control of Eurasian Watermilfoil

The ecological approaches that have been evaluated for the biological control of Eurasian watermilfoil include resource competition by other native macrophyte species, interference competition (allelopathy), both exotic and native insects, exotic herbivorous fish, and endemic pathogens. Use of exotic pathogens is also under consideration.

### *Competition*

Competing aquatic macrophyte species can be planted to control nuisance species (Yeo and Fisher 1970; van Zon 1974). Also, phytoplankton blooms can be simulated to block the light reaching lake sediments (Surber 1948). To evaluate the potential of resource competition by native macrophytes to control Eurasian watermilfoil, an area dominated by Eurasian watermilfoil was treated with the herbicide

Endothall, then two native macrophyte species were planted in the area (Smart 1992). Two submerged macrophytes, *Vallisneria americana* and *Potamogeton nodosus,* were planted in single species stands and together, and Eurasian watermilfoil invasion into the plots was followed (Smart 1992). Unfortunately, herbivores invaded the system, overwhelming the other variables in the experiment (Smart 1992). Thus, this approach needs more evaluation.

Native aquatic plants have been screened to identify species possibly allelopathic to Eurasian watermilfoil. When Eurasian watermilfoil was grown with extracts from fresh plants of twelve native macrophyte species, three submersed species, *Ceratophyllum demersum, P. nodosus,* and *Vallisneria americana,* inhibited Eurasian watermilfoil increase in length (Jones 1993). Biomass of Eurasian watermilfoil was significantly lower than controls with three species, *Ceratophyllum, Vallisneria,* and *Nymphaea oderata* (Jones 1993).

Use of competition by native species to control Eurasian watermilfoil may prove effective alone or in combination with other techniques. The two studies cited are continuing, generating important results.

### *Exotic Insects*

Many of the investigations into potential controls have focused on insects. Insects are well suited as biological controls for weeds because many are species specific (Huffaker et al. 1984; Strong et al. 1984), and in some cases their intrinsic rate of increase is high enough to be able to track the target plant.

Following the classical biological control methods, potential insect agents of biological control of Eurasian watermilfoil have been investigated in Europe and Asia. Gaevskaya (1969) published an inventory of animals associated with aquatic macrophytes, including a number of species found in association with Eurasian watermilfoil. Lekic (1970) identified a possible natural enemy of Eurasian watermilfoil. Caterpillars of *Parapoynx stratiotata,* a pyralid moth, fed on Eurasian watermilfoil in Yugoslavia (Lekic 1970). In the lab, *P. stratiotata* caterpillars fed on stem tissue, built day refugia on the plant, and made pupal chambers along a plant stem (Lekic 1970). Adults were nonfeeding; after emergence the moths mated, females quickly deposited eggs, and the males and females died within two days after emergence.

The results from a series of feeding experiments with *P. stratiotata* suggested that while the caterpillars did have a significant negative effect on *M. spicatum,* they also feed on another watermilfoil, *Myriophyllum verticillatum* and *Stratiodes aloides* (Lekic 1970). *Parapoynx stratiotata* was successfully reared on plants representing five families (Lekic 1970). In further studies, in addition to *M. spicatum, P. stratiotata* fed on water chestnut (*Trapa natans*), a number of pondweeds (*Potamogeton* spp.), and *Ceratophyllum* (Spencer and Lekic 1974). Habeck (1983) also evaluated use of *P. stratiotata.* He found caterpillars on a variety of species in the field in Italy. In feeding trials, they fed on a wide range of plants, and it was concluded that *P. stra-*

*tiotata* was too broad a generalist to be employed as a control for Eurasian watermilfoil (Habeck 1983).

Spencer and Lekic (1974) summarized the herbivorous insects found on Eurasian watermilfoil in Yugoslavia and Pakistan that might be candidates for agents of biological control of Eurasian watermilfoil. They found (number of species in Yugoslavia [Y], number of species in Pakistan [P]) Lepidoptera: Pyralidae 6Y, 2P; Gelechiidae 1P; Noctuidae 2P; Coleoptera: weevils 6Y, 2P; and Diptera: midges *Cricotopus* spp. in each country.

A chrysomelid *Haemonia appendiculata,* was found associated with declining populations of a pondweed *Potamogeton pectinatus,* and *M. spicatum* beds in France (Grillas 1988). The larvae fed on both plant species and could cause local damage, but Grillas (1988) suggested that it was unlikely that it could control Eurasian watermilfoil.

Herbivorous insects feeding on Eurasian watermilfoil in China and Korea have been collected and in some cases reared (Balciunas 1990, 1991; Buckingham 1992, 1993). A *Phytobius* species (Curculionidae) was found on Eurasian watermilfoil and collected (Buckingham 1992). Another species of weevil, *Eubrychius* sp., was found on Eurasian watermilfoil, but it was also found on another watermilfoil, *Myriophyllum verticillatum,* a species common in North America (Buckingham 1993). *Bagous* new species is a watermilfoil stem-boring weevil that was collected in China (Buckingham 1992) and transported to a quarantine facility in Tallahassee, Florida, USA, and is currently being cultured (Bennett 1993). Some *Bagous* n. sp. $F_2$'s have been reared, although the number of adults is few (Bennett 1993). Biologists at the Sino-American Biological Control Laboratory at the Chinese Academy of Agricultural Sciences in Beijing, are looking for and culturing insects found on Eurasian watermilfoil and *Hydrilla verticillata,* another submersed weed (Bennett 1993; Buckingham 1993; C. Bennett, personal communication).

Another pyralid, *Acentria nivea* (= *Acentropus niveus*), has been found both in Europe (Gaevskaya 1969; Spencer and Lekic 1974) and in North America (e.g., Batra 1977) on Eurasian watermilfoil. The life history of *A. nivea* is very similar to that described for *P. stratiotata,* including the formation of day refugia and a nonfeeding adult phase (Batra 1977). *Acentria nivea* had a significant negative effect on Eurasian watermilfoil (Batra 1977). Buckingham and Ross (1981) collected *A. nivea* from a native watermilfoil, *M. sibiricum* (= *M. exalbescens*), and carried out a series of feeding trials with native plants. They found that *A. nivea* is not a species-specific feeder (Buckingham and Ross 1981; Buckingham et al. 1981).

### *Endemic Insects*

Most texts suggest going to the native range of a nuisance species to find a potential biological control agent (e.g., Harley and Forno 1992), but finding a native agent of biological control in the exotic range of the pest may be a faster or more cost-effective means of biological control. In North America, Balciunas (1982) inventoried invertebrates associated with Eurasian watermilfoil and found many species of moths, along with weevils, caddisflies (Trichoptera), and chironomids (Diptera). A group of these native or naturalized insects has been investigated. Painter and McCabe (1988) reported that *A. nivea* may have been responsible for a Eurasian watermilfoil decline in a number of Ontario lakes, but others have suggested that *A. nivea* is not sufficiently species-specific to be used as an agent of biological control (Spencer and Lekic 1974; Batra 1977; Buckingham and Ross 1981; Buckingham et al. 1981).

In addition to the pyralid *A. nivea,* found in Europe and North America, other pyralid species, *Parapoynx allionealis* and *P. badiusalis,* have been found feeding on Eurasian watermilfoil in North America (Apperson and Axtell 1981; Balciunas 1982; Spencer and Lekic 1974). There has not been much work on these two *Parapoynx* species. *Parapoynx badiusalis* seems not to be a species-specific feeder; I have found caterpillars feeding and making day refugia on two native submerged macrophytes, *Heteranthera dubia* and *Potamogeton praelongus* (personal observation). There appear to be no ongoing studies investigating use of *Parapoynx* spp. as agents of biological control of Eurasian watermilfoil.

There are a number of midge species in the genus *Cricotopus* (Chironomidae) that feed on Eurasian watermilfoil (Balciunas 1982; Kangasniemi and Oliver 1983; Kangasniemi et al. 1993). In feeding trails *C. myriophylli* can have an extensive negative impact on *M. spicatum* (Kangasniemi et al. 1993). *Cricotopus myriophylli* acts by destroying meristems, which hinders meristematic development and retards shoot elongation. At high chironomid densities (> 500/ $m^2$) *M. spicatum* cannot reach the water surface (Kangasniemi et al. 1993). In areas of the Okanagan Valley in British Columbia, Canada, densities greater than 650 larvae/$m^2$ have been recorded (Kangasniemi et al. 1993). Host range tests show that *C. myriophylli* is quite specific; of the twelve native species tested, the midge fed or built cases only on another watermilfoil (*M. sibiricum*), a pondweed (*Potamogeton natans*), and *Ranunculus aquatilis* (MacRae et al. 1990). Because of the high densities of midges needed and the relative phenologies of *C. myriophylli* and Eurasian watermilfoil, Kangasniemi et al. (1993) suggest mass rearing larvae in the laboratory and releasing them in large numbers in the spring. *C. myriophylli* shows potential as a biological control agent of Eurasian watermilfoil; this research is still in progress (Kangasniemi and Newroth, Ministry of Environment, Victoria, British Columbia, Canada).

A number of weevil species have been found associated with Eurasian watermilfoil in North America: *Phytobius leucogaster* (= *Litodactylus leucogaster*; see Buckingham and Bennett 1981), *Euhrychiopsis lecontei* (Creed and Sheldon 1993), *Bagous bituberous* (e.g., Apperson and Axtell 1981), and *Perenthis vestitus* (e.g., Apperson and Axtell 1981; Balciunas 1982). Of these weevils, most studies have been done with *P. leucogaster* and *E. lecontei.*

*Phytobius leucogaster,* a holarctic weevil, has been found in North America on *M. spicatum* along the northern tier of states, and in California and Georgia (Buckingham and Bennett 1981). *Phytobius* larvae and adults fed on both vegetative plant tissue and inflorescences (Buckingham and Bennett 1981). *Phytobius* caused damage to Eurasian watermilfoil flowers, but because it is likely that most reproduction in Eurasian watermilfoil is vegetative, it is thought that *Phytobius* will not have a negative impact on Eurasian watermilfoil significant enough to control the weed (Buckingham et al. 1981).

The aquatic weevil *Euhrychiopsis lecontei* was found associated with a population of Eurasian watermilfoil that had undergone a decline (Sheldon 1990). Like *Phytobius, E. lecontei* larvae and adults both feed on *M. spicatum* (Creed and Sheldon 1993; Sheldon and Creed 1995). Weevil eggs are found on *M. spicatum,* primarily on the apical meristem or other meristems. Larval feeding destroys meristems, and stem vascular tissue is removed as a result of tunneling; adults eat both leaves and stems (Sheldon and Creed 1995).

In laboratory feeding trials, weevils had a significant negative impact on *M. spicatum* growth (Creed and Sheldon 1993). *Euhrychiopsis lecontei* is a very species-specific grazer. In feeding trials with ten native macrophyte species including the native northern watermilfoil, *M. sibiricum,* it did not have a significant effect on any native plant species (Sheldon and Creed 1995). When placed in enclosures with Eurasian watermilfoil at two lakes that did not have any *E. lecontei,* there was significantly less *M. spicatum* biomass in enclosures with weevils compared to control enclosures or to open water (Sheldon and Creed 1995). The Eurasian watermilfoil decline seen in a northeastern lake in Vermont (Sheldon 1990) is clearly associated with damage by weevils (Creed et al. 1992; Creed and Sheldon 1995). Eurasian watermilfoil populations in nine lakes with high densities of weevils are declining and being replaced by native plant populations.

*Euhrychiopis lecontei* is a native weevil that appears to have increased its host range to include Eurasian watermilfoil. It is likely that northern watermilfoil, *M. sibiricum,* was the original host of this aquatic weevil. In Alberta, Canada, where *M. spicatum* has never been found, *E. lecontei* were found on *M. sibiricum* (Creed and Sheldon 1994). The effects of weevils on northern watermilfoil appear to be different from their effects on Eurasian watermilfoil. We have found *E. lecontei* and weevil damage on the northern watermilfoil in three lakes in Vermont, but the weevils do not appear to have a significant negative effect on northern watermilfoil.

Kangasniemi et al. (1993) looked at the impact of a combination of three Eurasian watermilfoil grazers, *C. myriophylli, E. lecontei,* and a caddisfly (*Triaenodes tarda*), and found that the midge and weevil had greater negative effects on the plants than the caddisfly did.

### *Exotic Herbivorous Fish*

Two exotic herbivorous fishes, grass carp (*Ctenopharyngodon idella*) and *Tilapia* spp., are being used for aquatic plant control (Opuszynski 1972). Grass carp have been introduced for plant control in the southern United States, particularly in Florida and the Tennessee Valley Authority reservoir system. Sterile triploid grass carp had a significant negative impact on the submersed weed *Hydrilla verticillata* (Leslie et al.

1983; Chappelear et al. 1991) and on other aquatic weeds (Webb 1990), but Eurasian watermilfoil is not a preferred plant (Fowler and Robson 1978). *Tilapia* spp. have been introduced into irrigation canals in California for general aquatic plant control. *Tilapia* will eat Eurasian watermilfoil, but only after other preferred native macrophyte species, such as *Potamogeton pectinatus,* are depleted (Legner and Fischer 1980).

### *Endemic Pathogens*

Pathogens of other aquatic weeds have been identified but have not yet been extensively used for aquatic plant control (Zettler and Freeman 1972). Populations of Eurasian watermilfoil have declined at least two times in association with apparent increases in pathogens. In neither case—the "Lake Venice" disease in Maryland (Elser 1969) and "Northwest disease" (Bayley et al. 1968)—could a causative agent be identified. There appears to be no current work on these possible pathogens.

Native fungal pathogens collected from *M. spicatum,* often at sites where plants appeared moribund, have been studied. These include, for example, *Mycoleptodiscus terrestris* (Gunner et al. 1990), *Acremonium curvulum* (Andrews et al. 1982; Smith et al. 1989a), *Aureobasidium pullulans* (Smith et al. 1989a), *Cladosporum herbarum* (Smith et al. 1989a), *Colletrichum gloeosporiodes* (Smith et al. 1989a, 1989b; Kees and Theriot 1991), *Fusarium sporotrichiodes* (Andrews and Hecht 1981), *Macrophomena phaseolina* (Kees and Theriot 1991), and *Paecilomyces* sp. (Smith et al. 1989a). Of these, *M. terrestris* has undergone the most extensive studies.

Under laboratory conditions, *M. terrestris* had a significant negative effect on Eurasian watermilfoil (Smith et al. 1989a; Gunner et al. 1990; Stack 1990; Winfield 1990; Kees and Theriot 1991; Smith and Winfield 1991). In some pond experiments, the results from *M. terrestris* addition led to infection of 90 percent of the Eurasian watermilfoil plants (Stack 1990), although there were no significant negative effects on Eurasian watermilfoil in other open water trials (Smith and Winfield 1991; Shearer 1993). The lack of Eurasian watermilfoil response may have been due to the high temperature in the ponds (Smith and Winfield 1991). In follow-up lab trials, *M. terrestris* was found to be pathogenic at 20°C and 25°C, but did not have an impact on Eurasian watermilfoil at 15°C (Shearer 1992). Shearer (1992) estimated the upper lethal temperature of *M. terrestris* to be 28°C.

In studies with the other fungal pathogens, pathogen addition resulted in plant necrosis (Andrews and Hecht 1981; Andrews et al. 1982; Smith et al. 1989b), chlorosis (Andrews and Hecht 1981; Andrews et al. 1982), and a decrease in plant biomass compared to controls (Smith et al. 1989b), but researchers concluded that the fungi were not sufficiently pathogenic to warrant further development (Andrews and Hecht 1981; Andrews et al. 1982; Smith et al. 1989a, 1989b). In another set of trials with potential Eurasian watermilfoil pathogens, *Macrophomena phaseolina* and *Colletrichum gloeosporiodes* had a negative effect on Eurasian watermilfoil, but again results did not motivate further follow-up research (Kees and Theriot 1991).

Given the difficulties that have been encountered in using North American pathogens to control Eurasian watermilfoil, there is interest in taking the more classical approach of going to the native range of Eurasian watermilfoil and collecting pathogens there (Theriot et al. 1993).

## Biological Control of Other Aquatic Weeds

To date two aquatic weeds, *Salvinia molesta* (= water fern, = Kariba weed) and alligatorweed (*Althernanthera philoxeroides*), have been controlled by the introduction of a number of exotic insect species from the original ranges of the plants. In some locations waterlettuce, *Pistia stratiotes,* has been controlled. Other releases are in progress.

*Salvinia molesta* is a floating plant originally from South America. *Salvinia* was introduced in Sri Lanka, then spread to India, Africa, Southeast Asia, and Australia (Room 1990). Three South American insects—a weevil, a moth, and a grasshopper—were found feeding on *Salvinia.* All three insect species were released in Africa, but they did not control it (Julien 1992). A new species of weevil (*Cryptobagous salviniae*) was found on *Salvinia* in Brazil and was introduced to a number of areas. In Australia the weevil very effectively controlled *Salvinia;* subsequently the weevil has been introduced extensively in Africa and Asia and *Salvinia* has been controlled (Room 1990).

In North America, alligatorweed was the first aquatic weed controlled with biological control agents (Coulsen 1977). Like *Salvinia,* alligatorweed is a floating weed, originally from South America. A series of insects—a flea beetle, a thrips, and a moth larva—has been introduced and the plant controlled (Cofrancesco 1991). In eighteen years, alligatorweed decreased from 40,000 "problem" hectares to less than 400 hectares (Cofrancesco 1988).

In Australia, waterlettuce, *Pistia stratiotes,* has been controlled by a weevil from South America (Harley et al. 1984). In some U.S. locations waterlettuce has been controlled by the South American weevil, and a moth from Thailand was introduced in 1991 (Grodowitz and Freedman 1990; Grodowitz 1991; Dray and Center 1993). Control of water hyacinth (*Eichhornia crassipes*) has been more problematic. Water hyacinth, originally from South America, was intentionally introduced as a decorative plant (Sanders et al. 1985). To date, two weevils, a moth, and a pathogen have been released, but the plant has not yet been effectively controlled (Cofrancesco 1991).

Andres and Bennett (1975) divided aquatic weeds into two categories, submersed weeds and emergent/floating weeds, and suggested that it may be more difficult to find biological control agents for submersed plants. All of the examples of successful biological control of aquatic plants have been for floating plants using insects from the original range of the plants. Both *Hydrilla verticillata* and Eurasian watermilfoil are submerged species that are expanding in distribution throughout North America. The search for biological controls for both species has been extensive (see

Cofrancesco 1993 for a review). For biological control of *H. verticillata,* two weevils and two ephydrid flies from Pakistan and Australia have been identified, tested, cultured, and released (Center and Dray 1990; Cofrancesco 1993). The effects of native pathogens on *H. verticillata* have also been evaluated (Joye 1990; Shearer 1992, 1993). *Hydrilla verticillata* has not yet been controlled, although grass carp feed on the plant and decrease its densities (Leslie et al. 1983).

## Use of Native Insects to Control an Introduced Weed

Classical biological control, which involves collection of "natural enemies" from the original range of the pest (DeBach 1964) has been successfully used to control both weeds and herbivorous insects (for example, see Huffaker 1971; van den Bosch et al. 1982; Wood and Way 1988; Harley and Forno 1992). However, getting to the phase where a classical biological agent can be released is expensive and time consuming; it involves extensive work in quarantine (e.g., Harley and Forno 1992 for procedures). Despite this difference, use of native organisms for biological control is less common (Julien 1992). Many North American macrophytes, fungi, and insect species have been evaluated as potential controls of Eurasian watermilfoil. Two native insects, a fly, *C. myriophylli* (Kangasniemi et al. 1993), and a weevil, *E. lecontei* (Creed and Sheldon 1993), have been identified as possible biological control agents of Eurasian watermilfoil. A potential advantage in using endemic control agents is that there may be a decrease in the potential undesired effects on nontarget species. Presumably, native plants and their herbivores have coevolved. In situations when a nuisance exotic weed is related to endemic plants, researchers may want to consider identification and evaluation of native enemies of the native, related plant. It might be possible to induce a native enemy to use the exotic weed. If so, the native insect, possibly in conjunction with other biological controls, may be an important part of an integrated pest management program.

## References

Aiken, S. G., P. R. Newroth, and I. Wile. 1979. The biology of Canadian weeds. 34. *Myriophyllum spicatum* L. *Canadian Journal of Plant Science* 59:201–15.

Andres, L. A., and F. D. Bennett. 1975. Biological control of aquatic weeds. *Annual Review of Entomology* 20:31–46.

Andrews, J. H., and E. P. Hecht. 1981. Evidence for pathogenicity of *Fusarium sporotrichioides* to Eurasian water milfoil, *Myriophyllum spicatum. Canadian Journal of Botany* 59: 1069–77.

Andrews, J. H., E. P. Hecht, and S. Bashirian. 1982. Association between the fungus *Acremonium curvulum* and Eurasian water milfoil, *Myriophyllum spicatum. Canadian Journal of Botany* 60:1216–21.

Apperson, C. S., and R. C. Axtell. 1981. Arthropods associated with shoreline deposits of Eurasian watermilfoil in the Currituck Sound, North Carolina. *Journal of the Georgia Entomological Society* 16:53–59.

Arlington, A. H., and D. S. Mitchell. 1986. Aquatic invading species. In *Ecology of biological invasions,* ed. R. H. Groves and J. J. Burden, 34–53. Cambridge: Cambridge University Press.

Balciunas, J. K. 1982. *Insects and other macroinvertebrates associated with Eurasian watermilfoil in the United States.* Technical Report A-82-5. Vicksburg, Miss.: U.S. Army Corps of Engineers Waterways Experiment Station.

______. 1990. Biocontrol agents from temperate areas of Asia. In *Proceedings, 24th Annual Meeting, Aquatic Plant Control Research Program,* 25–33. Miscellaneous Paper A-90-3. Vicksburg, Miss.: U.S. Army Corps of Engineers Waterways Experiment Station.

______. 1991. Progress during 1990 in the search for biological control agents in temperate Asia. In *Proceedings, 25th Annual Meeting, Aquatic Plant Control Research Program,* 166–71. Miscellaneous Paper A-91-3. Vicksburg, Miss.: U.S. Army Corps of Engineers Waterways Experiment Station.

Batra, S. W. T. 1977. Bionomics of the aquatic moth *Acentropus niveus,* a potential biological control agent for Eurasian water-milfoil and hydrilla. *Journal of the New York Entomological Society* 85:143–52.

Bayley, S., H. Rabin, and C. H. Southwick. 1968. Recent decline in the distribution and abundance of Eurasian watermilfoil in the Chesapeake Bay. *Chesapeake Science* 9:173–81.

Bennett, C. A. 1993. Quarantine biocontrol operations. In *Proceedings, 27th Annual Meeting, Aquatic Plant Control Research Program,* 88–92. Miscellaneous Paper A-93-2. Vicksburg, Miss.: U.S. Army Corps of Engineers Waterways Experiment Station.

Bronmark, C. 1989. Interactions between epiphytes, macrophytes and freshwater snails: A review. *Journal of Molluscan Studies* 55:299–311.

Buckingham, G. R. 1992. Temperate biocontrol insects for eurasian watermilfoil and Hydrilla. In *Proceedings, 26th Annual Meeting, Aquatic Plant Control Research Program,* 222–25. Miscellaneous Paper A-92-2. Vicksburg, Miss.: U.S. Army Corps of Engineers Waterways Experiment Station.

______. 1993. Foreign research on insect biocontrol agents. In *Proceedings, 27th Annual Meeting, Aquatic Plant Control Research Program,* 85–87. Miscellaneous Paper A-93-2. Vicksburg, Miss.: U.S. Army Corps of Engineers Waterways Experiment Station.

Buckingham, G. R., and C. A. Bennett. 1981. Laboratory biology and behavior of *Litodactylus leucogaster,* a Ceutorhynchine weevil that feeds on water milfoils. *Annals of the Entomological Society of America* 74:451–58.

Buckingham, G. R., C. A. Bennett, and B. M. Ross. 1981. *Investigation of two insect species for control of Eurasian watermilfoil.* Technical Report A-81-4. Vicksburg, Miss.: U.S. Army Corps of Engineers Waterways Experiment Station.

Buckingham, G. R., K. H. Haag, and D. Habeck. 1986. Native insect enemies of aquatic macrophytes. *Aquatics* 8:28–34.

Buckingham, G. R., and B. M. Ross. 1981. Notes on the biology and host specificity of *Acentria nivea = Acentropus niveus. Journal of Aquatic Plant Management* 19:32–36.

Campbell, J. M., and W. J. Clark. 1983. Effects of microhabitat heterogeneity on the spatial dispersion of small plant-associated invertebrates. *Freshwater Invertebrate Biology* 2: 180–85.

Carpenter, S. R. 1980. The decline of *Myriophyllum spicatum* in a eutrophic Wisconsin lake. *Canadian Journal of Botany* 58:527–535.

Carpenter, S. R., and D. M. Lodge. 1986. Effects of submersed macrophytes on ecosystem processes. *Aquatic Botany* 26:341–70.

Center, T. D., and F. A. Dray Jr. 1990. Release, establishment and evaluation of insect biocontrol agents for aquatic plant control. In *Proceedings, 24th Annual Meeting, Aquatic Plant Control Research Program,* 39–49. Miscellaneous Paper A-90-3. Vicksburg, Miss.: U.S. Army Corps of Engineers Waterways Experiment Station.

Chappelear, S. J., J. W. Foltz, K. T. Chavis, J. P. Kirk, and K. J. Kilgore. 1991. Movements and habitat utilization of triploid grass carp in Lake Marion, South Carolina. In *Proceedings, 25th Annual Meeting, Aquatic Plant Control Research Program,* 157–65. Miscellaneous Paper A-91-3. Vicksburg, Miss.: U.S. Army Corps of Engineers Waterways Experiment Station.

Cofrancesco, A. F. Jr. 1988. *Alligatorweed survey of ten southern states.* Miscellaneous Paper A-88-3. Vicksburg, Miss.: U.S. Army Corps of Engineers Waterways Experiment Station.

______. 1991. A history and overview of biocontrol technology. In *Proceedings, 25th Annual Meeting, Aquatic Plant Control Research Program,* 117–23. Miscellaneous Paper A-91-3. Vicksburg, Miss.: U.S. Army Corps of Engineers Waterways Experiment Station.

______. 1993. Biological control. In *Proceedings, 27th Annual Meeting, Aquatic Plant Control Research Program,* 71–78. Miscellaneous Paper A-93-2. Vicksburg, Miss.: U.S. Army Corps of Engineers Waterways Experiment Station.

Cooke, G. D. 1986. Sediment surface covers for macrophyte control. In *Lake and reservoir restoration,* ed. G. D. Cooke, E. B. Welch, S. A. Peterson, and P. R. Neworth, 349–60. Boston, Butterworths.

Cooney, J. C. 1990. Aquatic vegetation-mosquito production, Guntersville Reservoir. In *Proceedings, 24th Annual Meeting, Aquatic Plant Control Research Program,* 150–52. Miscellaneous Paper A-90-3. Vicksburg, Miss.: U.S. Army Corps of Engineers Waterways Experiment Station.

Couch, R., and E. Nelson. 1986. *Myriophyllum spicatum.* In *Proceedings, First International Symposium on Watermilfoil (Myriophyllum spicatum) and Related Haloragaceae Species,* 8–18. Vicksburg, Miss.: Aquatic Plant Management Society.

Coulson, J. R. 1977. Biological control of alligatorweed, 1959–72: A review and evaluation. *USDA Technical Bulletin* 1547.

Creed, R. P. Jr., and S. P. Sheldon. 1993. The effect of feeding by a North American weevil, *Euhrychiopsis lecontei,* on Eurasian watermilfoil (*Myriophyllum spicatum*). *Aquatic Botany* 45:245—56.

______. 1994. Aquatic weevils (Coleoptera: Curculionidae) associated with Northern watermilfoil (*Myriophyllum sibiricum*) in Alberta, Canada. *Entomological News* 105:98–102

______. 1995. Weevils and milfoil: Did a North American herbivore cause the decline of an exotic plant? *Ecological Applications* 5:1113–21.

Creed, R. P. Jr., S. P. Sheldon, and D. M. Cheek. 1992. The effect of herbivore feeding on the buoyancy of Eurasian watermilfoil. *Journal of Aquatic Plant Management* 30:75–76.

Crowder, L. B., and W. E. Cooper. 1982. Habitat structural complexity and the interactions between bluegills and their prey. *Ecology* 63:1802–13.

Cyr, H., and J. A. Downing. 1988. The abundance of phytophilous invertebrates on different species of submerged macrophytes. *Freshwater Biology* 20:365–74.

DeBach, P. 1964. *Biological control by natural enemies.* Cambridge: Cambridge University Press.

Dray, F. A., and T. D. Center. 1993. Biocontrol of waterlettuce (*Pistia stratiotes*): An annual report. In *Proceedings, 27th Annual Meeting, Aquatic Plant Control Research Program,* 118–23. Miscellaneous Paper A-93-2. Vicksburg, Miss.: U.S. Army Corps of Engineers Waterways Experiment Station.

Dvorak, J., and E.P.H. Best. 1982. Macro-invertebrate communities associated with the macrophytes of Lake Vechten: Structural and functional relationships. *Hydrobiologia* 95:115–26.

Elser, H. J. 1969. Observations on the decline of the watermilfoil and other aquatic plants, Maryland 1962–1967. *Hyacinth Control Journal* 8:52–60.

Engel, S. 1987. The impact of submerged macrophytes on largemouth bass and bluegills. *Lake Reservoir Management* 3:227–34.

Fowler, M. C., and T. O. Robson. 1978. The effects of the food preferences and stocking rates of grass carp (*Ctenopharyngodon idella*) on mixed plant communities. *Aquatic Botany* 5:261–76.

Gaevskaya, N. S. 1969. *The Role of Higher Aquatic Plants in the Nutrition of the Animals of Freshwater Basins.* Translated by D. Muller. Yorkshire, England: National Lending Library of Science and Technology.

Getsinger, K. D., E. G. Turner, and J. D. Madsen. 1993. Field evaluation of triclopyr: One year posttreatment. In *Proceedings, 27th Annual Meeting, Aquatic Plant Control Research Program,* 190–93. Miscellaneous Paper A-93-2. Vicksburg, Miss.: U.S. Army Corps of Engineers Waterways Experiment Station.

Grillas, P. 1988. *Haemonia appendiculata* Panzer (Chrysomelidae, Dolanaciinae) and its implication on *Potamogeton pectinatus* L. and *Myriophyllum spicatum* L. beds in the Camargue (France). *Aquatic Botany* 31:347–53.

Grodowitz, M. J. 1991. Biological control of waterlettuce using insects: Past, present and future. In *Proceedings, 25th Annual Meeting of the Aquatic Plant Control Research Program,* 148–56. Miscellaneous Paper A-91-3. Vicksburg, Miss.: U.S. Army Corps of Engineers Waterways Experiment Station.

Grodowitz, M. J., and J. E. Freedman. 1990. Relationships between population dynamics and nutritional profile of waterhyacinth and *Neochetina eichhorniae.* In *Proceedings, 24th Annual Meeting of the Aquatic Plant Control Research Program,* 50–64. Miscellaneous Paper A-90-3. Vicksburg, Miss.: U.S. Army Corps of Engineers Waterways Experiment Station.

Guillarmod, A. J. 1977. *Myriophyllum,* an increasing water weed menace for South Africa. *South African Journal of Science* 73:89–90.

Gunner, H. B., Y. Limpa-amara, B. S. Bouchard, P. J. Weilstein, and M. E. Taylor. 1990. *Microbiological control of Eurasian watermilfoil.* Technical Report A-90-2. Vicksburg, Miss.: U.S. Army Corps of Engineers Waterways Experiment Station.

Habeck, D. H. 1983. The potential of *Parapoynx stratiotata* L. as a biological control agent for Eurasian watermilfoil. *Journal of Aquatic Plant Management* 21:26–29.

Harley, K.L.S., and I. W. Forno. 1992. *Biological control of weeds.* Melbourne, Australia: Inkata.

Harley, K.L.S., I. W. Forno, and D. P. A. Sands. 1984. Biological control of waterlettuce. *Journal of Aquatic Plant Management* 22:101–2.

Henderson, J. E. 1992. Economics and aquatic plant control. In *Proceedings, 26th Annual Meeting, Aquatic Plant Control Research Program,* 6–17. Miscellaneous Paper A-92-2. Vicksburg, Miss.: U.S. Army Corps of Engineers Waterways Experiment Station.

______. 1993. Economic evaluations of aquatic plant control. In *Proceedings, 27th Annual Meeting, Aquatic Plant Control Research Program,* 23–29. Miscellaneous Paper A-93-2. Vicksburg, Miss.: U.S. Army Corps of Engineers Waterways Experiment Station.

Huffaker, C. B. 1971. *Biological control.* New York: Plenum.

Huffaker, C. B., D. L. Dahlsten, D. H. Janzen, and G. G. Kennedy. 1984. Insect influences in the regulation of plant populations and communities. In *Ecological entomology,* ed. C. B. Huffaker and R. L. Rabb, 659–96. New York: Wiley.

Jones, H. L. 1993. The allopathic ability of three species of aquatic plants to inhibit the growth of *Myriophyllum spicatum.* In *Proceedings, 27th Annual Meeting, Aquatic Plant Control Research Program,* 124–27. Miscellaneous Paper A-93-2. Vicksburg, Miss.: U.S. Army Corps of Engineers Waterways Experiment Station.

Joye, G. F. 1990. Biocontrol of *Hydrilla verticillata* with the endemic fungus *Macrophomina phaesolina. Plant Disease* 74:1035.

Julien, M. H. 1992. *Biological control of weeds: A world catalogue of agents and their target weeds.* Oxon, U.K.: CAB International, Institute of Biological Control.

Kangasniemi, B. J. 1983. Observations on herbivorous insects that feed on *Myriophyllum spicatum* in British Columbia. *Proceedings of the Second Annual Conference, North American Lake Management Society,* 214–18.

Kangasniemi, B. J., and D. R. Oliver. 1983. Chironomid (Diptera) associated with *Myriophyllum spicatum* in Okanagan Valley lakes, British Columbia. *Canadian Entomologist* 115: 1545–46.

Kangasniemi, B., H. Speier, and P. Newroth. 1993. Review of Eurasian watermilfoil biocontrol by the milfoil midge. In *Proceedings, 27th Annual Meeting, Aquatic Plant Control Research Program,* 17–22. Miscellaneous Paper A-93-2. Vicksburg, Miss.: U.S. Army Corps of Engineers Waterways Experiment Station.

Keast, A. 1984. The introduced aquatic macrophyte, *Mryiophyllum spicatum,* as habitat for fish and their invertebrate prey. *Canadian Journal of Zoology* 62:1289–303.

Kees, S. L., and E. A. Theriot. 1991. Biotechnical approaches to aquatic plant management: Genetic engineering. In *Proceedings, 25th Annual Meeting, Aquatic Plant Control Research Program,* 138–42. Miscellaneous Paper A-91-3. Vicksburg, Miss.: U.S. Army Corps of Engineers Waterways Experiment Station.

Legner, E. F., and T. W. Fischer. 1980. Impact of *Tilapia zillii* (Gervais) on *Potamogeton pectinatus* L., *Myriophyllum spicatum* var. *exalbescens* Jepson, and mosquito reproduction in lower Colorado Desert irrigation canals. *Acta Oecologica* 1:3–14.

Lekic, M. 1970. Ecology of the aquatic insect species *Parapoynx stratiotata* L. (Pyraustidae, Lepidoptera). *Archiv Za Poljoprivredne Nauke* 23:49–62.

Leslie, A. J., L. E. Nall, and J. M. van Dyke. 1983. Effects of vegetation control by grass carp on selected water-quality variables in four Florida lakes. *Transactions of the Fisheries Society of America* 112:777–87.

Lodge, D. M. 1991. Herbivory on freshwater macrophytes. *Aquatic Botany* 41:195–224.

Lodge, D. M., J. W. Barko, D. Strayer, J. M. Howarth, R. W. Mittlebach, B. Menge, and J. E. Titus. 1988. Spatial heterogeneity and habitat interactions in lake communities. In *Complex interactions in lake communities,* ed. S. R. Carpenter, 181–208. New York: Springer-Verlag.

Lodge, D. M., and J. G. Lorman. 1987. Reductions in submersed macrophyte biomass and species richness by the crayfish *Orconectes rusticus. Canadian Journal of Fisheries and Aquatic Sciences* 44:591–97.

MacRae, I. V., N. N. Winchester, and R. A. Ring. 1990. Feeding activity and host preference of the milfoil midge, *Cricotopus myriophylli* Oliver (Diptera: Chironomidae). *Journal of the Aquatic Plant Management Society* 28:89–92.

Madsen, J. D., and C. W. Boylen. 1989. Eurasian watermilfoil seed ecology from an oligotrophic and eutrophic lake. *Journal of Aquatic Plant Management* 27:119–21.

Madsen, J. D., L. W. Eichler, and C. W. Boylen. 1988. Vegetative spread of Eurasian watermilfoil in Lake George, New York. *Journal of Aquatic Plant Management* 26:47–50.

Madsen, J. D., J. W. Sutherland, J. A. Bloomfield, L. W. Eichler, and C. W. Boylen. 1991. The decline of vegetation under dense Eurasian watermilfoil canopies. *Journal of Aquatic Plant Management* 29:94–99.

Miller, A.C., R. Peets, and D. C. Beckett. 1992. Value of aquatic macrophytes for invertebrates: Studies conducted in Lake Seminole, Florida and Georgia. In *Proceedings, 26th Annual Meeting, Aquatic Plant Control Research Program,* 163–68. Miscellaneous Paper A-92-2. Vicksburg, Miss.: U.S. Army Corps of Engineers Waterways Experiment Station.

Morrow, J. V., J. J. Hoover, and K. J. Killgore. 1992. Invertebrate foods and growth of juvenile large mouth bass in species of aquatic macrophytes. In *Proceedings, 26th Annual Meeting, Aquatic Plant Control Research Program,* 169–76. Miscellaneous Paper A-92-2. Vicksburg, Miss.: U.S. Army Corps of Engineers Waterways Experiment Station.

Newman, R. M. 1991. Herbivory and detritivory on freshwater macrophytes by invertebrates: A review. *Journal of the North American Benthological Association* 10:89–114.

Newroth, P. R. 1974. Studies on aquatic macrophytes. Part IV. A review of the ecology and control of some aquatic macrophyte species. Department of the Environment, Water Resources Service, Victoria, British Columbia. As cited by J. K. Balciunas, 1982. *Insects and other macroinvertebrates associated with Eurasian watermilfoil in the United States.* Technical Report A-82-5. Vicksburg, Miss.: U. S. Army Corps of Engineers Waterways Experiment Station.

Nichols, S. A. 1973. The effects of harvesting aquatic macrophytes on algae. *Transactions of the Wisconsin Academy of Natural Sciences, Arts and Letters* 59:107–19.

Nichols, S. A., and G. Cottam. 1972. Harvesting as a control for aquatic plants. *Water Research Bulletin* 8:1205–10.

Nichols, S. A., and B. H. Shaw. 1986. Ecological life histories of the three aquatic nuisance plants, *Myriophyllum spicatum, Potamogeton crispus* and *Elodea canadensis. Hydrobiologia* 131:3–21.

Opuszynski, K. 1972. Use of phytophagous fish to control aquatic plants. *Aquaculture* 1:67–74.

Painter, D. S., and K. J. McCabe. 1988. Investigation into the disappearance of Eurasian watermilfoil from the Kawartha Lakes. *Journal of Aquatic Plant Management* 26:3–12.

Reed, C. F. 1977. History and distribution of Eurasian watermilfoil in United States and Canada. *Phytologia* 36:417–36.

Room, P. M. 1990. Ecology of a simple plant-herbivore system: Biological control of *Salvinia. Trends in Ecology and Evolution* 5:74–79.

Sanders, D. R. Sr., E. A. Theriot, and P. Perfetti. 1985. *Large scale operations management test (LSOMT) of insects and pathogens for control of water hyacinth in Louisiana. Vol. 1: Results for 1979–1981.* Technical Report A-85-1. Vicksburg, Miss.: U.S. Army Corps of Engineers Waterways Experiment Station.

Savino, J. F., and R. A. Stein. 1982. Predator-prey interaction between largemouth bass and bluegills as influenced by simulated submersed vegetation. *Transactions of the American Fisheries Society* 111:255–66.

Shearer, J. F. 1992. Biological control of aquatic weeds using plant pathogens. In *Proceedings, 26th Annual Meeting, Aquatic Plant Control Research Program,* 253–57. Miscellaneous Paper A-92-2. Vicksburg, Miss.: U.S. Army Corps of Engineers Waterways Experiment Station.

______. 1993. Biocontrol of hydrilla and milfoil using plants. In *Proceedings, 27th Annual Meeting, Aquatic Plant Control Research Program,* 79–81. Miscellaneous Paper A-93-2. Vicksburg, Miss.: U.S. Army Corps of Engineers Waterways Experiment Station.

Sheldon, S. P. 1990. A sudden decline of a Eurasian watermilfoil population in Brownington Pond, Vermont. In *Proceedings, 24th Annual Meeting, Aquatic Plant Control Research Program,* 302–3. Miscellaneous Paper A-90-3. Vicksburg, Miss.: U.S. Army Corps of Engineers Waterways Experiment Station.

Sheldon, S. P., and R. P. Creed. 1995. Use of a native insect as a biological control for an introduced weed. *Ecological Applications* 5:1122–32.

Smart, R. M. 1992. Aquatic plant competition studies in Guntersville Reservoir. In *Proceedings, 26th Annual Meeting, Aquatic Plant Control Research Program,* 148–53. Miscellaneous Paper A-92-2. Vicksburg, Miss.: U.S. Army Corps of Engineers Waterways Experiment Station.

Smith, C. S., and J. W. Barko. 1990. Ecology of Eurasian watermilfoil. *Journal of Aquatic Plant Management* 28:55–64.

Smith, C. S., T. Chand, R. F. Harris, and J. H. Andrews. 1989a. Colonization of submersed aquatic plant, Eurasian water milfoil (*Myriophyllum spicatum*), by fungi under controlled conditions. *Applied and Environmental Microbiology* 55:2326–32.

Smith, C. S., S. J. Slade, J. H. Andrews, and R. F. Harris. 1989b. Pathogenicity of the fungus *Colletrorichum gloeosporiodes* (Penz.) Sacc. to Eurasian watermilfoil (*Myriophyllum spicatum* L.). *Aquatic Botany* 33:1–12.

Smith, C. S., and L. E. Winfield. 1991. Biological control of Eurasian watermilfoil. In *Proceedings, 25th Annual Meeting, Aquatic Plant Control Research Program,* 133–37. Miscellaneous Paper A-91-3. Vicksburg, Miss.: U.S. Army Corps of Engineers Waterways Experiment Station.

Smith, G. E. 1971. Resume of studies and control of Eurasian watermilfoil (*Myriophyllum spicatum* L.) in the Tennessee Valley from 1960 through 1969. *Hyacinth Control Journal* 9:23–25.

Smith, G. E., T. F. Hall, and R. A. Stanley. 1967. Eurasian watermilfoil in the Tennessee Valley. *Weeds* 15:95–98.

Spencer, N. R. 1974. Biological control of Eurasian water milfoil. In *Research Planning Conference on Integrated Systems of Aquatic Plant Control,* 75–83. Vicksburg, Miss.: U. S. Army Corps of Engineers Waterways Experiment Station.

Spencer, N. R., and M. Lekic. 1974. Prospects for biological control of Eurasian watermilfoil. *Weed Science* 2:401–4.

Stack, J. 1990. Development of the fungal agent *Mycoleptodiscus terrestris* for the biological control of Eurasian watermilfoil, *Myriophyllum spicatum* L. In *Proceedings, 24th Annual Meeting, Aquatic Plant Control Research Program,* 85–90. Miscellaneous Paper A-90-3. Vicksburg, Miss.: U.S. Army Corps of Engineers Waterways Experiment Station.

Strong, D. R., J. H. Lawton, and T.R.E. Southwood. 1984. *Insects on plants: Community patterns and mechanisms.* Oxford: Blackwell.

Surber, E. W. 1948. Fertilization of a recreational lake to control submerged plants. *Progressive Fish Culturist* 10:53–58.

Theriot, E. A., A. F. Cofrancesco Jr., and J. F. Shearer. 1993. Pathogen biocontrol for aquatic plant management. In *Proceedings, 27th Annual Meeting, Aquatic Plant Control Research*

*Program,* 82–84. Miscellaneous Paper A-93-2. Vicksburg, Miss.: U.S. Army Corps of Engineers Waterways Experiment Station.

van den Bosch, R., P. S. Messenger, and A. P. Gutierrez. 1982. *An introduction to biological control.* New York: Plenum.

van Zon, J. C. J. 1974. Studies on the biological control of aquatic weeds in the Netherlands. In *Proceedings of the Third International Symposium on Biological Control of Weeds,* ed. A. J. Wapshere, 31–38. Farnham Royal, England: Commonwealth Agricultural Bureaux.

Webb, D. H. 1990. Drawdowns and grass carp for aquatic weed control, Guntersville Reservoir. In *Proceedings, 24th Annual Meeting, Aquatic Plant Control Research Program,* 147–49. Miscellaneous Paper A-90-3. Vicksburg, Miss.: U.S. Army Corps of Engineers Waterways Experiment Station.

Winfield, L. E. 1990. Biological control of Eurasian watermilfoil. In *Proceedings, 24th Annual Meeting, Aquatic Plant Control Research Program,* 79–84. Miscellaneous Paper A-90-3. Vicksburg, Miss.: U.S. Army Corps of Engineers Waterways Experiment Station.

Wood, R. K. S., and M. J. Way. 1988. Biological control of pests, pathogens and weeds: Developments and prospects. *Philosophical Transactions of the Royal Society of London series B* 318:109–376.

Yeo, R. R., and T. W. Fisher. 1970. Progress and potential for biological weed control with fish, plants and snails. *FAO International Conference on Weed Control* 450–63.

Zettler, F. W., and T. E. Freeman. 1972. Plant pathogens as biocontrols of aquatic weeds. *Annual Review of Phytopathology* 10:455–70.

# 4

# Integrating Biological Control in IPM Systems

*David A. Andow*

Integrated pest management (IPM) as originally conceived was a methodology for integrating chemical and biological control (Stern et al. 1959). In its more practical incarnation, however, IPM was developed mainly as a means to optimize the use of chemical insecticides based on information about pest incidence. Natural control can be readily incorporated into this framework, and at least two contrasting approaches have been proposed. Ostlie and Pedigo (1987) propose using historical data on pest mortality to discount observed pest incidence data used to calculate the expected gain from chemical application. Because actual natural control will vary around the historical average, they suggest using a number that is lower than the mean historical value in discounting pest incidence. Nakasuji et al. (1973) propose using real-time estimates of natural control to discount observed pest incidence data. They suggest that natural enemy incidence data be collected concurrent with pest incidence data. Real-time estimates of natural control can be projected from a mechanistic understanding of the natural enemy–host pest relationship. Development and implementation of either idea in IPM has been slow.

For the past two decades there have been increasing calls to integrate biological control and IPM. Much of this has been aimed at incorporating inundative releases of natural enemies and methods to conserve natural enemies into contemporary pest management systems. Biological control of European corn borer in corn using inundative releases of *Trichogramma brassicae* has been implemented successfully in parts of Europe (Raynaud and Crouzet 1985; Bigler and Brunetti 1986; Hassan et al. 1986), and use of *T. dendrolimi* in control of Asian corn borer is common in China (Zhang 1986). Despite these successes, *Trichogramma* is used only sparingly in the United States against European corn borer. Several operational and biological factors constrain the use of *Trichogramma* against corn borer in the United States, and in this chapter I will discuss them from the IPM perspective, and provide an analysis of the technical constraints to their increased use.

## Corn IPM and *Trichogramma*

I use the framework of partial budget analysis for evaluating the potential use of *Trichogramma* as an inundative biological control agent in some U.S. corn commodities. Although corn production in the United States is predominantly oriented to grain production, other corn commodities require substantially different production processes that engender unique pest management systems. Some of the more significant of these include fresh market sweet corn (for wholesale or retail sale), sweet corn for processing (canning, freezing, or frozen whole cob), popcorn, and seed corn (field corn and sweet corn seed). Partial budget analysis allows comparison of pest management alternatives by comparing the relative benefits of alternative management tactics. In general, these relative benefits are calculated as:

$$B = P \times (Y - YL) - C,$$

where $B$ is the relative benefit, $P$ is the price of the yield, $Y$ is the yield, $YL$ is the yield loss, and $C$ is the costs associated with control. In this analysis I evaluate only direct costs. Indirect and external costs are only partly represented as constraints on pest management alternatives in this analysis.

Corn for grain has the lowest per acre gross value, and there are no insect pest standards that influence price prior to storage. Consequently, greater yield losses must be sustained before insect pest control can be economically justified. The key pest control decision maker is the grower, so the partial budget analysis must be taken from the grower's point of view. Relative benefits can be calculated from:

$$B_p = P \times Y \times (1 - YL_{\%} \times ECB_T \times (1 - E)) - C,$$

where $B_p$ are the benefits calculated from a partial budget analysis, $P$ is the price of yield $Y$ to the grower, $YL_{\%}$ is the percent yield loss per European corn borer tunnel per plant, $ECB_T$ is the density of tunnels per plant, $E$ is the efficacy of control, and $C$ is the cost of control. Values for this equation were taken from Bode and Calvin's (1990) yield loss relation for second-generation European corn borer, Ostlie's (1992) insecticide efficacy data, and two historical Minnesota densities of European corn borer, 0.75 and 3 tunnels per plant. An additional very high density of 5 tunnels per plant was included to illustrate the value of *Trichogramma* even under extremely high pest densities. *Trichogramma* efficacy was estimated from six of the published experiments in Prokrym et al. (1992) and Andow et al. (1995). Data from Super Sweet and Pinnacle in Prokrym et al. (1992) were excluded because pest densities were too low. Analysis of these data (Table 4.1) suggests that at current conditions, use of *Trichogramma* would cost a farmer between \$35 and \$45 per acre as a net operating loss. Indeed, even if efficacy of *Trichogramma* were improved substantially to 80 percent and costs were reduced by 40 percent to \$30 per application, use of *Trichogramma* would be a net operating loss. If corn borer densities or

Table 4.1. Field corn

| Cost of Control | Cost Application ($/acre) | Number Applications | Control Cost ($/acre) |
|---|---|---|---|
| Insecticide | 15 | 1 | 15 |
| Trichogramma | 50 | 1 | 50 |
| Nothing | 0 | 0 | 0 |

| | Yield (bu) | Price ($/bu) | Control Cost ($/acre) | Control Efficacy | Yield Loss (%/ECB tunnel) | ECB Density (Tunnel/plant) | Benefit | ECB Density (Tunnel/plant) | Benefit | ECB Density (Tunnel/plant) | Benefit |
|---|---|---|---|---|---|---|---|---|---|---|---|
| Insecticide | 150 | $2.50 | 15 | 0.65 | 0.028 | 0.75 | $357.24 | 3 | $348.98 | 5 | $341.63 |
| Trichogramma | 150 | $2.50 | 50 | 0.5 | 0.028 | 0.75 | $321.06 | 3 | $309.25 | 5 | $298.75 |
| Nothing | 150 | $2.50 | 0 | 0 | 0.028 | 0.75 | $367.13 | 3 | $343.50 | 5 | $322.50 |
| Improvements | | | | | | | | | | | |
| Success | 150 | $2.50 | 50 | 0.8 | 0.028 | 0.75 | $323.43 | 3 | $318.70 | 5 | $314.50 |
| Cost | 150 | $2.50 | 30 | 0.5 | 0.028 | 0.75 | $341.06 | 3 | $329.25 | 5 | $318.75 |
| Combination | 150 | $2.50 | 30 | 0.8 | 0.028 | 0.75 | $343.43 | 3 | $338.70 | 5 | $334.50 |

yield loss per tunnel were higher, then *Trichogramma* would become more competitive compared to doing nothing, but even under these conditions, use of *Trichogramma* would not be as profitable as using insecticides. For field corn, these results suggest that the use of *Trichogramma* is largely constrained by the prevailing economic conditions; technical improvements alone are not likely to faciliate increased use.

Sweet corn requires an entirely different approach to pest management because there are critical standards that the product must meet to obtain certain prices. For example, processors are very concerned about insect larvae ending up in the final canned or frozen product, and sweet corn entering a processing plant must have low enough insect infestations that the risk of product contamination is minimized sufficiently. Much of the processed corn is grown under contract with growers, but insect pest control remains the responsibility of the processor, who is the key decision maker. Although most processors would say that their pest-control goal is to eliminate all insect pests before the product enters the processing plant, this is impractical. Instead, the processing system and the costs associated with changing the system become a function of insect infestation. Prior to the availability of current effective insecticides, dozens of workers would staff inspection lines to remove European corn borer larvae from processed product. With the advent of more effective insecticides, these inspection lines have been largely eliminated, reducing costs and reducing the risk of insect contamination. The standards currently used by one commercial processor require that when 0.33 insect parts are found in 168.2 kg of finished, canned, cut sweet corn (assuming 37 percent recovery of product from 454.5 kg green uncut corn), then increased inspection begins along the processing lines of the plant and processing is slowed down. When 0.66 insect parts are found, then more serious measures, such as halting processing or downgrading product, are initiated (see Prokrym et al. 1992). These actions can incur significant costs to the processor and should be incorporated into the pest management decision-making process. The processing machinery removes insect parts with considerable efficiency during husk removal, washing, cutting, and flotation separation of debris (probably greater than 99.4 percent removal of large European corn borer larvae). Excellent control with insecticides results in about two larvae per 100 ears, and as many as four larvae per 100 ears can be acceptable for processing (Bartels and Hutchison 1996). Larval removal rates and product tolerances depend on the processing system, so the details of decision-making for pest management are likely to vary from processing plant to processing plant. Canned cut corn, canned cream corn, frozen cut corn, and frozen whole cob products have very different tolerances. In addition, different processing equipment used for the same product can alter significantly the removal rates of large larvae. Further research is needed to quantify the relationship between costs and pest density before realistic partial budgets can be constructed for processed sweet corn.

For fresh market sweet corn ears, the USDA (1992) has set three main standards: US Fancy, US #1, and US #2. To meet the standards for US Fancy, all ears must be untrimmed and on average less than 10 percent of the ears in any package (e.g., a 48- or 60-count bag) can have any of the following defects: kernel damage extending more than one quarter of the ear length from the tip of an untrimmed ear, presence of insect frass, or live larvae. Moreover, no package can have such defects in more than 15 percent of the ears. For US #1 with untrimmed ears, the same standards as those for US Fancy apply, except that the frass and larval standards do not apply. For trimmed US #1 ears, all must be at least five inches long, and on average less than 10 percent of these ears can have any kernel damage; no package can have more than 15 percent damaged ears. The standard for US #2 is on average less than 10 percent of the ears in any package with serious damage, and no package with more than 15 percent seriously damaged ears; all ears must be at least four inches long. An ear is seriously damaged if insect damage extends more than one-third of the length of the ear from the tip, if more than four kernels are damaged on other parts of the cob, or if insect damage extends more than one-fourth of an inch from the tip of a trimmed ear. A partial budget analysis of fresh market sweet corn is much more complex than field corn and includes a stochastic element related to the probability that the pest control tactics would succeed in producing yield that meets the USDA standards. Relative expected benefits for a risk-neutral producer are as follows:

$$B_p = P_F \times ((p_F \times Y) - C) + P_{\#1} \times ((p_{\#1} \times Y) - C) + P_{\#2} \times ((p_{\#2} \times Y) - C),$$

where $P_F$, $P_{\#1}$, and $P_{\#2}$ are the prices of US Fancy, US #1, and US #2 corn, respectively, and $p_F$, $p_{\#1}$, and $p_{\#2}$ are the probabilities that the insect control method will produce yield that can be graded US Fancy, US #1, or US #2 corn. All other symbols are defined earlier. The probabilities for the *Trichogramma* and the "do nothing" tactics were derived from the eight experiments published in Prokrym et al. (1992) and Andow et al. (1995) by applying the USDA standards to these results. The price for US Fancy was chosen on the low end of the price range paid to growers by wholesale buyers, and the price for US #1 was estimated from discussions with growers. US #2 corn was assumed to have a price of zero because no wholesaler or retailer was willing to handle it. Although use of *Trichogramma* is vastly superior to doing nothing, it falls far short of the benefits from using insecticides (Table 4.2). Technical improvements to *Trichogramma* could result in greater use. If the probability of producing Fancy grade corn increased to 70 percent, the probability of producing US #2 grade corn reduced to 5 percent, costs were reduced by 40 percent, and the producer were to receive a small price premium for producing sweet corn without insecticides, then use of *Trichogramma* would compete effectively against conventional insecticides. Thus, technical improvements in *Trichogramma* could result in greater use in U.S. wholesale fresh market sweet corn. Analysis of direct market retail (e.g., roadside stands) is likely to produce different results.

Seed corn production is usually grown under contract between growers and a seed corn company. The seed company is responsible for European corn borer control and

Table 4.2. Fresh market sweet corn

| Cost of Control | Cost/ Application ($/acre) | Number Application | Control Cost ($/acre) |
|---|---|---|---|
| Insecticide | $20.00 | 3 | $60.00 |
| Trichogramma | $50.00 | 2 | $100.00 |
| Nothing | $0.00 | 0 | $0.00 |

| | Yield (bags/acre) (60 ct bag) | Price Fancy ($/bag) | Control Cost ($/acre) | Prob ($p_F$) | Price #1 ($/bag) | Prob ($p\#1$) | Price #2 ($/bag) | Prob ($p\#2$) | Expected Benefit |
|---|---|---|---|---|---|---|---|---|---|
| Insecticide | 200 | $5.00 | $60.00 | 0.95 | $2.00 | 0.05 | $0.00 | 0 | $910.00 |
| Trichogramma | 200 | $5.00 | $100.00 | 0.5 | $2.00 | 0.25 | $0.00 | 0.25 | $500.00 |
| Nothing | 200 | $5.00 | $0.00 | 0.25 | $2.00 | 0.25 | $0.00 | 0.5 | $350.00 |
| Improvements | | | | | | | | | |
| Price | 200 | $6.00 | $100.00 | 0.5 | $2.50 | 0.25 | $0.00 | 0.25 | $625.00 |
| Success | 200 | $5.00 | $100.00 | 0.7 | $2.00 | 0.25 | $0.00 | 0.05 | $700.00 |
| Cost | 200 | $5.00 | $60.00 | 0.5 | $2.00 | 0.25 | $0.00 | 0.25 | $540.00 |
| Combination | 200 | $6.00 | $60.00 | 0.7 | $2.50 | 0.25 | $0.00 | 0.05 | $905.00 |

decides the timing and methods of control. Pest control decisions need to balance two critical objectives, the maximization of yield and the minimization of human exposure to toxic insecticides. Throughout much of the growing season, these objectives are not in great conflict, but care must be taken at all times to use insecticides with low human toxicity and low persistence. Seed corn is detasseled by hand over about a week. During this time, human exposure to insecticides must be minimized. Appropriate insecticides can be applied within two to three days of the start of detasseling, but this leaves a window of nearly ten days when insecticide control options are very limited. Moreover, this window of time corresponds to the period when maize is most attractive to ovipositing European corn borers. The use of *Trichogramma* during this period of approximately ten days may be economically competitive because insecticide options are so limited. From the seed company perspective, relative benefits can be calculated as:

$$B_p = P \times (Y - YL \times ECB_L \times (1 - E)) - C,$$

where $YL$ is the yield loss per acre per European corn borer larva per ear, $ECB_L$ is the density of larvae per ear, and all other symbols are defined as above. Insecticide efficacy was calculated for Sevin based on published data. Efficacy of *Trichogramma* was calculated from five of the experiments published in Prokrym et al. (1992) and Andow et al. (1995). Data from Sugar Buns and Pinnacle in Prokrym et al. (1992) were excluded because pest densities were too low, and data from Code 5 in 1990 in Andow et al. (1995) were excluded because the appropriate data on larval abundance in ears were missing. I assumed a yield loss relation of ten seeds per larva per ear, and a density of 1.0 larvae per ear. Under these assumptions, the use of *Trichogramma* appears to be nearly competitive with insecticides, even under current conditions, netting perhaps only $14 an acre less than insecticides (Table 4.3). With improvements in efficacy to 80 percent and a 40 percent reduction in costs, *Trichogramma* could net the seed company $60 an acre more than insecticides (Table 4.3). Technical improvements to *Trichogramma* could result in widespread use in the seed corn industry.

Technical improvements to *Trichogramma* could be aimed at improved efficacy, reduced cost, or both. Using the same partial budgets described earlier, I conducted a sensitivity analysis to relate the proportional increase in benefits to the proportional improvement in efficacy or cost (Table 4.4). Under prevailing economic and technical conditions, improvements in efficacy would result in fivefold to sevenfold greater increases in benefits than reduction in costs for both fresh market sweet corn and seed corn. Thus, improvements in efficacy should be the initial goal of applied research on *Trichogramma.*

## Improving Efficacy

Two approaches have been used for improving efficacy of *Trichogramma.* One has been to screen a variety of strains to select the most effective. Pak (1988) has

Table 4.3. Seed corn

| Cost of Control | Cost/ Application ($/acre) | Number Application | Control Cost ($/acre) |
|---|---|---|---|
| Insecticide | 15 | 2 | 30 |
| Trichogramma | 50 | 1 | 50 |
| Nothing | 0 | 0 | 0 |

| | Yield (bag/acre) | Price ($/bag) | Control Cost | Efficacy | Yield Loss (bags/ ECB/Ear) | ECB Density (Larvae/Ear) | Benefit |
|---|---|---|---|---|---|---|---|
| Insecticide | 55 | $95.00 | 30 | 0.6 | 3.125 | 1 | $5,076.25 |
| Trichogramma | 55 | $95.00 | 50 | 0.62 | 3.125 | 1 | $5,062.19 |
| Nothing | 55 | $95.00 | 0 | 0 | 3.125 | 1 | $4,928.13 |
| Improvements | | | | | | | |
| Success | 55 | $95.00 | 50 | 0.8 | 3.125 | 1 | $5,115.63 |
| Cost | 55 | $95.00 | 30 | 0.62 | 3.125 | 1 | $5,082.19 |
| Combination | 55 | $95.00 | 30 | 0.8 | 3.125 | 1 | $5,135.63 |

Table 4.4. Sensitivity Analysis of Technical Improvements to Trichogramma

Fresh Market Sweet Corn

| Control Cost | Prob ($p_F$) | Prob ($p$#1) | Prob (p#2) | Expected Benefit | Percent Change in Benefit |
|---|---|---|---|---|---|
| $100.00 | 0.5 | 0.25 | 0.25 | $500.00 | |
| $100.00 | 0.55 +10% | 0.25 | 0.2 | $550.00 | 10 |
| $100.00 | 0.6 +20% | 0.25 | 0.15 | $600.00 | 20 |
| $100.00 | 0.5 | 0.25 | 0.25 | $500.00 | |
| $90.00 -10% | 0.5 | 0.25 | 0.25 | $510.00 | 2 |
| $80.00 -20% | 0.5 | 0.25 | 0.25 | $520.00 | 4 |

Seed Corn

| Control Cost | Control Efficacy | Benefit | Percent Change In Benefit |
|---|---|---|---|
| $30.00 | 0.62 | $5,082.19 | |
| $30.00 | 0.682 +10% | $5,100.59 | 0.362049 |
| $30.00 | 0.744 +20% | $5,119.00 | 0.724294 |
| $30.00 | 0.62 | $5,082.19 | |
| $27.00 -10% | 0.62 | $5,085.19 | 0.05903 |
| $24.00 -20% | 0.62 | $5,088.19 | 0.118059 |

systematized this approach to identify strains that would appear to be more useful in cabbage production than those investigated previously. He suggested that tolerance of hot and cold temperatures, host selection, and host finding capacity were important elements to screen for effective strains. Bigler et al. (Chapter 15 in this volume) have conducted further research to suggest better screening methods for strain selection.

Another approach has been to analyze the limitations of a particular strain to identify the characteristics that need improvement. Andow et al. (1995) have evaluated some of the limitations of *Trichogramma nubilale* as a biological control agent against European corn borer in corn. They examined the relationship between release rates and parasitism rates and between parasitism rates and population suppression of European corn borer for a series of experimental releases conducted by Andow and Prokrym (1991), Prokrym et al. (1992), and Andow et al. (1995). At relatively low release rates, ten females per square meter of plant surface area, parasitism ranged from 20 to 90 percent (Figure 4.1a). At higher release rates, more than forty females per square meter of plant surface area, parasitism was consistently

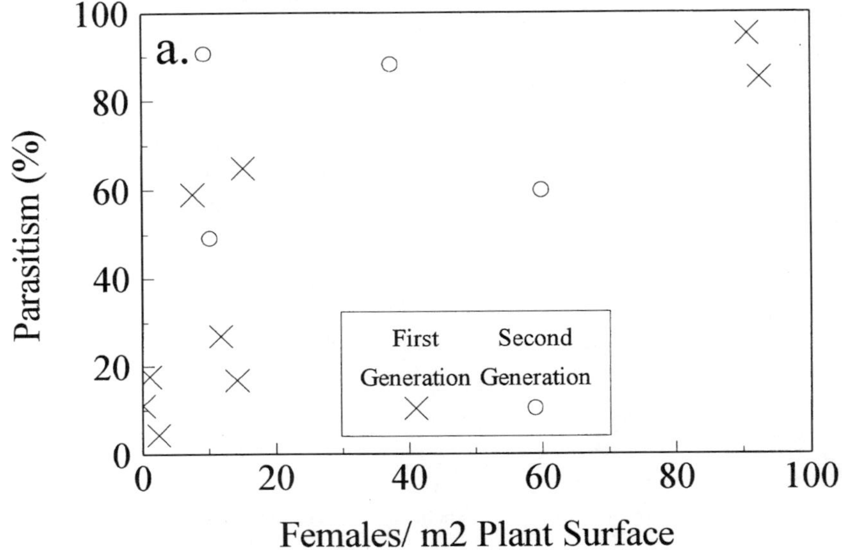


Figure 4.1.a. Relation between release rate (females/m$^2$ of plant surface area in the experimental area) of *Trichogramma nubilale* and egg mass parasitism rate in the field on sweet corn. *Source:* Redrawn from Andow et al. 1995.

high. This release rate, however, is equivalent to approximately 800,000 females per hectare and would be prohibitively expensive in most applications. Larval suppression was very closely related to egg parasitism rates (Figure 4.1b). Thus, the limitations of *T. nubilale* are associated with processes relating release numbers to parasitism rates, including variability in the quality of parasitoids produced, and environmental factors such as plant structure, weather, and availability of alternative food sources.

A closer examination of Figure 4.1a reveals a tendency for parasitism rates to be higher during the second-generation oviposition period of European corn borer than during the first-generation oviposition period. This might suggest either a seasonal effect on parasitism, with early season releases less effective than later season releases, or that releases in shorter, less mature corn are less effective than those in taller, more mature corn. When corn is short and the canopy has not closed, the winds that blow between rows and among plants may inhibit *T. nubilale* movement and reduce efficacy. In addition, the microclimate reaches greater temperature extremes in short corn, which could increase mortality and reduce efficacy.

To test the effects of time of season and plant height, two experiments were conducted during 1992, the first experiment during the first flight period and the second

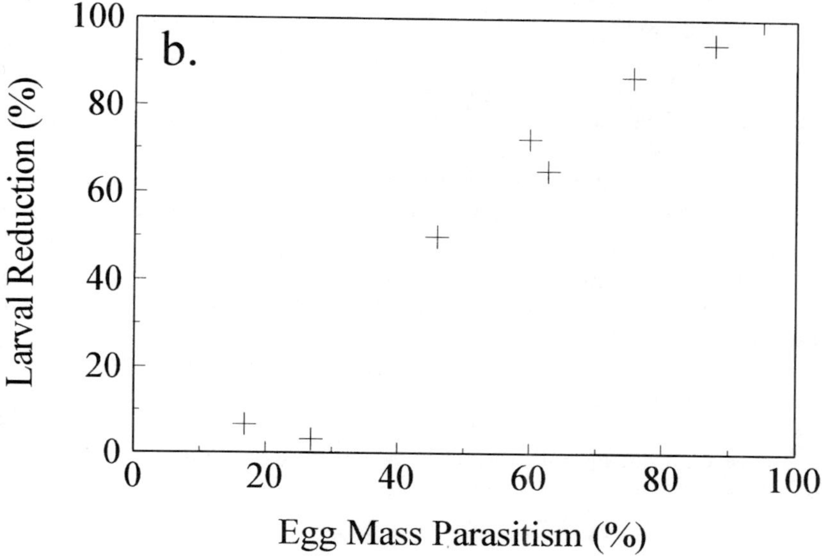


Figure 4.1.b. Relation between egg mass parasitism rate in the field on sweet corn and percent reduction in the density of larvae of European corn borer. *Note:* Data from first- and second-generation European corn borer are combined. *Source:* Redrawn from Andow et al. 1995.

one during the second flight period. Experiment 1 was conducted at the Werth Farm at LeSueur, Minnesota, at a site that did not receive insecticide applications. Three sweet corn varieties were planted to create three growth stages separated by 150 heat units. The growth stages were achieved by starting the first two plantings of each variety in peat pots in the greenhouse on April 21 and April 23. The third growth stage was planted by direct seeding on May 11. Transplanting was done on May 12. Each plot was three rows with nine plants per row, and a replicate set of all the variety by plant growth stage combinations consisted of nine plots. These nine plots were replicated four times in a randomized complete block design. The experiment was surrounded by a planting of sweet corn. Experiment 2 was conducted at the Ahlf Farm in LeSueur, Minnesota, at a site that did not receive insecticide applications. A different variety of sweet corn was planted on June 19 and June 30 to provide two distinct growth stages. Each plot was sixteen rows wide by forty feet long, and each planting was replicated four times. There was a buffer of fifteen feet separating each replicate set.

Mixed-age *T. nubilale* were released in jars for experiment 1 on June 15 and 22 (Andow and Prokrym 1991; Prokrym et al. 1992). Temperatures were below average for much of this period, but warmed up toward the end of the period. Few *T.*

*nubilale* probably emerged during the early part of this experiment, and in total an equivalent of 162,000 and 232,000 females per acre were released in the first and second release respectively. Egg masses of European corn borer were placed on five plants in each plot for two- to three-day intervals following methods described by Andow (1990). Egg masses were collected and the fate of eggs was determined in the laboratory. A similar release of *T. nubilale* was conducted for the second experiment on August 17. Temperatures were below 21°C during the first four days of this experiment, but warmed above 22°C during the end of the experiment. An equivalent of 46,500 females per acre were released. Forty egg masses were placed per plot for two- to three-day intervals, then were collected and examined in the laboratory. Data were analyzed by ANOVA for each date of collection of egg masses for both experiments.

There were no effects of sweet corn variety on parasitism by *T. nubilale* for any of the egg mass baiting periods, so parasitism was averaged across varieties (Fig. 4.2a). In general, parasitism rates were higher in the shorter, later-planted or direct-seeded corn during both the first- and second-generation oviposition periods (Fig. 4.2). Contrary to expectation, factors associated with corn maturity,

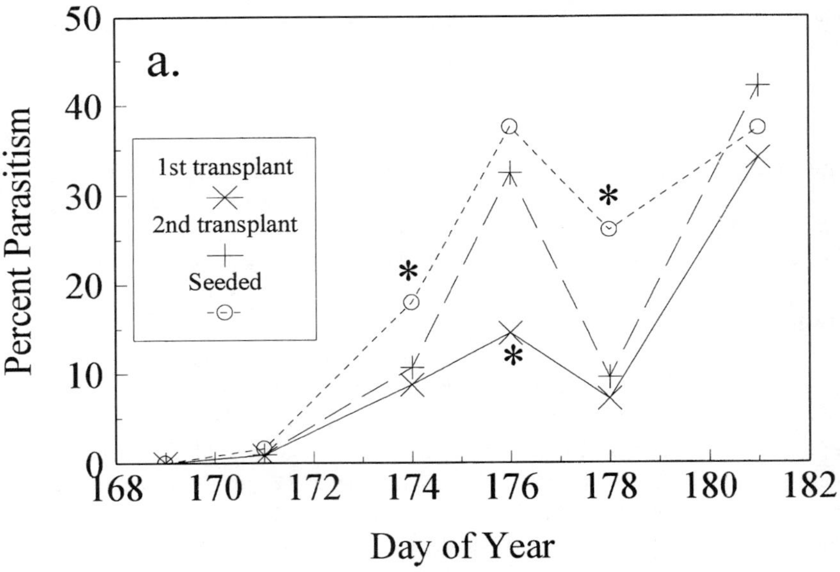


Figure 4.2.a. Percent parasitism of eggs at Werth Farm by *Trichogramma nubilale* during first-generation European corn borer. *Note:* Asterisks indicate a planting date significantly different from others on that day of the year.

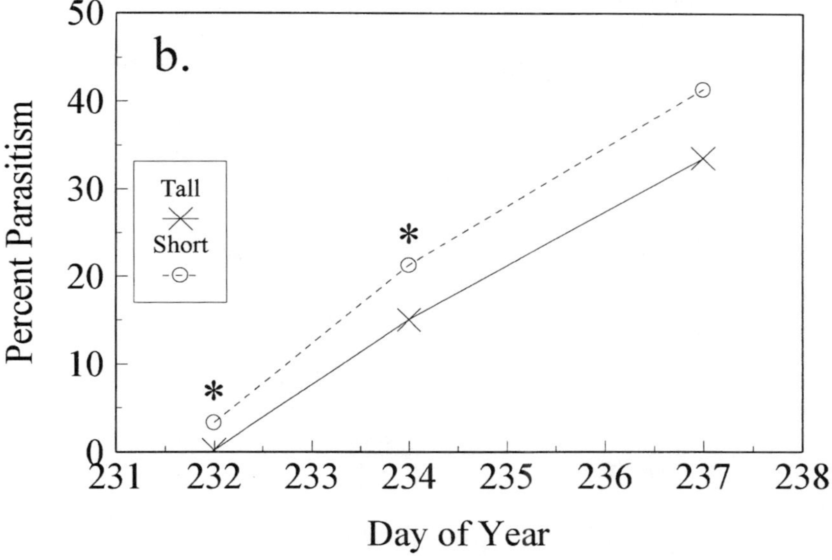


Figure 4.2.b. Percent parasitism of eggs at Ahlf Farm by *Trichogramma nubilale* during second-generation European corn borer. *Note:* Asterisks indicate significant differences between tall and short corn on that day of the year.

such as larger plant size, more complete canopy structure, less wind between rows, and reduced temperature extremes, did not increase efficacy. In addition, the range in observed parasitism was similar in both oviposition periods (Fig. 4.2), which suggests that there were no strong seasonal effects on efficacy. These results indicate that other factors may be affecting parasitism rates more than corn maturity or season. It is possible that the greater plant surface area of the more mature corn reduced parasitism rates (Need and Burbutis 1979).

Temperature is likely to affect parasitism by *T. nubilale* in the field. The relation between average temperature during an egg mass baiting interval and the calculated average daily percent parasitism for the two experiments is shown in Fig. 4.3. Although no clear relation is evident, these data might suggest that at the beginning of a release (the first two baiting intervals for the first-generation release and the first baiting interval for the second-generation release) adequate numbers of *T. nubilale* had not yet emerged. When adequate numbers had emerged, then parasitism might increase with temperature, perhaps about 10 percent for a 4 to 6°C increase. Further experimentation would be essential before quantitative conclusions regarding this relation could be drawn reliably.

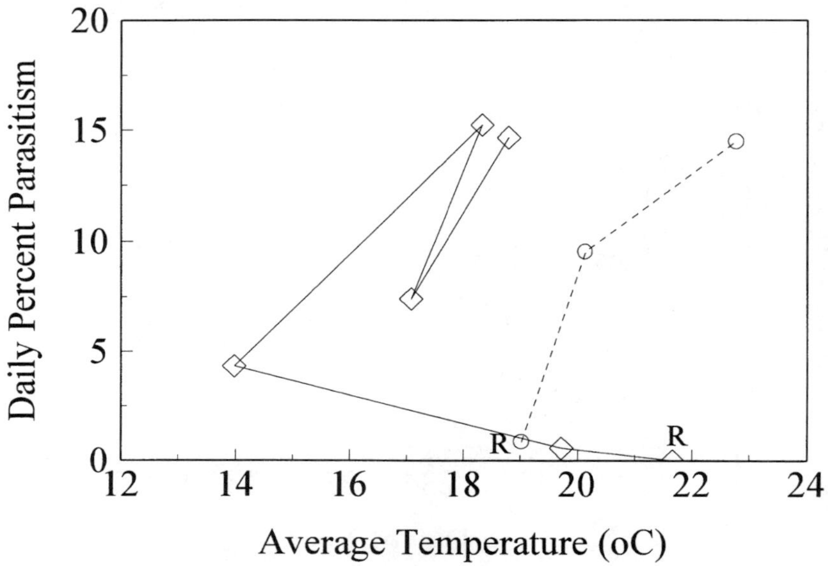


Figure 4.3. Temporal changes in estimated daily egg parasitism by *Trichogramma nubilale* in relation to average temperature. *Notes:* Each point is an average over a two- to three-day period. Consecutive periods are connected by lines. R indicates the first estimate after the field release of *T. nubilale.*

Even at the highest rate of parasitism observed in these experiments (about 15 percent daily parasitism), calculated parasitism over a five-day egg development period would result in only 66 percent parasitism. Thus, it is clear that considerable emphasis should be placed on improving existing strains of *T. nubilale* and screening and improving efficacy of additional species, such as *T. ostriniae* or *T. brassicae,* to identify more effective strains that can be used as inundative controls of European corn borer.

### Summary

Integration of inundative biological control agents in existing IPM systems requires an agent with sufficient efficacy and low enough cost that it will compete in the current pest management system. Using partial budget analysis and preliminary estimates of efficacy and cost, I evaluate the potential use of *Trichogramma* for controlling European corn borer in several corn commodities in the United States. Prevailing economic conditions constrain the use of *T. nubilale* in field corn. Even with substantial technical improvements, such as increases in efficacy and reductions in cost, *T. nubilale* is unlikely to be used in field corn. For fresh market sweet corn, *T. nubilale* is not now competitive, but with relatively modest improvements in effectiveness it would be nearly competitive with current control tactics. Use of *T. nubilale* in seed corn production is nearly competitive even under current conditions. Technical improvements could result in increases in net revenue of more than $60 an acre in this corn commodity. Although any technical improvement would improve the competitiveness of *T. nubilale,* increases in efficacy would result in fivefold to sevenfold greater increases in benefits than decreases in costs.

Two approaches have been used for improving efficacy of *Trichogramma.* Strains of *Trichogramma* can be screened to identify the most effective. Although screening methods are still being developed, current methods can identify better-performing strains for more intensive evaluation. The other approach has been to analyze the limitations of a particular strain to identify characteristics that need improvement. Artificial selection can then be used to improve the strain. Evaluation of data from several field releases reveals that limitations of *T. nubilale* are associated with processes relating release numbers to parasitism rates, including variability in the quality of parasitoids released, and environmental factors such as plant structure, weather, and availability of alternative food sources.

To further characterize some of the environmental components potentially limiting *T. nubilale,* the effects of time in the season, corn variety, and corn height were examined experimentally. Variety and time in the season had no effect on effectiveness of *T. nubilale,* and effectiveness was better in shorter corn than in taller corn, contrary to preliminary expectations. Other factors, such as temperature and quality of the released parasitoids, could have greater effects on effectiveness. Even at the highest observed parasitism rate, the expected reduction in European corn borer populations is low. Improvements to existing or additional strains or species is essential for making *Trichogramma* a competitive control tactic in the United States.

## References

Andow, D. A. 1990. Characterization of predation on egg masses of *Ostrinia nubilalis* (Hübner) *Annals of the Entomological Society of America* 83:482–86.

Andow, D. A., and D. R. Prokrym. 1991. Release density, efficiency and disappearance of *Trichogramma nubilale* for control of European corn borer. *Entomophaga* 36:105–13.

Andow, D. A., G. C. Klacan, D. Bach, and T. C. Leahy. 1995. Limitations of *Trichogramma nubilale* (Hymenoptera: Trichogrammatidae) as an inundative biological control of *Ostrinia nubilalis* (Lepidoptera: Crambidae). *Environmental Entomology* 24:1352–57.

Bartels, D. W., and W. D. Hutchison. 1996. Corn (Sweet), European corn borer. *Arthropod Management Tests* 21:107–8.

Bigler, F., and R. Brunetti. 1986. Biological control of *Ostrinia nubilalis* Hbn. by *Trichogramma maidis* Pint. et Voeg. on corn for seed production in southern Switzerland. *Journal of Applied Entomology* 102:303–8.

Bode, W. M., and D. Calvin. 1990. Yield-loss relationships and economic injury levels for European corn borer (Lepidoptera: Pyralidae) populations infesting Pennsylvania field corn. *Journal of Economic Entomology* 83:1595–603.

Hassan, S. A., E. Stein, K. Dannemann, and W. Reichel. 1986. Massenproduktion und Anwendung von *Trichogramma*: 8. Optimierung des Einsatzes zur Bekampfung des Maiszünslers *Ostrinia nubilalis* Hbn. *Zeitschrift für angewandte Entomologie* 101:508–15.

Nakasuji, F., H. Yamanaka, and K. Kiritani. 1973. Control of tobacco cutworm *Spodoptera litura* F., with polyphagous predators and ultra-low concentration of chlorophenamidine. *Japanese Journal of Applied Entomology and Zoology* 17:171–80.

Need, J. T., and P. P. Burbutis. 1979. Searching efficiency of *Trichogramma nubilale. Environmental Entomology* 8:224–27.

Ostlie, K. R. 1992. Corn (Field): European corn borer. *Insecticide and Acaricide Tests* 17: 215–16.

Ostlie, K. R., and L. P. Pedigo. 1987. Incorporating pest survivorship into economic thresholds. *Bulletin of the Entomological Society of America* 33:98–102.

Pak, G. A. 1988. *Selection of* Trichogramma *for inundative biological control.* Wageningen, The Netherlands: Grafisch bedrijf Ponsen en Looijen.

Prokrym, D. R., D. A. Andow, J. A. Ciborowski, and D. D. Sreenivasam. 1992. Suppression of *Ostrinia nubilalis* by *Trichogramma nubilale* in sweet corn. *Entomologia experimentalis et applicata* 64:73–85.

Raynaud, B., and B. Crouzet. 1985. Mais. La lutte contre la pyrale par les trichogrammes. *Phytoma* 36:17–18.

Stern, V. M., R. F. Smith, R. van den Bosch, and K. S. Hagen. 1959. The integration of chemical and biological control of the spotted alfalfa aphid: The integrated control concept. *Hilgardia* 29:81–101.

United States Department of Agriculture (USDA). 1992. United States standards for grades of sweet corn. USDA Agricultural Marketing Service. 7 CFR 51, Section 51.835–51.845.

Zhang, Z. L. 1986. *Trichogramma* spp. parasitizing the eggs of Asian corn borer *Ostrinia furnacalis* and its efficacy in Beijing suburbs. In *Trichogramma* and other egg parasites, Les Colloques de l'INRA 43:629–32.

# 5

# Foliar Pathogens in Weed Biocontrol: Ecological and Regulatory Constraints

*Roberte M.D. Makowski*

On a global scale, over half of all pesticides used are for weed control. In the United States, herbicides accounted for 65 percent of domestic pesticide sales in 1990 (Standard & Poor's 1991). It is a formidable task for biologicals to displace or complement some of this market in an integrated approach for weed control, even with the increasing public concern over the use of pesticides. The use of pathogens as biocontrol agents for weed control is relatively new in comparison to their use in insect control. Yet with little over two decades of research in this area, many advances have been made in both the classical and inundative strategies (TeBeest 1991). There are now three commercially available mycoherbicides in North America (Templeton 1982b; Bowers 1986; Kenney 1986; Makowski and Mortensen 1992) and approximately five or six others in practical use worldwide (Charudattan 1991). In addition, eight pathogens are in use worldwide as classical control agents (Watson 1991).

One of the first and most noteworthy examples of the classical approach was the successful introduction of *Puccinia chondrillina* Bubak & Syd. into Australia in 1971 for the control of rush skeletonweed (*Chondrilla juncea* L.) (Hasan 1972; 1981; Hasan and Wapshere 1973). It was also the first exotic plant pathogen intentionally introduced into North America in 1976 (Supkoff et al. 1988). Bioherbicide research was initiated in the United States in the late 1960s and early 1970s (Daniel et al. 1973), and in Canada in the mid-1970s (Watson 1975), with practical use not occurring until a decade later. Two bioherbicides have been registered for use in the United States: DeVine in 1981, consisting of chlamydospores of *Phytophthora palmivora* (Butler) Butler for control of strangler vine (*Morrenia odorata* Lindl.) in citrus groves (Burnett et al. 1974; Ridings et al. 1976; Kenney 1986; Ridings 1986); and *Colletotrichum gloeosporioides* (Penz.) Penz. & Sacc. f. sp. *aeschynomene* in 1982 as Collego for control of northern jointvetch (*Aeschynomene virginica* (L.) B.S.P.) in rice and soybean (Daniel et al. 1973; Templeton 1982a; TeBeest and Templeton 1985; Bowers 1986; Smith

1986). In 1992 BioMal, which consists of spores of *C. gloeosporioides* f. sp. *malvae,* was registered as the first bioherbicide in Canada for control of round-leaved mallow (*Malva pusilla* Sm.) in field crops (Mortensen 1988; Makowski and Mortensen 1992; Makowski 1995). Even with the few commercial products brought to market in the past decade, the interest in plant pathogens, especially with the bioherbicide approach, has increased dramatically worldwide. The number of projects using the classical approach has remained static (Evans and Ellison 1990; Watson 1991), while those involving the use of plant pathogens as bioherbicides for weed control more than doubled between 1982 and 1989 (Charudattan 1991).

For the purpose of this chapter, foliar pathogens refer to pathogens that act above ground, as opposed to soil-borne organisms (Chapters 10 and 14 in this volume). Recent in-depth reviews on biocontrol of weeds using plant pathogens have dealt with a number of ecological constraints that will only be highlighted here (Watson 1989, 1991; Hasan and Ayres 1990; Charudattan 1991; TeBeest 1991). Rather, this chapter will outline other ecological constraints that have been less commonly considered. The bioherbicide strategy will be discussed, with examples from the classical approach used for comparison purposes. One cannot discuss constraints to the successful use of bioherbicides without also dealing with regulatory issues. Prospects for overcoming these constraints and for future research will be suggested.

## Ecological Constraints

### *Common Constraints*

***Environment.*** The greatest constraint to effective use of pathogens as biocontrol agents of weeds is the environment (Templeton 1982b). The two components of the environment repeatedly identified as the most limiting are temperature and moisture (Greaves et al. 1989; TeBeest and Templeton 1985; TeBeest 1991). Each host-parasite system has its own set of requirements for length of dew period, temperature during dew, incubation temperature for optimal spore germination, penetration, and disease development. Whether these requirements present constraints is dependent on the type of organism and the biocontrol strategy. Rust fungi, which are more appropriate for the classical strategy, are wind-dispersed and require less moisture (Hasan and Wapshere 1973) than *Colletotrichum* spp., which are water-disseminated with spores contained in a slimy matrix (TeBeest 1991). The epidemiological and ecological requirements of fungal plant pathogens for use as weed biocontrol agents have been thoroughly reviewed (TeBeest 1991). TeBeest (1991) concluded that for most pathogens evaluated for use as bioherbicides, disease development was greatest at temperatures most favorable for spore germination and growth, and that disease severity increased with increasing length of dew periods up to a maximum before leveling off.

It is fairly easy to determine the environmental requirements for various host-pathogen systems under controlled environments, but it is much more difficult to evaluate this interaction in the field, and to extrapolate whether sufficient efficacy would be achieved under field conditions. Projects may not be pursued or carried to the field because of long dew requirements established in greenhouse trials; long period requirements of sixteen hours or more have been suggested to limit the practical field use of pathogens as biocontrol agents for weeds (Andersen and Walker 1985). However, many of the pathogens considered or used as bioherbicides have dew requirements close to and greater than sixteen hours, especially if other conditions are less than optimum (TeBeest et al. 1978; Walker 1981; Kirkpatrick et al. 1982; Andersen and Walker 1985; Makowski 1987; McRae and Auld 1988; Wymore et al. 1988; Chiang et al. 1989). This apparent limiting factor may be overcome by appropriate timing of application to take advantage of dew present in the field. This may be accomplished by spraying in the evening, immediately before or after a rain, in irrigated crops, or by formulation. A better initial choice of the organism selected for bioherbicide development would avoid some of these constraints and lead to more rapid progression toward a commercial product. A good understanding of the biology, infection process, and pathogenicity of the organisms selected is also essential.

*Colletotrichum* spp. offer tremendous potential and have been the most widely studied group of organisms as potential bioherbicides not only due to the successes of Collego and BioMal, but also because of the biology and cosmopolitan distribution of the genus (Templeton 1992). Contrary to expectation, the optimum dew requirement of twenty hours or repetitive dews of sixteen hours needed for establishment of *C. g. malvae* under greenhouse conditions (Makowski 1987) has not been a limiting factor in its development. In the eight years of field testing prior to registration, it was only in two drought years when hot and dry conditions prevailed after inoculation that less than 80 percent plant kill occurred (Mortensen 1988; unpublished data). Disease symptoms were not observed for over one month after inoculation, and complete mortality of *C. g. malvae* spores was suspected. However, when moisture conditions improved later in the growing season disease developed rapidly, resulting in complete plant kill by harvest. Appressoria are involved in infection and penetration in many fungi, but have also been reported as potential dormant and survival structures in *Colletotrichum* spp. (Hepperley et al. 1980; Muirhead 1981; Parbery 1981). Appressoria of *C. g. malvae* formed on glass coverslips can remain viable for up to six months (Holmström-Ruddick and Mortensen 1992). Even under the semi-arid climate of the Canadian prairies, *C. g. malvae* is efficaceous against *M. pusilla,* partly because the appressoria can serve as dormant structures until conditions become more favorable. The decision to reject or pursue more detailed studies on a biocontrol agent must be based on field results. For *C. g. malvae,* the potential dormancy and protection against unfavorable conditions were not predicted from the initial greenhouse trials, and the project would have been dropped if the decision had been based solely on greenhouse data. More commonly, pathogens show excellent potential under controlled conditions, but once they are brought to the field, their efficacy is greatly reduced, as is the case for *C. g. malvae* on *Abutilon theophrasti* Medic. (Mortensen 1988).

***Survival.*** Many of the plant pathogens being developed as bioherbicides are endemic and return to relatively low levels of disease incidence after overwintering (TeBeest 1991). To overcome this constraint, the strategy is to inundate the desired target weed with inoculum at the appropriate time for an epidemic to develop. With the reduction in allowing registration of persistent chemicals in Canada, regulatory agencies would have concerns if a potential bioherbicide was persistent and had survival structures, such as sclerotia of *Colletotrichum coccodes* (Blakeman and Hornby 1966). However, for a farmer, some residual effect would be beneficial: the product would not have to be applied every year. Moreover, high levels of survival are essential for the classical strategy. The pathogen must establish, build up, and perpetuate on its own (Watson 1991).

***Dispersal.*** For classical biocontrol, pathogens must have the ability to disperse. If the target weed has a patchy distribution or populations are separated, the ability of the pathogen to disperse widely is critical. Often the target weed is a widespread and dominant species capable of forming dense infestations in pastures and rangelands, and consequently more localized dispersal ability is required of the pathogen. To date, rust fungi have been the organisms used almost exclusively for the classical approach, and dispersal has not been a problem as they readily disseminate by wind (Cullen et al. 1973; Phatak et al. 1983; Watson 1991).

Bioherbicides have the potential to overcome these constraints (Templeton et al. 1990; Watson 1991), because the pathogen is simply applied to the target weed when and where it is a problem, at the appropriate time, with the appropriate amount of inoculum. If suitable environmental conditions prevail, successful control can be achieved.

***Specificity.*** Host specificity of the pathogenic organisms used for classical biocontrol is absolutely essential. The regulations and guidelines relating to introductions and release, as well as the theory and the procedures for conducting host-specificity tests, have been thoroughly reviewed (Zwölfer and Harris 1971; Wapshere 1974; Charudattan 1982; Klingman and Coulson 1982; Watson 1985; Weidemann 1991). Rigorous testing is conducted to ensure that the introduction of an exotic pathogen will not harm any beneficial plant species in the country of introduction or in neighboring ones (Leonard 1982; Watson 1985, 1991; Weidemann 1991).

Strict specificity is not a requirement for bioherbicides. Because of their similarity to chemicals, they can be site-directed with label restrictions so as not to endanger nontarget species, thus avoiding potential conflicts of interest (Leonard 1982; Weidemann 1991). Specificity may actually present a major constraint to commercialization of bioherbicides. Farmers and industry are interested in one application that will control all weed problems in a given field, not an application for each species. To date, bioherbicides are restricted to small niche markets because of their high specificity (i.e., Collego, DeVine, and BioMal). It may be difficult to attract industrial partners for such products because of the low profit potential of these small markets in relation to the cost of registration, large-scale production, fermentation, and formulation (Templeton 1992). Genetic manipulation may help overcome strict host specificity (Kistler 1991) and allow for improvement of biocontrol agents (Charudattan 1985; Templeton 1990; Kistler 1991). Genetic manipulation of the broad spectrum pathogen *Sclerotinia sclerotiorum* (Lib.) de Bary has been accomplished, resulting in a number of virulent mutants that have restricted host range or decreased survival (nonsclerotial mutant) or spread, and could have application as bioherbicides (Miller et al. 1989a, 1989b, 1992; Sands et al. 1990). To increase the spectrum of weeds controlled, two pathogens, *C. g. aeschynomene* and *C. gloeosporioides* f. sp. *jussiaeae,* have been tank-mixed to control northern jointvetch and winged waterprimrose (*Jussiaea decurrens* (Watt.) DC.) (Boyette et al. 1979). Two pathogens could also be combined to control one difficult weed. However, this approach has not been successful when tank-mixing *C. g. malvae* with *C. coccodes* for control of velvetleaf, as these two *Colletotrichum* spp. were found to be antagonistic (Watson, personal communication). Tank-mixing of two pathogens would only reduce costs of application, not costs of development, since both pathogens would still have to be registered individually.

### *Other Considerations*

***Colonization Potential.*** In the past two decades, there has been considerable research on microorganisms in the phyllosphere, with much of the focus on the potential for biocontrol of plant pathogens (Preece and Dickinson 1971; Dickinson and Preece 1976; Blakeman 1981; Windels and Lindow 1985; Fokkema and Van den Heuvel 1986; Andrews and Hirano 1991). The colonization potential of pathogens used for biocontrol of weeds and their interaction with other microorganisms in the phyllosphere has been virtually unexplored. The failure or lack of efficacy of many pathogens for biocontrol of weeds in the field may be due in part to the complexity of the phyllosphere community rather than plant barriers, low virulence, or lack of optimum environmental conditions (Fokkema 1991). The colonization potential of a biocontrol agent can be affected by competition, antagonism, or synergism by other microorganisms in the phyllosphere. The microbial community on the leaf surface results from immigration, emigration, growth, and death of microorganisms (Kinkel et al. 1989). The biocontrol agent must enter and interact in this complex community, and the ability to invade is dependent on what is already present (see Chapter 6 in this volume). Little is known of the phyllosphere of crop plants and virtually nothing of that of weed species. Understanding the interactions in the phyllosphere is key to developing more effective formulations and products, especially with the general belief that antagonism of saprophytic microflora against pathogens reduces incidence of disease in the field, acting as natural biocontrol agents of pathogens (Blakeman 1985). In the bioherbicide approach, enhancement of disease is the goal. The growth of fungi and bacteria on plant surfaces is affected not only by the phylloplane community but also by the physical and chemical characteristics of the leaf surface, extracellular materials of fungal and bacterial origin, and their interaction with the plant cuticle (Allen et al. 1991). Surface-active *Pseudomonas* strains were found to increase leaf wetability and thus affect the availability of water to microorganisms as well as the redistribution of microorganisms and nutrients on the plant surface (Bunster et al. 1989). Bioherbicides could be formulated with such surface-active strains to improve efficacy. Preliminary data indicate that *C. coccodes* in mixture with certain strains of *Pseudomonas* is a more effective velvetleaf control (Fernando et al. 1992). Much additional research is needed to help choose the appropriate organism or formulation that could improve efficacy in the phyllosphere.

***Nature of the Weed and the Biocontrol Agent.*** The nature of the target weed and the type of biocontrol agent selected can give an indication of the level of control that can be expected and ultimately may help us predict if the system has a chance of success. According to the recent review by Charudattan (1991) on the status of bioherbicide projects, the majority were aimed at herbaceous angiosperms and weeds in crops, followed by grasses, trees, and vines, with equal emphasis on annuals and perennials. The successful products in commercial and practical use are for broad-leaved weeds, whereas none have been employed against important monocot weeds (Winder and Van Dyke 1990). Grasses pose a difficult challenge not only because of the difficulty in reaching the growing point with a foliar-applied pathogen, but also because so many grassy weeds are closely related to crop, pasture, and lawn grass species (Wapshere 1990). Two leaf-spotting pathogens, *Bipolaris setariae* (Saw.) and *Piricularia grisea* (Cke.) Sacc., caused severe damage to goosegrass (*Eleusine indica* (L.) Gartner), but showed slight infections on a few crop cultivars (Figliola et al. 1988). Most often only leaf injury occurs, with limited or no plant kill. Chiang et al. (1989) investigated the potential of four pathogens for control of johnsongrass, *Sorghum halepense* (L.) Pers. *Exserohilum turcicum* (Pass.) Leonard & Sugg. caused more than 90 percent leaf injury, but no seedlings were killed by any of the fungi tested. Damage to johnsongrass by two *Bipolaris* strains also caused limited damage (Winder and Van Dyke 1990). The use of the rust *Puccinia canaliculata* (Schw.) Lagerh. in an augmentative strategy to control the weed yellow nutsedge (*Cyperus esculentus* L.) in crops has been demonstrated and is expected to be commercially available in the near future (Phatak et al. 1983, 1987). However, rusts are obligate parasites that typically cause less damage to their host than facultative parasites or saprophytes (Quimby 1982) and are difficult to mass produce. A better strategy for grasses may be the use of rhizobacteria (Chapter 10 in this volume).

Weeds with a high rate of vegetative growth would also be more difficult to control since they can outgrow disease more readily (Charudattan 1991), as is the case with *Cercospora rodmanii* Conway for control of water hyacinth, *Eichhornia crassipes* (Mart.) Solms (Conway 1976; Charudattan et al. 1985). Velvetleaf inoculated with *C. g. malvae* outgrew the disease under conditions where round-leaved mallow was completely killed (Makowski 1987; Mortensen 1988). Velvetleaf is controlled by *C.*

*coccodes* at the very young seedling stage but outgrows the disease at older growth stages (Wymore et al. 1988).

Pathogens that infect stems may have a better chance of killing plants than leaf pathogens do. *Colletotrichum coccodes* is a leaf pathogen and requires large amounts of inoculum applied at the cotyledon stage to control velvetleaf (Wymore et al. 1988). With older plants, infected leaves may senesce rapidly and fall, but the growing point may remain intact and the plant can outgrow the disease readily. On the other hand, infections of the stem pathogens *C. g. aeschynomene* on northern jointvetch and *C. g. malvae* on round-leaved mallow can girdle the stem, with everything above the lesion wilting and dying, resulting in rapid plant kill (Daniel et al. 1973; Makowski 1987; Mortensen 1988).

***Level of Control.*** The general requirements for selection of a potential bioherbicide include the ability of the pathogen to produce abundant, durable, and genetically stable inoculum in artificial culture; specificity to the target weed; and the ability to infect and kill a high proportion of the weed under broad environmental conditions (Daniel et al. 1973). A high level of control and consistency similar to that of chemical herbicides has also become a desired trait for bioherbicides. However, the level of control acceptable from an agricultural perspective may need to be reconsidered from an ecological perspective by taking competing vegetation into account. Complete plant kill may not be required to prevent crop losses if applications of nonlethal mycoherbicides early in the season can be as effective at suppressing weeds because of the subsequent effects of competition, as suggested by Hasan and Ayres (1990) and Paul et al. (1992). Paul and Ayres (1986, 1987) have shown that disease can reduce host competitive fitness in intra- as well as interspecific mixtures with nonhost plants. The rust *Puccinia lagenophorae* Cooke reduced the ability of groundsel (*Senecio vulgaris* L.) to compete with noninfected plants of the same species as well as with lettuce and petty spurge (*Euphorbia peplus* L.) (Paul and Ayres 1987; Ayres and Paul 1990). Nutrient and water stress may affect the interactions between infection and competition (Paul and Ayres 1990; Paul et al. 1990; Ayres 1991). These interactions do not necessarily apply to lethal bioherbicides such as *C. g. malvae,* where plant kill was high regardless of the level of competition (Makowski and Mortensen 1990). Nevertheless, pathogens that do not kill weeds could be effective biocontrol agents when infection and abiotic stress act additively, through the mechanism of competition, to damage the weed (Hasan and Ayres 1990). To my knowledge, including competition as part of a weed control strategy has been accomplished only with rusts. This should be investigated in other systems because many pathogens have been rejected as potential bioherbicides because their level of control is considered poor. The more difficult challenge may be to convince the agricultural community that crop yields can be improved without killing weeds.

The use of a dual-pathogen tactic has recently been reported as a method for increasing the level of control by an otherwise nonlethal pathogen. In the groundsel system just described, the rust may predispose or render the plant more susceptible to invasion by a second necrotrophic fungus (*Botrytis cinerea* Pers.) that is not normally a pathogen, but the combination of the two fungi resulted in plant kill (Hallett et al. 1990; Hallett and Ayres 1992).

***Metabolites.*** The mode of action of many plant pathogens is through the production of phytotoxins (Strobel et al. 1990). Phytotoxins can serve as templates for the development of biorational herbicides that can be applied with conventional technology, and the genetic information governing the production of such compounds can be cloned for more efficient production or delivery. If they are used alone as biorational herbicides, the advantages in the registration process of biologicals over chemicals would be lost, as these would be considered chemicals. However, these compounds may replace older, more toxic chemicals and may alleviate some of the environmental concerns associated with more conventional molecules. Knowledge of the metabolites of bioherbicides would alleviate the concern of regulators regarding the potential production of mycotoxins and could provide a rapid and efficient screening method for detection of potential residues of bioherbicides in food. The role of metabolites alone or as a component of a microbial product has recently been reviewed (Hoagland 1990). An understanding of the metabolites and their role in pathogenicity could serve as a first step in directing future genetic manipulation of potential bioherbicides. Reduced efficacy on marginal hosts may be overcome by selection of strains that overproduce the phytotoxins responsible for plant kill (Makowski and Miller 1995).

***Weed Management Systems.*** Integration of biocontrol of weeds with existing weed management systems is not a new concept but is essential if biocontrol is to be successful (Watson and Wymore 1990; Smith 1991). To date, bioherbicides have been specific to a single target weed; however, a complex community of weeds usually infests most crops. Integration with chemicals and other existing weed control strategies is necessary (Watson 1989; Charudattan1991; Smith 1991). Tank-mixing of chemicals and bioherbicides has been successful in a few cases, but chemicals, especially fungicides, can also adversely affect viability, germination, and growth of the pathogens. *Colletotrichum g. aeschynomene, C. g. malvae,* and other potential bioherbicides can be tank-mixed with certain herbicides and adjuvants, while others can adversely affect bioherbicide activity (Klerk et al. 1985; Smith 1986, 1991; Grant et al. 1990a, 1990b). Appropriately timed, sequential applications have been used successfully to overcome these adverse antagonistic interactions and allow for integration into the existing cropping systems (Klerk et al. 1985; Grant et al. 1990b; Watson and Wymore 1990; Smith 1991). Pesticide-resistant strains can also be developed to counteract antagonism. Benomyl-tolerant strains of *C. g. aeschynomene* and *C. g. malvae* have been produced (TeBeest 1984; Holmström-Ruddick and Mortensen 1992). Combinations of bioherbicides and chemicals or plant growth regulators can be used to increase the spectrum of weeds controlled and can also interact synergistically to increase efficacy or overcome plant host barriers. The mixture of thidiazuron, a plant growth regulator, and *C. coccodes* resulted in increased plant mortality and effective control of velvetleaf (Wymore et al. 1987). Continued research will be required to evaluate the effect of new, improved chemical pesticides and bioherbicides as well as genetically altered strains with pesticide tolerance in each weed-crop management system (Smith 1991).

## Regulatory Constraints

Two decades ago, when research in the field of biological herbicides was just beginning, some scientists envisaged that applications of genetic engineering would be achieved readily over the ensuing few decades and that these advances could resolve many of the world's problems. What was not predicted was the public concern over this new technology. The public has become very skeptical of politicians, government, and, to a lesser extent, scientists. With the increasing concerns over the safety of chemical products (whether valid or not), coupled with a general "pathophobia" over plant pathogens, the stage has been set for the restrictive regulatory atmosphere that now reigns in Canada and the United States. In Canada, bioherbicides are considered microbial pest control products and are regulated under the Pest Control Products Act. There were no guidelines prior to the first submission of the data package for registration of BioMal in October 1987. Guidelines now in place cover the following main areas of data requirements: agent specification, manufacturing methods and quality control, human health safety testing, food and feed residue studies, environmental fate, environmental toxicology, and efficacy. These are similar to the U.S. requirements, but more restrictive (Charudattan 1982). The data requirements and the registration process in Canada have been reviewed (Makowski 1995). In Canada, scientists have had the opportunity to comment on draft regulatory guidelines, but the influence of such comments on the regulation of microbials has been limited to date. More recently, scientists and specialists in industry have been involved in actually drafting regulations for the commercial use of insects, mites, and nematodes for biocontrol. This is a step forward that will, it is hoped, prevent overregulation and unreasonable requirements.

Regulations are one of the main constraints to the successful development and commercialization of pathogens as weed biocontrol agents. BioMal was registered for use in Canada in 1992 (Makowski and Mortensen 1992) for control of round-leaved mallow and is only the third commercially available bioherbicide. The two others were registered in the United States more than a decade ago. Unfortunately, the regulatory climate is becoming increasingly restrictive. Thus, many potential agents are not developed because of the high cost of meeting regulatory requirements, which has discouraged private sector involvement. Templeton (1992) has coined the term "orphaned mycoherbicides" for these products. The only reason BioMal and the other commercial bioherbicides (Templeton et al. 1990) were registered was heavy public sector support and involvement. Industry considerations for involvement in the development of this control strategy have been reviewed (Bowers

1982; Auld 1991). The role of the public and private sectors in the development and commercialization of bioherbicides is presented in Table 5.1. This partnership will remain the same for the foreseeable future, until a major change in the rate of return on investment occurs.

## Future Prospects

Bioherbicides can overcome many of the constraints on the classical use of pathogens for weed control (Watson 1991). Probability of successful development of a bioherbicide can be increased if decisions are made from scientific and economic or market-driven perspectives, many of which have been presented here. The organisms pathogen and target weed can be appropriately chosen to facilitate the achievement of a commercial product. The environmental and specificity constraints can be addressed through a better understanding of infection processes and phyllosphere ecology, which can lead to improved formulations, and through genetic manipulation and use of phytotoxic metabolites. Integration into existing weed management systems can be achieved through continued efforts in determining compatibility with chemicals or using chemicals to increase efficacy. As Paul et al. (1992) so eloquently put it: "[We can also seek] to increase the effectiveness of mycoherbicides by exploiting our developing understanding of ecology, rather than the biochemistry of plant diseases."

The track record of bioherbicides is relatively good. Three products are commercially available of the approximately 200 projects to date (Charudattan 1991). In comparison, only one chemical compound is registered for every 10,000 compounds screened (Auld 1991). However, to obtain industrial commitment in this method of weed control, the profitability of this approach must be proven. Technical feasibility has been proven, but not economic feasibility (Auld 1991; Templeton 1990, 1992). Smaller companies may be more successful than the larger companies in returning a profit because of their lower overhead. Suggestions include choosing microbial candidates for more important weeds in important crops (Watson 1989; Auld 1991; Charudattan 1991), weeds not readily controlled by chemicals, or weeds that have developed resistance to chemicals (Hasan and Ayres 1990; Templeton 1992). Another suggestion that is still being actively debated is whether it is truly necessary to commercialize a product and have it registered, with all the costs involved, or whether the product could be provided directly to the farming community in the public interest from a public institution. Whatever the outcome, continued involvement and funding from the public sector is imperative for the future of biocontrol (Auld 1991). This is especially true because biologicals are expected to provide pest control solutions, in response to the growing concern about the safety of present chemical control strategies (Hoagland 1990). Public/private multidisciplinary teams are required for the development of bioherbicides to place this technology in the hands of farmers in the most timely manner (Templeton et al. 1990; Templeton 1990, 1992).

Table 5.1 Role of the Public and Private Sectors in the Development and Commercialization of Bioherbicides

| **Public Sector** | **Private Sector** |
|---|---|
| Discovery | |
| Isolation | |
| Identification | |
| Host range | |
| Biology (dew, temperature, etc.) | |
| Efficacy (greenhouse, field) | |
| Patenting | Patenting |
| | Fermentation |
| | Scale-up |
| | Stability---Shelf Life |
| | Registration |
| | Marketing and sales |

The public is increasingly concerned about the use of chemical pesticides. Whether these concerns are well-founded or not, it has forced regulators to be more restrictive, often slowing newer, safer products in coming to market. Biologicals, generally perceived to be safer, offer an alternative or complementary tool for pest control. Nevertheless, the safety of bioherbicides, naturally occurring or genetically altered, to the environment and to health must be proven. Only through education can the public be won over and help influence the regulatory atmosphere to allow for development of these products. Most of the ecological constraints can be overcome by appropriately focused research; the regulatory constraints, however, can be overcome only by public support, understanding, and a high level of public and regulatory comfort with this developing technology.

## Summary

Constraints to development and successful use of plant pathogens as biocontrol agents of weeds are mainly ecological and regulatory. The ecological contraints usually investigated are moisture and temperature requirements, survival, dispersal, and specificity of the pathogenic organisms. Less commonly considered is the colonization potential of a biocontrol agent affected by competition and antagonism in the phyllosphere. Understanding the infection process at this level is key to developing more effective formulations and products. The nature of the target weed and the type of biocontrol agent can indicate the level of control that may be achieved. The level of control acceptable from an agricultural perspective may need to be reconsidered from an ecological perspective by taking competing vegetation into account.

Regulations are one of the major constraints to the successful development and commercialization of pathogens as weed biocontrol agents. BioMal was registered for use in

Canada in 1992 for control of round-leaved mallow and is only the third bioherbicide available commercially. The two others were registered in the United States more than a decade ago. Unfortunately, the regulatory climate is becoming increasingly restrictive. Thus, many potential agents are not developed because of the high cost of meeting regulatory requirements, which is discouraging private sector involvement.

## References


Allen, E. A., H. C. Hoch, J. R. Steadman, and R. J. Stavely. 1991. Influence of leaf surface features on spore deposition and the epiphytic growth of phytopathogenic fungi. In *Microbial ecology of leaves,* ed. J. H. Andrews and S. S. Hirano, 87–110. New York: Springer-Verlag.

Andersen, R. N., and H. L. Walker. 1985. *Colletotrichum coccodes*: A pathogen of eastern black nightshade (*Solanum ptycanthum*). *Weed Science* 33:902–5.

Andrews, J. H., and S. S. Hirano. 1991. *Microbial ecology of leaves.* New York: Springer-Verlag.

Auld, B. A. 1991. Economic aspects of biological weed control with plant pathogens. In *Microbial control of weeds,* ed. D. O. TeBeest, 262–73. New York: Chapman and Hall.

Ayres, P. G. 1991. Growth responses induced by pathogens and other stresses. In *Response of plants to multiple stresses,* ed. H. A. Mooney, W. E. Winner, E. J. Pell, and E. Chu, 227–48. San Diego, Calif.: Academic Press.

Ayres, P. G., and N. D. Paul. 1990. The effects of disease on interspecific plant competition. *Aspects of Applied Biology* 24:155–62.

Blakeman, J. P. 1981. *Microbial ecology of the phylloplane.* London: Academic Press.

______. 1985. Ecological succession of leaf surface microorganisms in relation to biological control. In *Biological control on the phylloplane,* ed. C. E. Windels and S. E. Lindow, 6–30. St. Paul, Minn.: American Phytopathological Society.

Blakeman, J. P., and D. Hornby. 1966. The persistence of *Colletotrichum coccodes* and *Mycosphaerella ligulicola* in soil, with special reference to sclerotia and conidia. *Transactions of the British Mycological Society* 48:227–40.

Bowers, R. C. 1982. Commercialization of microbial biological control agents. In *Biological control of weeds with plant pathogens,* ed. R. Charudattan and H. L. Walker, 157–73. New York: Wiley.

______. 1986. Commercialization of Collego: An industrialist's view. *Weed Science* 34: 24–25.

Boyette, C. D., G. E. Templeton, and R. J. Smith Jr. 1979. Control of winged waterprimrose (*Jussiaea decurrens*) and northern jointvetch (*Aeschynomene virginica*) with fungal pathogens. *Weed Science* 27:497–501.

Bunster, L., N. J. Fokkema, and B. Schippers. 1989. Effect of surface-active *Pseudomonas* spp. on leaf wettability. *Applied and Environmental Microbiology* 55:1340–45.

Burnett, H. C., D.P.H. Tucker, and W. H. Ridings. 1974. Phytophthora root and stem rot of milkweed vine. *Plant Disease Reporter* 58:355–57.

Charudattan, R. 1982. Regulation of microbial weed control agents. In *Biological control of weeds with plant pathogens,* ed. R. Charudattan and H. L. Walker, 175–88. New York: Wiley.

______. 1985. The use of natural and genetically altered strains of pathogens for control. In *Biological control in agricultural IPM systems,* ed. M. A. Hoy and D. C. Herzog, 347–72. New York: Academic Press.

______. 1991. The mycoherbicide approach with plant pathogens. In *Microbial control of weeds,* ed. D. O. TeBeest, 24–57. New York: Chapman and Hall.

Charudattan, R., S. B. Linda, M. Kluepfel, and Y. A. Osman. 1985. Biocontrol efficacy of *Cercospora rodmanii* on waterhyacinth. *Phytopathology* 75:1263–69.

Chiang, M. Y., C. G. Van Dyke, and W. S. Chilton. 1989. Four foliar pathogenic fungi for controlling seedling Johnsongrass (*Sorghum halepense*). *Weed Science* 37:802–9.

Conway, K. E. 1976. Evaluation of *Cercospora rodmanii* as a biological control of waterhyacinth. *Phytopathology* 66:914–17.

Cullen, J. M., P. F. Kable, and M. Catt. 1973. Epidemic spread of a rust imported for biological control. *Nature* 244:462–64.

Daniel, J. T., G. E. Templeton, R. J. Smith Jr., and W. T. Fox. 1973. Biological control of northern jointvetch in rice with an endemic fungal disease. *Weed Science* 21:303–7.

Dickinson, C. H., and T. F. Preece. 1976. *Microbiology of aerial plant surfaces.* London: Academic Press.

Evans, H. C., and C. A. Ellison. 1990. Classical biological control of weeds with microorganisms: Past, present, prospects. *Aspects of Applied Biology* 24:39–49.

Fernando, W. G. D., A. K. Watson, and T. C. Paulitz. 1992. Interaction of phylloplane bacteria in the enhancement of disease caused by the bioherbicide *Colletotrichum coccodes* on velvetleaf. *Phytopathology* 82:1081.

Figliola, S. S., N. D. Camper, and W. H. Ridings. 1988. Potential biological control agents for goosegrass (*Eleusine indica*). *Weed Science* 36:830–35.

Fokkema, N. J. 1991. The phyllosphere as an ecologically neglected milieu: A plant pathologist's point of view. In *Microbial ecology of leaves,* ed. J. H. Andrews and S. S. Hirano, 3–18. New York: Springer-Verlag.

Fokkema, N. J., and J. Van Den Heuvel. 1986. *Microbiology of the phyllosphere.* Cambridge: Cambridge University Press.

Grant, N. T., E. Prusinkiewicz, R.M.D. Makowski, B. Holmström-Ruddick, and K. Mortensen. 1990a. Effect of selected pesticides on survival of *Colletotrichum gloeosporioides* f. sp. *malvae,* a bioherbicide for round-leaved mallow (*Malva pusilla*). *Weed Technology* 4:701–15.

Grant, N. T., E. Prusinkiewicz, K. Mortensen, and R.M.D. Makowski. 1990b. Herbicide interactions with *Colletotrichum gloeosporioides* f. sp. *malvae* a bioherbicide for round-leaved mallow (*Malva pusilla*) control. *Weed Technology* 4:716–23.

Greaves, M. P., J. A. Bailey, and J. A. Hargreaves. 1989. Mycoherbicides: Opportunities for genetic manipulation. *Pesticide Science* 26:9–101.

Hallett, S. G., and P. G. Ayres. 1992. Invasion of rust (*Puccinia lagenophorae*) aecia on groundsel (*Senecio vulgaris*) by secondary pathogens: Death of the host. *Mycological Research* 96:142–44.

Hallett, S. G., N. D. Paul, and P. G. Ayres. 1990. *Botrytis cinerea* kills groundsel (*Senecio vulgaris*) infected by rust (*Puccinia lagenophorae*). *New Phytologist* 114:105–9.

Hasan, S. 1972. Specificity and host specialization of *Puccinia chondrillina. Annals of Applied Biology* 72:257–63.

______. 1981. A new strain of the rust fungus *Puccinia chondrillina* for biological control of skeleton weed in Australia. *Annals of Applied Biology* 99:119–24.

Hasan, S., and P. G. Ayres. 1990. Tansley Review No. 23. The control of weeds through fungi: Principles and prospects. *New Phytologist* 115:201–22.

Hasan, S., and A. J. Wapshere. 1973. The biology of *Puccinia chondrillina* a potential biological control agent of skeletonweed. *Annals of Applied Biology* 74:325–32.

Hepperly, P. R., B. L. Kirkpatrick, and J. B. Sinclair. 1980. *Abutilon theophrasti,* wild host for three fungal parasites of soybean. *Phytopathology* 70:307–10.

Hoagland, R. E. 1990. Microbes and microbial products as herbicides: An overview. In *Microbes and microbial products as herbicides,* ed. R. E. Hoagland, 2–52. Washington, D.C.: American Chemical Society.

Holmström-Ruddick, B., and K. Mortensen. 1992. Various factors affecting infectivity of a benomyl-resistant strain of the mycoherbicide agent *Colletotrichum gloeosporioides* f. sp. *malvae. Canadian Journal of Plant Pathology* 14:239.

Kenney, D. S. 1986. DeVine—the way it was developed: An industrialist's view. *Weed Science* 34:15–16.

Kinkel, L. L., J. H. Andrews, and E. V. Nordheim. 1989. Fungal immigration dynamics and community development on apple leaves. *Microbial Ecology* 18:45–58.

Kirkpatrick, T. L., G. E. Templeton, D. O. TeBeest, and R. J. Smith Jr. 1982. Potential of *Colletotrichum malvarum* for biological control of prickly sida. *Plant Disease* 66:323–25.

Kistler, H. C. 1991. Genetic manipulations of plant pathogenic fungi. In *Microbial control of weeds,* ed. D. O. TeBeest, 152–70. New York: Chapman and Hall.

Klerk, R. A., R. J. Smith Jr., and D. O. TeBeest. 1985. Integration of a microbial herbicide into weed and pest control programs in rice (*Oryza sativa*). *Weed Science* 33:95–99.

Klingman, D. L., and J. R. Coulson. 1982. Guidelines for introducing foreign organisms into the U.S. for biological control of weeds. *Plant Disease* 66:1205–9.

Leonard, K. J. 1982. The benefits and potential hazards of genetic heterogeneity in plant pathogens. In *Biological control of weeds with plant pathogens,* ed. R. Charudattan and H. L. Walker, 99–112. New York: Wiley.

Makowski, R.M.D. 1987. The evaluation of *Malva pusilla* Sm. as a weed and its pathogen *Colletotrichum gloeosporioides* (Penz.)Sacc. f. sp. *malvae* as a bioherbicide. Ph.D. dissertation, University of Saskatchewan, Saskatoon.

______. 1995. Regulating microbial pest control agents in Canada: The first mycoherbicide. In *Proceedings VIII International Symposium on Biological Control of Weeds,* ed. E. S. Delfosse and R. R. Scott, 641–2. Canterbury, New Zealand: Lincoln University.

Makowski, R.M.D., and J. D. Miller. 1995. Phytotoxic metabolites of *Colletotrichum gloeosporioides* f. sp. *malvae,* a mycoherbicide for round-leaved mallow control. In *Proceedings VIII International Symposium on Biological Control of Weeds,* ed. E. S. Delfosse and R. R. Scott, 703. Canterbury, New Zealand: Lincoln University.

Makowski, R.M.D., and K. Mortensen. 1990. *Colletotrichum gloeosporioides* f. sp. *malvae* as a bioherbicide for round-leaved mallow (*Malva pusilla*): Conditions for successful control in the field. In *Proceedings VII International Symposium on Biological Control of Weeds,* ed. E. S. Delfosse, 513–22. Rome: Istituto Sperimentate per la Patologia Vegetale, Ministero dell'Agricoltura e delle Forestale.

______. 1992. The first mycoherbicide in Canada: *Colletotrichum gloeosporioides* f. sp. *malvae* for round-leaved mallow control. *Proceedings of the First International Weed Control Congress* 2:298–300.

McRae, C. F., and B. A. Auld. 1988. The influence of environmental factors on anthracnose of *Xanthium spinosum. Phytopathology* 78:1182–86.

Miller, R. V., E. J. Ford, and D. C. Sands. 1989a. A nonsclerotial pathogenic mutant of *Sclerotinia sclerotiorum. Canadian Journal of Microbiology* 35:517–20.

Miller, R. V., E. J. Ford, N. J. Zidack, and D. C. Sands. 1989b. A pyrimidine auxotroph of *Sclerotinia sclerotiorum* for use in biological weed control. *Journal of General Microbiology* 135:2085–91.

Miller, R. V., K. A. Glass, M. K. McCarthy, and E. J. Ford. 1992. Efficacies of *Sclerotinia sclerotiorum* mutants for biological control of weeds. *Phytopathology* 82:1081.

Mortensen, K. 1988. The potential of an endemic fungus, *Colletotrichum gloeosporioides,* for biological control of round-leaved mallow (*Malva pusilla*) and velvetleaf (*Abutilon theophrasti*). *Weed Science* 36:473–78.

Muirhead, I. F. 1981. The role of appressorial dormancy in latent infection. In *Microbial ecology of the phylloplane,* ed. J. P. Blakeman, 155–67. London: Academic Press.

Parbery, D. G. 1981. Biology of anthracnose on leaf surfaces. In *Microbial ecology of the phylloplane,* ed. J. P. Blakeman, 135–54. London: Academic Press.

Paul, N. D., and P. G. Ayres. 1986. Interference between healthy and rusted groundsel (*Senecio vulgaris* L.) within mixed populations of different densities and proportions. *New Phytologist* 104:257–69.

______. 1987. Effects of rust infection of *Senecio vulgaris* on competition with lettuce. *Weed Research* 27:431–41.

______. 1990. Effects of interactions between nutrient supply and rust infection of *Senecio vulgaris* L. on competition with *Capsella bursa-pastoris* (L.) Medic. *New Phytologist* 114: 667–74.

Paul, N. D., P. G. Ayres, and S. G. Hallett. 1992. Making biological herbicides more effective. *Journal of Biological Education* 26:94–99.

Paul, N. D., K. A. Laxmi, and P. G. Ayres. 1990. Responses of rust (*Puccinia lagenophorae* Cooke) to nutrient supply in groundsel (*Senecio vulgaris* L.) and effects of infection on host nutrient relations. *New Phytologist* 115:99–106.

Phatak, S. C., M. B. Callaway, and C. S. Vavrina. 1987. Biological control and its integration in weed management systems for purple and yellow nutsedge (*Cyperus rotundus* and *C. esculentus*). *Weed Technology* 1:84–91.

Phatak, S. C., D. R. Sumner, H. D. Wells, D. K. Bell, and N. C. Glaze. 1983. Biological control of yellow nutsedge with the indigenous rust fungus *Puccinia canaliculata. Science* 219:1446–47.

Preece, T. F., and C. H. Dickinson. 1971. *Ecology of leaf surface micro-organisms.* London: Academic Press.

Quimby, P. C. Jr. 1982. Impact of diseases on plant populations. In *Biological control of weeds with plant pathogens,* ed. R. Charudattan and H. L. Walker, 47–60. New York: Wiley.

Ridings, W. H. 1986. Biological control of stranglervine in citrus—a researcher's view. *Weed Science* 34:31–32.

Ridings, W. H., D. J. Mitchell, C. L. Schoulties, and N. E. El-Gholl. 1976. Biological control of milkweed vine in Florida citrus groves with a pathotype of *Phytophthora citrophthora.* In *Proceedings IV International Symposium on Biological Control of Weeds,* ed. E. Freeman, 224–40. Gainesville: University of Florida.

Sands, D. C., E. J. Ford, and R. V. Miller. 1990. Genetic manipulation of broad host-range fungi for biological control of weeds. *Weed Technology* 4:471–74.

Smith, R. J. Jr. 1986. Biological control of northern jointvetch in rice and soybeans—a researcher's view. *Weed Science* 34:17–23.

______. 1991. Integration of biological control agents with chemical pesticides. In *Microbial control of weeds,* ed. D. O. TeBeest, 189–208. New York: Chapman and Hall.

Standard and Poor's Corp. 1991. *Standard & Poor's Industry Surveys.* New York: Standard & Poor's.

Strobel, G., A. Stierle, S. H. Park, and J. Cardellina. 1990. A host-specific phytotoxin from *Alternaria alternata* on spotted knapweed. In *Microbes and microbial products as herbicides,* ed. R. E. Hoagland, 54–62. Washington, D.C.: American Chemical Society.

Supkoff, D. M., D. B. Joley, and J. J. Marois. 1988. Effect of introduced biological control organisms on the density of *Chondrilla juncea* in California. *Journal of Applied Ecology* 25: 1089–95.

TeBeest, D. O. 1984. Induction of tolerance to benomyl in *Colletotrichum gloeosporioides* f. sp. *aeschynomene* by ethyl methanesulfonate. *Phytopathology* 74:864.

______. 1991. Ecology and epidemiology of fungal plant pathogens studied as biological control agents of weeds. In *Microbial control of weeds,* ed. D. O. TeBeest, 97–114. New York: Chapman and Hall.

TeBeest, D. O., and G. E. Templeton. 1985. Mycoherbicides: Progress in the biological control of weeds. *Plant Disease* 69:6–10.

TeBeest, D. O., G. E. Templeton, and R. J. Smith Jr. 1978. Temperature and moisture requirements for development of anthracnose on northern jointvetch. *Phytopathology* 68: 389–93.

Templeton, G. E. 1982a. Status of weed control with plant pathogens. In *Biological control of weeds with plant pathogens,* eds. R. Charudattan and H. L. Walker, 29–44. New York: Wiley.

______. 1982b. Biological herbicides: Discovery, development, deployment. *Weed Science* 30:430–33.

______. 1990. Weed control with pathogens: future needs and directions. In *Microbes and microbial products as herbicides,* ed. R. E. Hoagland, 321–29. Washington, D.C.: American Chemical Society.

______. 1992. Use of *Colletotrichum* strains as mycoherbicides. In *Colletotrichum: Biology, pathology and control,* ed. J. A. Bailey and M. J. Jeger, 358–80. Oxon, U.K.: CAB International.

Templeton, G. E., R. J. Smith Jr., and D. O. TeBeest. 1990. Perspectives on mycoherbicides two decades after discovery of the Collego pathogen. In *Proceedings VII International Symposium on Biological Control of Weeds,* ed. E. S. Delfosse, 553–58. Rome: Istituto Sperimentate per la Patologia Vegetale, Ministero dell'Agricoltura e delle Forestale.

Walker, H. L. 1981. Factors affecting biological control of spurred anoda (*Anoda cristata*) with *Alternaria macrospora. Weed Science* 29:505–7.

Walker, H. L., and C. D. Boyette. 1986. Influence of sequential dew periods on biocontrol of sicklepod (*Cassia obtusifolia*) by *Alternaria cassiae. Plant Disease* 70:962–63.

Wapshere, A. J. 1974. A strategy for evaluating the safety of organisms for biological weed control. *Annals of Applied Biology* 77:201–11.

______. 1990. Biological control of grass weeds in Australia: An appraisal. *Plant Protection Quarterly* 5:62–75.

Watson, A. K. 1975. The potential of a nematode, *Paranguina picridis* Kirjanova & Ivanova, as a biological control agent of Russian knapweed (*Acroptilon repens* (L.) DC). Ph.D. thesis, University of Saskatchewan, Saskatoon.

______. 1985. Host specificity of plant pathogens in biological weed control. In *Proceedings VI International Symposium on Biological Control of Weeds,* ed. E. S. Delfosse, 577–86. Ottawa: Canada Government Publication Center.

______. 1989. Current advances in bioherbicide research. *Proceedings of the Brighton Crop Protection Conference—Weeds,* 987–96.

______. 1991. The classical approach with plant pathogens. In *Microbial control of weeds,* ed. D. O. TeBeest, 3–23. New York: Chapman and Hall.

Watson, A. K., and L. A. Wymore. 1990. Biological control, a component of integrated weed management. In *Proceedings VII International Symposium on Biological Control of Weeds,* ed. E. S. Delfosse, 101–6. Rome: Istituto Sperimentate per la Patologia Vegetale, Ministero dell'Agricoltura e delle Forestale.

Weidemann, G. J. 1991. Host-range testing: Safety and science. In *Microbial control of weeds,* ed. D. O. TeBeest, 83–96. New York: Chapman and Hall.

Windels, C. E., and S. E. Lindow. 1985. *Biological control on the phylloplane.* St. Paul, Minn.: American Phytopathological Society.

Winder, R. S., and C. G. Van Dyke. 1990. The pathogenicity, virulence, and biocontrol potential of two *Bipolaris* species on Johnsongrass (*Sorghum halepense*). *Weed Science* 38: 89–94.

Wymore, L. A., C. Poirier, and A. K. Watson. 1988. *Colletotrichum coccodes,* a potential bioherbicide for control of velvetleaf (*Abutilon theophrasti*). *Plant Disease* 72:534–38.

Wymore, L. A., A. K. Watson, and A. R. Gotlieb. 1987. Interaction between *Colletotrichum coccodes* and thidiazuron for control of velvetleaf (*Abutilon theophrasti*). *Weed Science* 35: 377–83.

Zwölfer, H., and P. Harris. 1971. Host specificity determination of insects for biological control of weeds. *Annual Review of Entomology* 16:159–78.

PART TWO

# Development and Implementation

# 6

# Antibiosis and Beyond: Genetic Diversity, Microbial Communities, and Biological Control

*Jocelyn Milner, Laura Silo-Suh, Robert M. Goodman, and Jo Handelsman*

Healthy plants grown in normal environments, such as the field or a greenhouse, are virtually never free of microbes. Anyone who has attempted to take plant tissues from the open environment into axenic culture, a technology commonly used in commercial horticulture, can attest to the perversity of microbial "contamination." Some of these organisms cause disease and some provide a known benefit to the plant. Many plant-associated microorganisms may have no effect on the plant or may influence plant health in important ways that have not yet been discovered. One goal of plant health management is to minimize the impacts of the detrimental microorganisms and maximize the effects of beneficial ones. A specific application of this concept is microbial biological control, which we define as the use of added microorganisms or organisms already in the soil as part of the crop production system to keep losses from disease and pests below economically acceptable thresholds. An extension of the concept of microbial biological control is the potential for managing communities of microorganisms to sustain plant health.

## The Challenge of Microbial Biological Control for Plant Disease

Conceptually, there are numerous possible approaches to developing microbial biological control strategies (Campbell 1994; Jacobsen and Backman 1993; Gutterson 1990; Deacon and Barry 1993; Fravel 1988; Weller 1988; O'Sullivan and O'Gara 1992). Most biological control research has started with either a chance observation of, or a deliberate attempt to find, specific microorganisms that suppress a disease or exhibit antibiosis toward a pathogen. Although there are numerous examples of microorganisms that inhibit growth and development of pathogens in the laboratory, few of these antagonists have translated into effective products or strategies for management of crop disease in the field (Jacobsen and Backman 1993; Powel and Jutsum 1993). The reality is that the agroecosystem is much more complex than any system we construct in the laboratory. One essential variable is the crop plant itself. Many laboratory studies have selected potential biological control organisms based on their ability to inhibit growth of the pathogen on culture medium, thereby avoiding the variability introduced by the plant (see Fravel 1988; Weller 1988; Gutterson 1990; O'Sullivan and O'Gara 1992, and references therein). However, the ability to inhibit growth of pathogens in a simple petri plate system is a poor predictor of disease suppression on a plant.

Even organisms that suppress disease on plants in the laboratory often do not suppress disease in the field. This may be explained in part by the variability introduced by the environment in the field. Plant growth is quite different in soil versus potting mixes, under the high light intensity of natural sunlight versus artificial light, and under conditions of constant humidity and air flow versus the variable wind, rain, and dew conditions of the field. Furthermore, to be a successful product, a microbe used for disease suppression must control many strains of the pathogen on many varieties of the crop. The genetic variability of the pathogen or the host plant is rarely considered in initial identification of microbes for biological control in the laboratory. The complexity of the physical, chemical, and biological environment in which a biological control agent must function requires that we examine biological control with a wide range of scientific hypotheses, approaches, and tools.

Consideration of these issues has led us to take a strongly ecological and mechanistic approach in our work to understand disease suppression and antibiosis by a biological control agent. In this article we will discuss our ideas about how research directed to improved biological control strategies might contribute to long-term progress toward more ecologically based crop protection strategies. These ideas are illustrated with examples from our own work, placed in the context of the wider literature. We make no attempt to review all of the literature on biological control of plant disease by bacteria. Many excellent reviews have been published elsewhere (Campbell 1994; Jacobsen and Backman 1993; O'Sullivan and O'Gara 1992; Gutterson 1990; Fravel 1988; Weller 1988) or appear as other chapters in this book.

## Isolation of *Bacillus cereus* Strain UW85 and Its Potential as a Biological Control Agent

Our work focuses on a strain of *Bacillus cereus,* designated UW85, that suppresses root diseases of alfalfa and soybeans. UW85 was identified from a collection of root-associated bacteria isolated from healthy alfalfa plants grown in field sites in which alfalfa damping-off was a common problem. Our rationale was that the roots of healthy plants that thrived under disease-producing conditions may host bacteria that suppress disease and promote plant health. We developed a laboratory bioassay in which we screened hundreds of bacterial isolates for the ability to suppress damping-off disease caused by the oomycete pathogen, *Phytophthora medicaginis,* on alfalfa seedlings (Handelsman et al. 1990). This bioassay allowed us to screen directly for disease suppression under conditions that are conducive to *P. medicaginis* infection: *P. medicaginis* was applied as zoospores, the major infective propagule for *Phytophthora* spp. in soil, and the alfalfa seedlings were kept moist for the duration of the assay. The isolate designated UW85, identified as a strain of *B. cereus,* was the only isolate that consistently suppressed alfalfa damping-off in the laboratory bioassay (Handelsman et al. 1990).

Two advantageous characteristics of *B. cereus* that may contribute to its potential as a successful biological control agent are evident. First, UW85 is a facultative aerobe, which might contribute to its ability to grow or persist under the water-saturated soil conditions that are conducive to damping-off. Second, its ability to sporulate might enable *B. cereus* UW85 to survive on seeds during storage, in the soil, and under many adverse conditions. These advantages apply equally well to other *Bacillus* spp., including commercially available strains of *B. subtilis* (Turner and Backman 1991; Mahafee and Backman 1993; Jacobsen and Backman 1993). Because the biology of *Bacillus* spp. differs from that of *Pseudomonas, Agrobacterium,* and other gram-negative bacteria that currently dominate the biological control literature, our studies with *B. cereus* offer a new perspective on the behavior of biological control agents. For example, unlike pseudomonads, aggressive colonization does not appear to be a salient feature of UW85's behavior in the field (Halverson et al. 1993a; 1993b). We anticipate that continued research with *Bacillus* spp. will offer alternative viewpoints on basic concepts that shape our understanding and perceptions of biological control of plant disease.

Studies with UW85 have demonstrated that it effectively suppresses plant diseases caused by several oomycete pathogens in the laboratory and in the field. It is effective against strains of four species of *Phytophthora,* two species of *Pythium,* and one species of *Aphanomyces.* In addition to suppressing damping-off of alfalfa caused by *P. medicaginis* in a laboratory bioassay (Handelsman et al. 1990), it suppresses infection of tobacco seedlings by *P. parasitica* var. *nicotianae,* the causative agent of black-shank (Handelsman et al. 1991). UW85 also suppresses seedling disease of alfalfa, tomato, maize, and tobacco caused by *Pythium* spp. (our unpublished results). UW85 also can suppress postharvest fruit disease: coapplication of UW85 and zoospores of *Pythium aphanidermatum* on cucumber wounds results in significantly less rotting than in cucumbers inoculated with zoospores alone (Smith et al. 1993).

UW85 provided consistent yield benefits on soybeans in a field trial conducted in Wisconsin over five consecutive growing seasons: UW85 improved soybean yields as much as did metalaxyl, the fungicide registered and used to control seedling and root rot diseases (Osburn et al. 1995). Similar results were obtained at several other field sites in the Midwest (Smith, Osburn, and Handelsman, unpublished). UW85 provided impressive protection against sclerotinia blight of peanut caused by *Sclerotinia minor* (Phipps 1992). We are continuing to conduct field trials to explore the range of crop plants on which UW85 may contribute to disease suppression and plant health.

Despite those demonstrations of efficacy, variability plagues the performance of UW85 on certain crops such as alfalfa. It has failed completely to improve the emergence or yield of soybeans at field sites and with the cultivars grown in the southern United States. Moreover, farmers are likely to demand that biologicals provide a clear advantage over chemicals before they will break with their reliance on chemical control and switch to biological control. Therefore, much of our work has focused on finding out how UW85 suppresses disease and why it fails under some conditions. Understanding the organism and its interaction with the plant, the pathogen, and the environment is likely to suggest strategies for enhancing and stabilizing its disease-suppressive ability. Thus, an important aspect of our work centers on the mechanisms of biological control by UW85.

## Antibiosis as a Mechanism of Biological Control

Competition, parasitism, and antibiosis are mechanisms of biological control that involve direct antagonism between the biological control agent and the plant pathogen, while the stimulation of plant growth and the induction of host resistance are mechanisms that do not involve direct antagonism between the biological control agent and the plant pathogen. In antibiosis, the biological control agent inhibits the growth or function of the pathogen by the production of toxic metabolites including antibiotics, lytic enzymes, and volatile substances (Fravel 1988).

Four distinct approaches to the study of antibiosis as a mechanism of disease suppression by the biological control agent are commonly taken. One approach is to generate mutants of the biological control agent that do not produce the antibiotic and determine whether there is an associated loss of biological control activity (see Fravel 1988; Gutterson 1990; Weller 1988; O'Sullivan and O'Gara 1992, and references therein). Another approach used to implicate an antibiotic in biological control is to demonstrate that the purified antibiotic suppresses disease (Howell and Stipanovic 1979, 1980). It may be difficult to isolate sufficient quantities of antibiotic to conduct such an analysis or purified antibiotic may be unstable in or adsorb to soil. The third approach is to demonstrate that the biological control agent does not suppress disease caused by isolates of the pathogen that are insensitive to the antibiotic (Kerr and Htay 1974). The last approach requires the transfer of the genes for antibiotic biosynthesis to an isolate that does not normally synthesize the antibiotic to determine whether the ability to suppress disease follows the antibiotic biosynthesis genes (Ellis and Kerr 1978). The use of more than one of these approaches strengthens the case for the role of an antibiotic in biological control.

Multiple antibiotics produced by the same biological control agent may contribute to the suppression of plant disease. Pseudomonads are recognized for producing a multitude of inhibitory compounds that contribute to disease suppression. However, not all mechanisms are effective against all pathogens and on all hosts. For example, *Pseudomonas fluorescens* strain CHAO produces several compounds with antifungal activity including hydrogen cyanide, the antibiotics pyoluteorin and 2, 4-diacetylphloroglucinol, and the siderophore, pyoverdine (Keel et al. 1992; Maurhofer et al. 1994; Voisard et al. 1989). The production of hydrogen cyanide and 2, 4-diacetylphloroglucinol by strain CHAO contributes to the suppression of tobacco root rot caused by *Thielaviopsis basicola,* while 2, 4-diacetylphloroglucinol contributes to the suppression of take-all of wheat caused by *Gaeumannomyces graminis,* and pyoluteorin contributes to the suppression of damping-off disease of cress caused by *Pythium ultimum.* In another system, *P. fluorescens* strain PF-5 produces two antibiotics, pyrrolnitrin and pyoluteorin. On cotton seedlings, pyrrolnitrin contributes to the suppression of *Rhizoctonia solani* and pyoluteorin contributes to the suppression of the oomycete pathogen *P. ultimum* (Howell and Stipanovic 1979, 1980).

A third example of a biological control agent that utilizes multiple toxic compounds is *Pseudomonas fluorescens* strain 2-79. It produces three compounds that have antifungal properties: phenazine–1-carboxylic acid, a pyoverdine siderophore, and an additional antifungal factor (Hamdan et al. 1991). The phenazine antibiotic has been demonstrated to contribute strongly to the suppression of take-all of wheat caused by *Gaeumannomyces graminis,* while the antifungal factor plays a minor role in the suppression of this disease. However, mutants that did not produce either the phenazine antibiotic or antifungal factor suppressed disease above the nontreated control levels, suggesting that strain 2-79 suppresses disease by another mechanism. Likewise, *Pseudomonas aureofaciens* strain 30-84 produces three distinct phenazine antibiotics, all of which contribute to the suppression of take-all of wheat caused by *Gaeumannomyces graminis* (Pierson and Thomashow 1992).

One of the most effective and commercially successful biological control agents to date is *Agrobacterium radiobacter* strain 84, used to control crown gall formation caused by *A. tumefaciens.* Both the biological control agent and the pathogen are gram-negative bacteria, which are amenable to genetic manipulation (Kerr 1980). Antibiosis has been identified as a contributing mechanism of biological control (Kerr and Htay 1974). *Agrobacterium radiobacter* strain 84 contains a plasmid, pAgK84, that codes for the production of and immunity to the antibiotic agrocin 84, while pathogenic *Agrobacterium* strains contain a tumor-inducing plasmid, pTi, that codes for sensitivity to agrocin 84 and for the induction of crown gall formation. However, mutants of *A. radiobacter* strain 84 that do not produce agrocin 84 suppress crown gall infection under certain conditions, which suggests that additional mechanisms contribute to suppression of crown gall formation (Cooksey and Moore 1982). A second antibiotic, agrocin 434, which is not inhibitory to strains of *A. tumefaciens,* may contribute to suppression of gall formation caused by *Agrobacterium* strains of biovar 2, including strains that are not inhibited by agrocin 84 (Donner et al. 1993). Competitive exclusion also has been proposed to play a role in control of crown gall, although conclusive data to support that hypothesis are lacking. Continuing work on *A. radiobacter* strain 84 has demonstrated that, even for successful biological control systems, their reliability may be improved and their spectrum of protected hosts may be broadened by an improved understanding of the mechanisms of action (Burr et al. 1993; Hendson et al. 1983; Chen and Xiang 1986; Webster et al. 1986; Shim et al. 1987).

The production of multiple antibiotics may make biological control agents formidable antagonists, especially when the compounds are directed at different target sites or different stages of the pathogen's life cycle. Pathogens that elude one mechanism may be overcome by another. The use of multiple mechanisms decreases the probability of the pathogen's developing resistance to a biological control agent. This suggests that biological control agents may provide a longer duration of control for plant diseases in the field compared with chemical fungicides, for which the development of resistance by the pathogen is an ever increasing problem.

## Antibiosis: Role in Disease Suppression by *B. cereus* UW85

For UW85, by-products of its metabolic activity that are antagonistic to oomycetes accumulate in culture supernatants (Handelsman et al. 1990; Gilbert et al. 1990; Silo-Suh et al. 1994). A portion of that inhibitory effect is attributable to sequestering of calcium and the production of large amounts of ammonium, which results in a concurrent increase in the pH of the culture medium (Gilbert et al. 1990). This ability to increase the ammonium to calcium ratio, a characteristic of *B. cereus,* causes zoospore lysis. Since *Bacillus* isolates generally lyse zoospores, but few protect plants as effectively as UW85, this characteristic does not account for the majority of UW85's unusual ability to suppress disease.

Some of the disease-suppressive activity of UW85 can be attributed to several metabolites produced by UW85 that have antifungal activity. The chemical structure of one of the antibiotics, zwittermicin A, has been determined (Silo-Suh et al. 1994; He et al. 1994). It is a linear aminopolyol, and it appears to represent a new class of antibiotics. Zwittermicin A has a broad target range, inhibiting the growth of fungi and bacteria, including many plant pathogens (Silo-Suh 1994; Silo-Suh et al. 1994; Osburn et al. 1995). It reversibly inhibits germ tube elongation of *Phytophthora medicaginis* and is bactericidal to certain enteric bacteria, including *Escherichia coli.* Zwittermicin A also potentiates the insecticidal activity of *B. thuringiensis* and Bt toxin, although zwittermicin A itself has no independent insecticidal activity (Manker et al. 1994). Preliminary results suggest that a second antibiotic, given the provisional designation antibiotic B, is an aminoglycoside. It has a narrower target range than does zwittermicin A, but it also inhibits both fungi and bacteria (Silo-Suh 1994). These antifungal metabolites are associated with the water soluble, cationic fraction of UW85 culture supernatants. They do not account for all of the antifungal activity in culture supernatants of UW85, suggesting that UW85 produces additional metabolites with antifungal activity.

We have demonstrated that zwittermicin A and antibiotic B are associated with disease suppression by UW85 in the laboratory. The antibiotics accumulate in culture supernatants following the onset of sporulation when the disease suppressiveness of cultures also increases (Handelsman et al. 1990; Milner et al. 1995). We identified several mutants of UW85 reduced in the ability to accumulate zwittermicin A and antibiotic B and less effective than UW85 in disease suppression (Silo-Suh et al. 1994). There is a significant correlation between the amount of each of the two antibiotics that the mutants produce and the ability of the mutants to suppress alfalfa damping-off in the laboratory. Furthermore, the addition of either or both of the two purified antibiotics restores the ability of the UW85 mutants to suppress disease (Silo-Suh 1994; Silo-Suh et al. 1994). However, antibiotics are unlikely to account for all the disease suppressiveness of UW85. For example, certain mutants that do not accumulate either antibiotic retain some disease-suppressive activity (Silo-Suh et al. 1994).

Further work is required to determine whether zwittermicin A and antibiotic B are also important under field conditions. If the antibiotics contribute to disease suppression in the field, then our knowledge of regulation of antibiotic production may suggest methods for improving biological control. For example, ferric iron increases accumulation of zwittermicin A and antibiotic B in culture and increases the disease suppressiveness of UW85, whereas phosphate reduces the accumulation of antibiotics and disease suppressiveness (Milner et al. 1995). Thus, conditions that are highly variable in soils have a substantial impact on the metabolic activity of UW85 and may influence the efficacy of biological control. We do not know how antibiotic production is regulated in the spermosphere, the rhizosphere, or the soil. Our laboratory experiments suggest that plant tissues have an impact on the physiology of UW85: alfalfa root exudates stimulate antibiotic production, whereas seed exudates inhibit growth of UW85 in minimal medium (Milner et al. 1995).

## Approaches to Improve Biological Control Through the Manipulation of Antibiosis

Antibiosis is a mechanism of biological control that can be readily manipulated and exploited in order to enhance disease suppression by biological control agents. An analysis of the regulation and biosynthesis of the antibiotic may suggest approaches to enhance the level of antibiotic synthesized or to improve the pattern of antibiotic synthesis. For example, the synthesis of many antibiotics is influenced by specific nutrients (Martin and Demain 1980; Gutterson 1990; Fravel 1988; Slininger and Jackson 1992; Thomashow et al. 1990; Clarke et al. 1992), which could be provided in the form of soil amendments. Soil amendments can serve the additional role of supporting growth of the microbial biological control agent in the rhizosphere and may even be exploited to favor growth of specific microbes (Colbert et al. 1993).

Once the genes for antibiotic biosynthesis have been cloned, alternative approaches to improve and enhance antibiotic synthesis become available. For example, the synthesis of a regulated antibiotic could be made constitutive by placing the biosynthetic genes under the control of a constitutive promoter. Alternatively, antibiotic synthesis may be enhanced by placing the biosynthetic genes on a multicopy plasmid. However, enhancing antibiotic synthesis does not always result in enhanced biological control. *Pseudomonas fluorescens* strain CHAO produces the antifungal antibiotic pyoluteorin, to which *Pythium ultimum* is extremely sensitive (Maurhofer et al. 1994). Unfortunately, enhanced expression of pyoluteorin appears to be phytotoxic to cress and to sweet corn (Maurhofer et al. 1992).

One approach to enhance disease suppression is to induce synthesis of the antibiotic under conditions in which it may not normally be synthesized. For example, the production of phenazine antibiotics by *P. aureofaciens* strain 30-84 appears to be regulated in part by cell density signals (Pierson et al. 1994). This may explain the need for pseudomonads to aggressively colonize the rhizosphere of wheat in order to suppress take-all effectively (for example, Bull et al. 1991). Altered regulation of phenazine antibiotics may be accomplished by supplying the autoinducer as a soil amendment or by placing the biosynthesis of the phenazines under the expression of a constitutive promoter. Elucidation of the regulation of the *phzR* gene, which responds to cell density and activates phenazine biosynthesis, may suggest further approaches to alter the regulation of antibiotic synthesis and enhance biological control activity of strain 30-84.

The *gacA* gene is a global regulator of antibiotic production in *P. fluorescens* strain CHAO and is required for effective disease suppression of tobacco root rot (Laville et al. 1992; Nasch et al. 1994). Regulation by *gacA* may limit the production of antibiotics to stationary or restricted phases of growth of CHAO. Manipulation of antibiotic regulation by *gacA* may result in more consistent disease suppression by CHAO.

## Exploiting Microbial Genetic Diversity

We can also use an understanding of mechanisms by which biological control agents suppress disease to develop strategies to search for better agents from the rich variety of organisms in the soil. Screening or selecting for specific traits known to be important for biological control activity can be used to isolate related strains from a wide range of environments. We have approached this angle of biological control through the discovery that many strains of *B. cereus* produce zwittermicin A, the antibiotic produced by UW85. These findings unveiled a potential wealth of genetic diversity for biological control activity in *B. cereus* strains isolated from around the world.

We have developed two screening methods by which zwittermicin A–producing isolates of *B. cereus* with biological control activity can be presumptively identified. First, a rapid screening method, based on phage typing, arose from the fortuitous discovery that a phage that infects *B. cereus* was frequently lytic on zwittermicin A–producing *B. cereus* strains and rarely lytic on nonproducing strains (Stabb et al. 1994). A second method is based on PCR amplification of *zmaR*, a zwittermicin A resistance gene from UW85 (Milner, Stohl, Raffel, and Handelsman, unpublished). The rationale for this approach is that zwittermicin A–producing strains are predicted to contain determinants of resistance to ensure their survival under zwittermicin A–producing conditions. Using the PCR-based method, we detected *zmaR* in zwittermicin A–producing, but not in nonproducing, isolates of *B. cereus.* By these methods, we have isolated zwittermicin A–producing strains from every soil sample we tested from environments of diverse geographic origin on three continents (Stabb et al. 1994; Raffel and Handelsman, unpublished).

Zwittermicin A–producing strains generally suppress damping-off disease more effectively than nonproducing strains in the laboratory bioassay (Stabb et al. 1994). However, we identified a number of strains that suppress disease effectively but do not produce zwittermicin A or antibiotic B, the antifungal metabolites for which we routinely assay. Therefore, this panglobal collection of *B. cereus* isolates may provide the basis for discovery of new mechanisms of disease suppression.

This discovery of zwittermicin A–producing *B. cereus* strains in every soil sample we tested suggested that the naturally occurring genetic diversity of a related group of bacteria could be exploited to improve biological control. We hypothesize that the efficacy of biological control could be increased and the associated variability decreased by using specific bacterial strains that are both adapted to the environment in which they would be applied and also share at least a subset of mechanisms of disease suppression with related and well-characterized strains. Such an approach has been applied successfully in the development of *B. thuringiensis* for the biological control of insects (Feitelson et al. 1992; Carlton 1993). A second approach would be to make mixtures or combinations of related bacterial strains that have overlapping mechanisms of biological control and distinct environmental adaptive histories. Two powerful advantages of making use of strains that are closely related to well-characterized biological control agents, either for application in new environments or for use in mixtures, is that all of the strains need not be well characterized and the formulation of the product for application may be simplified since the existing technology for the related strains will suffice.

There is evidence that combinations of like biological control strains may suppress plant disease more effectively than single isolates (Kloepper 1983; Sivasithamparam and Parker 1978; Weller and Cook 1983; Pierson and Weller 1994). In a careful study of the suppression of take-all of wheat caused by *Gaeumannomyces graminis* var. *tritici,* various combinations of *Pseudomonas* strains were shown to be more effective than single isolates in controlling disease (Pierson and Weller 1994). However, no single combination of strains was superior in all field trials. Surprisingly, many of the *Pseudomonas* strains that comprised effective combinations inhibited or were inhibited by other strains in the combination in culture. Thus, compatibility among biological control strains may not be required for effective disease suppression. Pierson and Weller (1994) propose several thoughtful explanations for the superior disease suppressiveness of combinations of *Pseudomonas* strains. The genetic diversity in a combination of strains establishes a more stable rhizosphere community that is better equipped to respond to environmental variations than are single isolates of bacteria. Furthermore, the greater variety of biological control mechanisms and differences in temporal expression of these mechanisms provide greater opportunities for suppressing plant pathogens.

Thus, the fundamental principle—taking advantage of genetic diversity of microorganisms to meet the challenges of diverse environments—may have broad applications in processes that are driven by microorganisms, such as biological control of plant diseases, biological control of pests other than pathogens, or plant growth promotion.

## Exploiting the Genetic Diversity of Host Plants

Our work on biological control is also directed to understanding the role that the plant might play, either directly or indirectly, in the biological control equation. There is a great deal of evidence suggesting that plants have genetic mechanisms that influence microorganisms associated with their roots. We envision the possibility of breeding or engineering plants to be better hosts for biological control agents, or for the ability to recruit specific organisms, or to influence the composition of microbial communities in the rhizosphere to enhance suppression of pathogens.

Relatively little contemporary basic research focuses on the contribution of the plant to noninvasive beneficial associations between plants and microorganisms (Bliss 1991). Previous work has clearly established that plant roots alter the physical and chemical properties of the soil immediately adjacent to them, thereby affecting the microbial communities that develop on and around them (Rovira 1965; Bowden and Rovira 1991). The effect of the root on the microbial flora in the soil is referred to as the "rhizosphere effect" (Lochhead 1940; Rovira 1965).

Many studies have shown that the plant genotype influences the nature and extent of the rhizosphere effect (Lochhead 1940; Lochhead et al. 1940; Neal et al. 1970; Peterson and Rouatt 1967; Timonin 1940, 1946; West and Lochhead 1940; Hebbar et al. 1992). Interpretation of much of this work, however, is limited by the small number of plant genotypes examined and the lack of genetic analysis. More rigorous studies relied on near-isogenic lines that differ in their infectibility by pathogens (Neal et al. 1970, 1973) or nitrogen-fixing symbionts (Elkan 1962) and also differ in the microbial communities that develop on their roots. These studies provide significant support for the hypothesis that a relationship exists between genetic resistance of plants to disease and the structure of microbial communities on the roots of those plants. However, most of the evidence suggesting that microbial community structure influences disease resistance is corollary.

Some recent studies are directed to the idea of improving crop cultivars to support biological control associations with microorganisms (Bliss 1991; Vakili and Bailey 1989; Vakili 1992). Studies of the effect of corn genotype on the efficacy of mycopathogenic biological control agents, including *Gliocladium roseum,* led Vakili to propose that both the host plant and the mycopathogen be selected as a unit of biological control against a given pathogen (Vakili and Bailey 1989; Vakili 1992).

We have begun studies to test the hypothesis that plants have genes that specifically influence associations with (noninvasive) beneficial microbes in the rhizosphere. This work is directed to characterizing microbial communities associated with healthy plants and to a genetic analysis, in alfalfa and tomato, of the contribution of plant genotype to biological control by a specific agent, UW85. Since the ecological parameters that determine successful biological control are still largely unknown, we are taking approaches requiring as few assumptions as possible about the functional roles of the genes we seek.

In tomato, we are taking advantage of the superior genetics, including molecular mapping and a long history of introgression from wild relatives, of this diploid crop. We have screened a range of tomato cultivars for differences in their response to inoculation with UW85 in the presence of virulent isolates of *Pythium* spp. The results indicate that the cultivars within the genus *Lycopersicon esculentum* differ in their responsiveness to the activity of the biological control agent (Smith, Handelsman, and Goodman, unpublished). On the basis of these and other preliminary results, we are now mapping quantitative trait loci (QTLs) in two populations, including recombinant inbred lines derived from a cross between *L. esculentum* and *L. cheesmanii* and a backcross population from a cross between *L. esculentum* and *L. pennellii.*

Cultivated alfalfa is a highly cross-pollinating tetraploid in which commercial cultivars are derived from interbreeding many clones selected by the breeder for various desirable characteristics. The result is a population, referred to by plant breeders as a "synthetic," in which each individual is a unique genotype. Within such populations, which resemble humans in their genetic heterogeneity, we have found genotypes that differ in their responsiveness to the biological control activity of UW85 when they are coinoculated with virulent isolates of *Pythium* spp. (Johnson, Handelsman, Grau, and Goodman, unpublished). When the populations are inoculated with *Pythium* spp. zoospores alone, without UW85, 100 percent of the population develops disease. Based on this observation, we set out to utilize (and compare) both divergent selection and mass selection to develop new alfalfa populations with increased responsiveness to biological control by UW85. These studies are in progress; based on early assessment of the performance of the selected populations in comparison with controls, the evidence is consistent with a role for plant genotype in biological control and suggestive of its heritability at a useful level (Johnson, Handelsman, Grau, and Goodman, unpublished).

During the course of the studies on alfalfa and biological control, we stumbled across an unexpected result that illuminates some of the potential for interactions between plant genotype and associative microbes such as the biological control agent UW85. In several different cultivars (populations) of alfalfa, we have found seeds that have, and others that do not have, a diffusible inhibitory material that is active against UW85 in an agar diffusion assay (Milner et al. 1995; Blackson, Milner, Olivares, and Handelsman, unpublished). The frequency of seeds with the inhibitory activity varies among the alfalfa cultivars tested. This inhibitory "spermosphere" activity, which among the rhizosphere microorganisms tested so far is quite specific to *B. cereus* strains, could influence biological control in several ways. For example, if the activity is bactericidal, it could prevent biological control in the root zones of those plants arising from seeds with the activity. In this case, breeding might be directed toward reducing or eliminating its frequency in alfalfa cultivars. Alternatively, a bacteriostatic material that is highly diffusible might enhance biological control, for example by delaying UW85 spore germination or population growth during the early phases of seed germination when biological control activity might be less important than thereafter, when the root is elongating and susceptible to infection by pathogens. In either case, components of seed diffusates or root exudates that modulate microbial population dynamics or community structure are attractive targets for breeding to improve the plant-microbial partnership in biological control with microorganisms.

With UW85, information about its requirements for growth and for antibiotic production (Milner et al. 1995) suggests an approach to altering host plants so as to better support the activity of UW85 as a biological control agent. UW85 requires exogenous threonine for growth. For soil-borne UW85, root exudates may be a source of threonine, and may be the source on which UW85 depends. We propose that root exudates enriched with threonine may alter, perhaps improve, the rhizosphere environment for UW85. We are testing this idea using tobacco plants engineered to produce excess free threonine (Karchi et al. 1993). If these plants also have increased threonine in their root exudates, or if such plants could be developed, we may have an opportunity to test the idea that rhizospheres can be nutritionally biased by genetic engineering to favor biological control.

## Beyond Antibiosis: Microbial Community Ecology

Throughout our investigations of mechanisms of biological control by the biological control organism, the contribution of the genetic diversity of the population of the biological control agent and the contribution of the plant and plant genotypic diversity, we continually return to the question of what is going on with the microbial community and how is it perturbed by practices associated with biological control applications.

To begin to examine this question, we studied the effect of UW85 on the microbial community that develops on emerging soybean roots in the field (Gilbert et al. 1993). That work demonstrated that treatment with UW85 can lead to the development of a consortium of other bacteria that resemble communities in soil more than do the communities that are typical of roots. We do not yet know whether this change in the community is responsible for biological control, but the results are consistent with a large body of older research showing that treatments that lead to reduced root disease influence the microbial communities on roots in a manner that makes them resemble soil more than root microflora (Gilbert et al. 1994 and references therein).

It is remarkable that UW85 can exert such strong effects on the composition of the microbial community and on plant health, although it does not establish itself as a dominant member of the heterotrophic, culturable community that develops on soybean roots grown from UW85-treated seeds (Halverson et al. 1993a, 1993b). UW85 persists on seeds and on roots from planting to harvest, but it makes up less than 0.1 percent of culturable bacteria recovered from roots as early as fourteen days after planting. Not only does UW85 suppress disease without being a dominant member of the population, but it also promotes nodulation of soybeans, which may contribute to plant health (Halverson and Handelsman 1991). The lack of relationship between efficacy of UW85 and dominance in the rhizosphere is in marked contrast to the observations made with other biological control agents, particularly *Pseudomonas* spp., for which persistent, aggressive colonization of roots is critical for efficacy (Weller 1988; Bull et al. 1991).

The results from our work coupled with previous findings led us to develop the "camouflage hypothesis" (Gilbert et al. 1994), which might provide a new model for thinking about mechanisms of biological control. The camouflage hypothesis proposes that roots whose associated microbial communities resemble the communities found in soil will be disguised, or camouflaged, so that the pathogen will not detect the roots and therefore they will be protected from disease. Although the camouflage hypothesis has not yet been tested, it provides an appealing and somewhat radical model on which to base the design of experiments.

The camouflage hypothesis and the effect that UW85 has on microbial communities raise some important issues about the concepts that have traditionally guided research in biological control. Many researchers have presented models of biological control that conjure up a vision of mechanisms of antibiosis acting as "magic bullets" directed toward the plant pathogen. In this model, a successful "hit" results in disease suppression. This model, derived largely from the agrichemical approach to disease control, reflects general thinking about how conventional pesticides operate. In reality, biological control organisms and the antibiotics they produce may play much more subtle and complex roles in suppressing disease. More research is needed to examine the effects of antibiotics on microbial communities on roots and to determine whether antibiotics directly affect the plant and its ability to defend itself against the pathogen.

## Toward an Understanding of the Microbial Ecology of the Rhizosphere

The extent of microbial biodiversity in the environment is truly astounding. The number of bacterial cells in a gram of soil has been estimated to be between $10^8$ and $10^{10}$. For many natural environments, the proportion of microbes that can be detected by culturing is less than 1 percent of the viable bacteria that can be seen in the microscope. Soil is a dilute and nutrient-poor environment by comparison with a laboratory culture medium. Therefore, many soil organisms are in a quiescent state and are refractory to culturing (Rollins and Colwell 1986; Byrd et al. 1991; Ward et al. 1990). But the numbers tell only part of the story. Even more remarkable than the total number of bacteria is the number of different kinds of microbes in simple environmental samples. One study of forest soil indicated a complexity of DNA sequence consistent with the presence of more than 4,000 distinct bacteria (Torsvik et al. 1990a, 1990b; Wilson 1992). This is more than the total number of bacterial species that have been described since the beginning of microbiology in the nineteenth century! And from the same samples, using plate culturing techniques, Torsvik recovered only about 200 (or four percent) of the different bacteria present. For one rhizosphere community, we find that the culturable population is in the range of five percent of the total bacterial cells observed microscopically for the same samples (Quirino and Goodman, unpublished). So both in terms of numbers and in terms of diversity, the bacterial world that exists in the soil, and in other microbial habitats, is truly an unexplored world.

Molecular studies of environments that are less complex than soil indicate that microbial communities are far more diverse and complex than those inferred from organisms that could be cultured from those environments (Mills and Wassel 1980; Olsen et al. 1986; Stahl et al. 1988; Zelles and Bai 1993; DeLong et al. 1989). The environments studied include hot springs, marine sediments, the picoplankton of the open ocean, the ruminant intestine, hydrothermal vents on the ocean floor, and groundwater (see Angert et al. 1993 and references therein; Giovannoni et al. 1990; Ward et al. 1990).

Our work is directed to extending molecular approaches to analyses of microbial communities in agricultural soils and plant rhizospheres, which, in the long term, will provide knowledge and methods useful for any environmental situation where soils are involved, including biological control. Our approach to the study of microbial communities in soils and rhizospheres is based on the use of ribosomal RNA (rRNA) sequences (Olsen et al. 1986; Ward et al. 1990; Porteous and Armstrong 1991; Torsvik 1980). Total DNA or RNA isolated directly from the environmental sample is amplified using the polymerase chain reaction (PCR). The amplified DNA is used to prepare libraries containing rRNA sequences, from which the DNA sequence can be determined. Many environments have been sampled by this general approach (for some examples see Angert et al. 1993; Giovannoni et al. 1990; Ward et al. 1990), and databases now exist for over 2,000 16S rRNA gene sequences and about 200 23S rRNA sequences (Neefs et al. 1993; Gutell 1993; Gutell et al. 1993; Larsen et al. 1993).

We anticipate that a 16S rRNA gene-based approach to devising a census of microbial communities in soil or the rhizosphere will reveal that the communities are highly complex and contain rRNA sequences, and therefore bacteria, that have never before been recorded. A goal of our work is to develop and test methods in which labeled 16S RNA sequences, amplified from environmental DNA samples, will be hybridized to large arrays of oligonucleotides that have been designed to detect and discriminate between specific 16S RNA sequences in complex mixtures. The hybridization arrays will be designed to reveal patterns that will be diagnostic of the structure of the microbial community from which the DNA was amplified. The ability to quantify changes in microbial communities following perturbations of and additions to the soil environment, such as biological control inoculants, would provide new and needed insights into microbial ecosystems.

## Conclusions

We predict that a significant part of the answer to the effectiveness of UW85 and other biological control agents will be the influence the agent has on the composition and dynamics of microbial communities. We are also expecting that the identification of functions provided by the plant that support effective biological control will point to some further ecological principles on which a better understanding of systems can be built. Thus, we are expecting that it will be the integrative effects on communities of micro- and macroorganisms that will truly explain biological control.

We envision the kind of work described here as being the infancy of new approaches in which biological control will be based on mimicking the workings of nature and be more responsive to management of the soil, perhaps not even requiring amendment or supplementation with nonindigenous organisms. In limited ways such approaches, termed organic or biodynamic, have worked empirically. Can we understand these approaches well enough at the ecological level that the salient features of their successes could be implemented in the mainstream? We need to learn about the microbial world around us and gain a better understanding of its dynamics and regulation. We must find new tools with which to "see" the microbial world and study its intricacies and learn how to employ emerging tools in agricultural research. We must be able to address fundamental questions about microbes in the agroecosystem. Which organisms are the important ones in supporting healthy plant growth? What in their repertoire of biological activities should be encouraged and what discouraged in a productive agricultural system? How do plants and rhizosphere microflora communicate? An understanding of microbial populations and communities offers tremendous potential for developing knowledge from which we will be able to develop new approaches to disease control for the future.

## Addendum

We have identified recently antibiotic B as the aminoglycoside, kanosamine (Milner et al. 1996a). A method for detecting zwittermicin A–producing isolates, based on PCR-amplification of a zwittermicin A resistance gene (Milner et al. 1996b) has successfully identified zwittermicin A–producing isolates from diverse soils (Raffel et al. 1996). Studies with tomato have demonstrated cultivar differences in responsiveness to biological control in the presence of *Pythium* (Smith et al. 1996). Research by other workers published after April 1995 has not been incorporated into the text.

## Acknowledgments

We are indebted to Greg Gilbert for conceptualizing the camouflage hypothesis and enriching our thinking about microbial community ecology, to Eric Stabb for shaping our thinking about genetic diversity of microorganisms, and to Craig Grau for contributing to our thinking about the role of the host genotype in biological control. This work was supported by National Science Foundation grant DCB-8819401, the McKnight Foundation, the Midwest Plant Biotechnology Consortium, and Hatch funding from the University of Wisconsin–Madison College of Agricultural and Life Sciences.

## References

Angert, E. R., K. D. Clemments, and N. R. Pace. 1993. The largest bacterium. *Nature* 362: 239–41.

Anonymous. 1987. *Our common future.* World Commission on Environment and Development. Oxford: Oxford University Press.

Bliss, F. A. 1991. Breeding plants for enhanced beneficial interactions with soil microorganisms. In *Plant breeding for the 1990s,* ed. H. T. Stalker and J. P. Murphy, 251–77. Oxon, U.K.: CAB International.

Bowen, G. D., and A. D. Rovira. 1991. The rhizosphere: The hidden half of the hidden half. In *Plant roots: The hidden half,* ed. Y. Waisel, A. Eshel, and U. Kafkafi, 641–69. New York: Marcel Dekker.

Bull, C. T., D. M. Weller, and L. S. Thomashow. 1991. Relationship between root colonization and suppression of *Gaeumannomyces graminis* var. *tritici* by *Pseudomonas fluorescens* strain 2–79. *Phytopathology* 81:954–59.

Burr, T. J., C. L. Reid, B. H. Katz, M. E. Tagliati, C. Bazzi, and D. I. Breth. 1993. Failure of *Agrobacterium radiobacter* strain K-84 to control crown gall on raspberry. *HortScience* 28:1017–19.

Byrd, J. J., H. S. Xu, and R. R. Colwell. 1991. Viable but nonculturable bacteria in drinking water. *Applied and Environmental Microbiology* 57:875–78.

Campbell, R. 1994. Biological control of soil-borne diseases: Some present problems and different approaches. *Crop Protection* 13:4–13.

Carlton, B. C. 1993. Development of improved bioinsecticides based on *Bacillus thuringiensis.* In *Pest control with enhanced environmental safety,* ed. S. O. Duke, J. J. Menn, and J. R. Plimmer, 258–66. Washington, D.C.: American Chemical Society.

Chen, X., and W. Xiang. 1986. A strain of *Agrobacterium radiobacter* inhibits growth and gall formation by biotype III strains of *A. tumefaciens* from grapevine. *Acta Microbiologica Sinica* 26:193–99.

Clarke, G.R.G., J. A. Leigh, and C. J. Douglas. 1992. Molecular signals in the interactions between plants and microbes. *Cell* 71:191–99.

Colbert, S. F., M. N. Schroth, A. R. Weinhold, and M. Hendson. 1993. Enhancement of population densities of *Pseudomonas putida* PpG7 in agricultural ecosystems by selective feeding with the carbon source salicylate. *Applied and Environmental Microbiology* 59:2064–70.

Cooksey, D. A., and L. W. Moore. 1982. Biological control of crown gall with an agrocin mutant of *Agrobacterium radiobacter. Phytopathology* 72: 919–21.

Deacon, J. W., and L. A. Barry. 1993. Biocontrol of soil-borne plant pathogens: Concepts and their application. *Pesticide Science* 37:417–26.

DeLong, E. F., G. S. Wickham, and N. R. Pace. 1989. Phylogenetic strains: Ribosomal RNA-based probes for the identification of single cells. *Science* 243:1360–63.

Donner, S. C., D. A. Jones, N. C. McClure, G. M. Rosewarne, M. E. Tate, A. Kerr, N. N. Fajardo, and B. G. Clare. 1993. Agrocin 434, a new plasmid encoded agrocin from the biocontrol *Agrobacterium* strains K84 and K1026, which inhibits biovar 2 agrobacteria. *Physiological and Molecular Plant Pathology* 42:185–94.

Elkan, G. H. 1962. Comparison of rhizosphere microorganisms of genetically related nodulating and non-nodulating soybean lines. *Canadian Journal of Microbiology* 8:79–87.

Ellis, J. G., and A. Kerr. 1978. Developing biological control agents for soil-borne pathogens. In *Proceedings of the Fourth International Conference on Plant Pathogenic Bacteria,* ed. M. Ride, 245–50. Angers: Station de pathologie vegetable et phytobacteriologie, INRA.

Feitelson, J. S., J. Payne, and L. Kim. 1992. *Bacillus thuringiensis:* Insects and beyond. *Biotechnology* 10:271–75.

Fravel, D. R. 1988. Role of antibiosis in the biocontrol of plant diseases. *Annual Review of Phytopathology* 26:75–91.

Gilbert, G. S., J. Handelsman, and J. L. Parke. 1990. Role of ammonia and calcium in lysis of zoospores of *Phytophthora cactorum* by *Bacillus cereus* strain UW85. *Experimental Mycology* 14:1–8.

______. 1994. Root camouflage and disease control. *Phytopathology* 84:222–25.

Gilbert, G. S., J. L. Parke, M. K. Clayton, and J. Handelsman. 1993. Effects of an introduced bacterium on bacterial communities on roots. *Ecology* 74:840–54.

Giovannoni, S. J., T. B. Britschgi, C. L. Moyer, and K. G. Field. 1990. Genetic diversity in Sargasso Sea bacterioplankton. *Nature* 345:60–63.

Gutell, R. R. 1993. Collection of small subunit 16S- and 16S-like ribosomal RNA structures. *Nucleic Acids Research* 21:3051–54.

Gutell, R. R., M. W. Gray, and M. N. Schnare. 1993. A compilation of large subunit 23S and 23S-like ribosomal RNA structures. *Nucleic Acids Research* 21:3055–74.

Gutterson, N. 1990. Microbial fungicides: recent approaches to elucidating mechanisms. *Critical Review of Biotechnology* 10:69–91.

Halverson, L. J., M. K. Clayton, and J. Handelsman. 1993a. Population biology of *Bacillus cereus* UW85 in the rhizosphere of field-grown soybeans. *Soil Biology and Biochemistry* 25:485–93.

______. 1993b. Variable stability of antibiotic resistance markers in *Bacillus cereus* UW85 in the soybean rhizosphere in the field. *Molecular Ecology* 2:65–78.

Halverson, L. J., and J. Handelsman. 1991. Enhancement of soybean nodulation by *Bacillus cereus* UW85 in the field and in a growth chamber. *Applied and Environmental Microbiology* 57:2767–70.

Hamdan, H., D. M. Weller, and L. S. Thomashow. 1991. Relative importance of fluorescent siderophores and other factors in biological control of *Gaeumannomyces graminis* var. *tritici* by *Pseudomonas fluorescens* 2-79 and M4-80R. *Applied and Environmental Microbiology* 57:3270–77.

Handelsman, J., W. C. Nesmith, and S. J. Raffel. 1991. Microassay for biological and chemical control of infection of tobacco by *Phytophthora parasitica* var. *nicotianae. Current Microbiology* 22:317–19.

Handelsman, J., S. Raffel, E. H. Mester, L. Wunderlich, and C. R. Grau. 1990. Biological control of damping-off of alfalfa seedlings with *Bacillus cereus* UW85. *Applied and Environmental Microbiology* 56:713–18.

He, H., L. A. Silo-Suh, J. Handelsman, and J. Clardy. 1994. Zwittermicin A, an antifungal and plant protection agent from *Bacillus cereus. Tetrahedron Letters* 35:2499–502.

Hebbar, K. P., A. G. Davey, J. Merrin, T. J. McLoughlin, and P. J. Dart. 1992. *Pseudomonas cepacia.* A potential suppressor of maize soil-borne diseases—seed inoculation and maize root colonization. *Soil Biology and Biochemistry* 24:999–1007.

Hendson, M., L. Askjaer, J. A. Thomson, and M. Montagu. 1983. Broad-host range agrocin of *Agrobacterium tumefaciens. Applied and Environmental Microbiology* 45:1526–32.

Howell, C. R., and R. D. Stipanovic. 1979. Control of *Rhizoctonia solani* on cotton seedlings with *Pseudomonas fluorescens* and with an antibiotic produced by the bacterium. *Phytopathology* 69:480–82.

______. 1980. Suppression of *Pythium ultimum*-induced damping-off of cotton seedlings by *Pseudomonas fluorescens* and its antibiotic, pyoluteorin. *Phytopathology* 70:712–15.

Jacobsen, B. J., and P. A. Backman. 1993. Biological and cultural plant disease controls: Alternatives and supplements to chemicals in IPM systems. *Plant Disease* 77:311–15.

Karchi, H., O. Shaul, and G. Galili. 1993. Seed-specific expression of a bacterial desensitized aspartate kinase increases the production of seed threonine and methionine in transgenic tobacco. *Plant Journal* 3:721–27.

Keel, C., U. Schnider, M. Maurhofer, C. Voisard, J. Laville, U. Burger, P. Wirthner, D. Haas, and G. Defago. 1992. Suppression of root diseases by *Pseudomonas fluorescens* CHAO: Importance of the bacterial secondary metabolite 2,4-diacetylphloroglucinol. *Molecular Plant-Microbe Interactions* 5:4–13.

Kerr, A. 1980. Biological control of crown gall through production of agrocin 84. *Plant Disease* 64:24–30.

Kerr, A., and K. Htay. 1974. Biological control of crown gall through bacteriocin production. *Physiological Plant Pathology* 4:37–44.

Kloepper, J. W. 1983. Effect of seed piece inoculation with plant growth–promoting rhizobacteria on populations of *Erwinia carotovora* on potato roots and in daughter tubers. *Phytopathology* 73:217–19.

Larsen, N., G. J. Olsen, B. L. Maidak, M. J. McCaughey, R. Overbeek, T. J. Macke, T. L. Marsh, and C. R. Woese. 1993. The ribosomal database project. *Nucleic Acids Research* 21: 3021–23.

Laville, J., C. Voisard, C. Keel, M. Maurhofer, G. Defago, and D. Haas. 1992. Global control in *Pseudomonas fluorescens* mediating antibiotic synthesis and suppression of black root rot of tobacco. *Proceedings of the National Academy of Science USA* 89:1562–66.

Lochhead, A. G. 1940. Qualitative studies of soil micro-organisms. III. Influence of plant growth on the character of the bacterial flora. *Canadian Journal of Research* 18:42–53.

Lochhead, A. G., M. I. Timonin, and P. M. West. 1940. The microflora of the rhizosphere in relation to resistance of plants to soil-borne pathogens. *Scientific Agriculture* 20:414–18.

Mahafee, W. F., and P. A. Backman. 1993. Effects of seed factors on spermosphere and rhizosphere colonization by *Bacillus subtilis* GB03. *Phytopathology* 83:1120–25.

Manker, D. C., W. D. Lidster, R. L. Starnes, and S. C. MacIntosh. 1994. Potentiator of *Bacillus* pesticidal activity. Patent Cooperation Treaty #W094/09630.

Martin, J. F., and A. L. Demain. 1980. Control of antibiotic biosynthesis. *Microbiological Reviews* 44:230–51.

Maurhofer, M., C. Keel, D. Haas, and G. Defago. 1994. Pyoluteorin production by *Pseudomonas fluorescens* strain CHAO is involved in the suppression of *Pythium* damping-off of cress but not of cucumber. *European Journal of Plant Pathology* 100:221–32.

Maurhofer, M., C. Keel, U. Schnider, C. Voisard, D. Haas, and G. Defago. 1992. Influence of enhanced antibiotic production in *Pseudomonas fluorescens* strain CHAO on its disease suppressive capacity. *Phytopathology* 82:190–95.

Mills, A. L., and R. A. Wassel. 1980. Aspects of diversity measurement for microbial communities. *Applied and Environmental Microbiology* 40:578–86.

Milner, J. L., S. Raffel, and J. Handelsman. 1995. Culture conditions that influence the accumulation of zwittermicin A. *Applied Microbiology and Biotechnology* 43:685–91.

Milner, J. L., L. A. Silo-Suh, J. C. Lee, H. He, J. Clardy, and J. Handelsman. 1996a. Production of kanosamine by *Bacillus cereus* UW85. *Applied and Environmental Microbiology* 62:3061–5.

Milner, J. L., E. A. Stohl, and J. Handelsman. 1996b. Zwittermicin A resistance gene from *Bacillus cereus. Journal of Bacteriology* 178:4266–72.

Nasch, A., C. Keel, H. A. Pfirter, D. Haas, and G. Defago. 1994. Contribution of the global regulator gene *gacA* to persistence and dissemination of *Pseudomonas fluorescens* biocontrol strain CHAO introduced into soil microcosms. *Applied and Environmental Microbiology* 60:2553–60.

Neal, J. L. Jr., T. G. Atkinson, and R. I. Larson. 1970. Changes in the rhizosphere microflora of spring wheat induced by disomic substitution of a chromosome. *Canadian Journal of Microbiology* 16:153–58.

Neal, J. L. Jr., R. I. Larson, and T. G. Atkinson. 1973. Changes in rhizosphere populations of selected physiological groups of bacteria related to substitution of specific pairs of chromosomes in spring wheat. *Plant and Soil* 39:209–12.

Neefs, J. M., Y. Van de Peer, P. De Rijk, S. Chapelle, and R. De Wachter. 1993. Compilation of small ribosomal subunit RNA structures. *Nucleic Acids Research* 21:3025–49.

Olsen, G. J., D. J. Lane, S. J. Giovannoni, N. R. Pace, and D. A. Stahl. 1986. Microbial ecology and evolution: A ribosomal RNA approach. *Annual Review of Microbiology* 40: 337–65.

Osburn, R. M., J. L. Milner, E. S. Oplinger, P. S. Smith, and J. Handelsman. 1995. Effect of *Bacillus cereus* UW85 on the yield of soybean at two field sites in Wisconsin. *Plant Disease* 79:551–56.

O'Sullivan, D. J., and F. O'Gara. 1992. Traits of fluorescent *Pseudomonas* spp. involved in suppression of plant root pathogens. *Microbiological Reviews* 56:662–76.

Peterson, E. A., and J. W. Rouatt. 1967. Soil microorganisms associated with flax root. *Canadian Journal of Microbiology* 13:199–203.

Phipps, P. M. 1992. Evaluation of biological agents for control of Sclerotinia blight of peanut, 1991. *Biological and Cultural Tests for Control of Plant Disease* 7:60.

Pierson, E. A., and D. M. Weller. 1994. Use of mixtures of fluorescent pseudomonads to suppress take-all and improve the growth of wheat. *Phytopathology* 84:940–47.

Pierson, L. S., V. D. Keppenne, and D. W. Wood. 1994. Phenazine antibiotic biosynthesis in *Pseudomonas aureofaciens* 30-84 is regulated by PhzR in response to cell density. *Journal of Bacteriology* 176:3966–74.

Pierson, L. S., and L. S. Thomashow. 1992. Cloning and heterologous expression of the phenazine biosynthetic locus from *Pseudomonas aureofaciens* 30-84. *Molecular Plant-Microbe Interactions* 5:330–39.

Porteous, L. A., and J. L. Armstrong. 1991. Recovery of bulk DNA from soil by a rapid, small-scale extraction method. *Current Microbiology* 22:345–48.

Powel, K. A., and A. R. Jutsum. 1993. Technical and commercial aspects of biocontrol products. *Pesticide Science* 37:315–21.

Raffel, S. J., E. V. Stabb, J. L. Milner, and J. Handelsman. 1996. Genotypic and phenotypic analysis of zwittermicin A–producing strains of *Bacillus cereus. Microbiology,* in press.

Rollins, D. M., and R. R. Colwell. 1986. Viable nonculturable stage of *Campylobacter jejuni* and its role in survival in the natural aquatic habitat. *Applied and Environmental Microbiology* 52:531–35.

Rovira, A. D. 1965. Interactions between plant roots and soil microorganisms. *Annual Review of Microbiology* 19:241–66.

Shim, J. S., S. K. Farrand, and A. Kerr. 1987. Biological control of crown gall: Construction and testing of new biocontrol agents. *Phytopathology* 77:463–66.

Silo-Suh, L. A. 1994. Biological activities of two antibiotics produced by *Bacillus cereus* UW85. Ph.D. thesis, University of Wisconsin–Madison.

Silo-Suh, L. A., B. J. Lethbridge, S. J. Raffel, H. He, J. Clardy, and J. Handelsman. 1994. Biological activities of two fungistatic antibiotics produced by *Bacillus cereus* UW85. *Applied and Environmental Microbiology* 60:2023–30.

Sivasithamparam, K., and C. A. Parker. 1978. Effects of certain isolates of bacteria and actinomycetes on *Gaeumannomyces graminis* var. *tritici* and take-all of wheat. *Australian Journal of Botany* 26:773–82.

Slininger, P. J., and M. A. Jackson. 1992. Nutritional factors regulating growth and accumulation of phenazine-1-carboxylic acid by *Pseudomonas fluorescens* 2-79. *Applied Microbiology and Biotechnology* 37:388–92.

Smith, K. P., M. J. Havey, and J. Handelsman. 1993. Suppression of cottony leak of cucumber with *Bacillus cereus* strain UW85. *Plant Disease* 77:139–42.

Smith, K. P., J. Handelsman, and R. M. Goodman. 1996. Modeling of dose-response relationships in biological control: Partitioning host responses to the pathogen and the biological control agent. *Phytopathology*, submitted.

Stabb, E. V., L. M. Jacobson, and J. Handelsman. 1994. Zwittermicin A–producing strains of *Bacillus cereus* from diverse soils. *Journal of Bacteriology* 60:4404–12.

Stahl, D. A., B. Flesher, H. R. Mansfield, and L. Montgomery. 1988. Use of phylogenetically based hybridization probes for studies of ruminal microbial ecology. *Applied and Environmental Microbiology* 54:1079–84.

Thomashow, L. S., D. M. Weller, R. F. Bonsall, and L. S. Pierson III. 1990. Production of the antibiotic phenazine-1-carboxylic acid by fluorescent *Pseudomonas* species in the rhizosphere of wheat. *Applied and Environmental Microbiology* 56:908–12.

Timonin, M. I. 1940. The interaction of higher plants and soil micro-organisms. II. Study of the microbial population of the rhizosphere in relation to resistance of plants of soil-borne diseases. *Canadian Journal of Research* 18:444–55.

______. 1946. Microflora of the rhizosphere in relation to the manganese-deficiency disease of oats. *Soil Science Society Proceedings* 10:284–92.

Torsvik, V. L. 1980. Isolation of bacterial DNA from soil. *Soil Biology and Biochemistry* 12:15–21.

Torsvik, V., J. Goksøyr, and F. L. Daae. 1990b. High diversity in DNA of soil bacteria. *Applied and Environmental Microbiology* 56:782–87.

Torsvik, V. L., K. Salte, R. Sørheim, and J. Goksøyr. 1990a. Comparison of phenotypic diversity and DNA heterogeneity in a population of soil bacteria. *Applied and Environmental Microbiology* 56:776–81.

Turner, J. T., and P. A. Backman. 1991. Factors relating to peanut yield increases after seed treatment with *Bacillus subtilis*. *Plant Disease* 75:347–52.

Vakili, N. G. 1992. Biological seed treatment of corn with mycopathogenic fungi. *Journal of Phytopathology* 134:313–23.

Vakili, N. G., and T. B. Bailey Jr. 1989. Yield response of corn hybrids and inbred lines to phylloplane treatment with mycopathogenic fungi. *Crop Science* 29:183–90.

Voisard, C., C. Keel, D. Haas, and G. Defago. 1989. Cyanide production by *Pseudomonas fluorescens* helps suppress black root rot of tobacco under gnotobiotic conditions. *EMBO Journal* 8:351–58.

Ward, D. M., R. Weller, and M. M. Bateson. 1990. 16S rRNA sequences reveal numerous uncultured microorganisms in a natural community. *Nature* 345:63–65.

Webster, J., M. Dos Santos, and J. A. Thomson. 1986. Agrocin-producing *Agrobacterium tumefaciens* strain active against grapevine isolates. *Applied and Environmental Microbiology* 52:217–19.

Weller, D. M. 1988. Biological control of soil-borne plant pathogens in the rhizosphere with bacteria. *Annual Review of Phytopathology* 26:379–407.

Weller, D. M., and R. J. Cook. 1983. Suppression of take-all of wheat by seed treatments with fluorescent pseudomonads. *Phytopathology* 73:463–69.

West, P. M., and A. G. Lochhead. 1940. Qualitative studies of soil micro-organisms. IV. The rhizosphere in relation to the nutritive requirements of soil bacteria. *Canadian Journal of Research* 18:129–35.

Wilson, E. O. 1992. *The diversity of life.* Cambridge, Mass.: Belnap Press.

Zelles, L., and Q. Y. Bai. 1993. Fractionation of fatty acids derived from soil lipids by solid phase extraction and their quantitative analysis by GC-MS. *Soil Biology and Biochemistry* 25:495–507.

# 7

# Microbial Competition and Plant Disease Biocontrol

*Linda L. Kinkel and Steven E. Lindow*

Competitive interactions among microbes on plant surfaces can influence disease development, and considerable efforts are now concentrated towards increasing the effectiveness of resource competition in reducing pathogen populations and disease. The practical use of competition as a strategy for biocontrol, however, is still very limited. A number of factors constrain the effective use of competition in biocontrol: (1) a lack of efficient or relevant strategies for screening effective competitors, (2) a limited understanding of the diversity and specificity of competitive interactions among epiphytic microbial populations, and (3) an absence of data on the habitat conditions under which competitive interactions may be most intense. In this chapter, we will review briefly systems in which competitive interactions are important in disease biocontrol. *Pseudomonas syringae* is used as a model to consider the types of ecological information needed to enhance competition-based biocontrol.

## Defining Competition

Competition between coexisting populations can be broken down into at least two distinct categories (Roughgarden 1979): competition for limiting resources (exploitative competition); and interaction through some form of physical or chemical attack (interference competition) (Arthur 1987). In addition, apparent competition, in which members of coexisting species appear to be negatively influenced by one another, can result from the activities of a predator consuming members of both coexisting populations (Arthur 1987). In biological control, interference competition is commonly the result of the production by microbial antagonists of antibiotics effective in inhibiting the pathogen population. This topic is addressed in Milner et al. (Chapter 6, in this volume). Even in cases where antibiotic production is the major mechanism by which pathogen populations and disease development are reduced, however, exploitative competitive ability may influence the outcome of the interaction (Handelsman and Parke 1989; Weller 1988). In this chapter, we will discuss competition as exploitative, or resource competition. The outcome of resource competition is the population reduction of coexisting species (Paulitz 1990; Roughgarden 1979).

## Examples of Competition in Biocontrol

The suggestion that competitive interactions among coexisting microbial populations may be useful in limiting disease development is not new (Hartley 1921; Leben 1965; Fokkema 1971; Last and Warren 1972). The importance of competition in community or population dynamics, however, is difficult to quantify (Lewin 1983a, 1983b; Connor and Simberloff 1986; Peters 1991). In general, a rigorous, direct, definitive strategy for determining that competitive interactions play a predominant role in biocontrol does not exist (Handelsman and Parke 1989). Rather, competitive interactions are usually inferred to be a primary mechanism by one or two forms of indirect evidence. Nutrient limitation and therefore competition have been inferred from observations that epiphytic or pathogenic microbial populations are enhanced following the addition of nutrients (Morris and Rouse 1985; Alabouvette et al. 1986; Fokkema et al. 1979, 1983). Additionally, competition has been inferred in cases where it has been shown that the presence of one population has a detrimental effect on the population size of a coexisting microbe and that antibiosis, predation, and parasitism do not play significant roles in the interactions (Lindow 1988). For example, for antibiosis, if non-antibiotic producing mutants of a potential biocontrol agent are as effective in reducing pathogen populations as the antibiotic-producing parent strains (Handelsman and Parke 1989), or if there is correlative evidence that antibiotic production is independent of biocontrol efficacy, then exploitative competition is inferred to be the mechanism by which biocontrol is achieved. Although both of these indirect approaches can provide evidence consistent with the explanation that resource competition limits pathogen populations, alternative explanations are possible (Morris and Rouse 1985; Handelsman and Parke 1989). For example, the demonstration that pathogenic populations are increased following nutrient applications to plants can suggest that the applied nutrient(s) are toxic or inhibitory to antagonistic microflora (Morris and Rouse 1985; Handelsman and Parke 1989). In addition, although antibiosis, parasitism, and predation may be ruled out through the use of mutants or microscopic examination of coexisting microbes on plants, either resource competition or microbial habitat alteration—such as surfactants, pH modification—remain as possible mechanisms of biocontrol. There is, unfortunately, relatively little information on the potential importance of habitat alteration on epiphytic microbial population dynamics, though habitat alteration has been suggested to play a role in control in some systems (Wilson and Lindow 1993). Experiments to distinguish habitat alteration from resource competition are difficult to conceive. In reality, resource competition as a mechanism for biocontrol exists along a continuum of microbial interactions (predation-parasitism-antibiosis-habitat alteration-resource competition) that are not always discrete or independent. Resource competition is often settled upon as a default explanation when other processes have been eliminated as potential explanations.

Despite the difficulties in definitively resolving mechanisms, competitive interactions have been implicated in biocontrol on a wide range of different plant parts. One of the earliest practical suggestions for biocontrol based upon microbial inoculations was for the use of *Peniophora gigantea* as a pine stump protectant against *Heterobasidion (Fomes) annosus* (Rishbeth 1957, 1963). *Peniophora gigantea* is a vigorous stump colonizer and was shown to be effective in competitively excluding the pathogen and also, in some cases, to replace *H. annosus* on stumps. Rishbeth (1957, 1963) attributed part of the success of competition in this system to the fact that the infection court for the pathogen is a highly selective "virgin substrate," thus eliminating much of the biological complexity encountered by biocontrol agents at the root or leaf surface.

Competitive interactions have also been shown to account for significant levels of disease control on both flowers (Wilson and Lindow 1993) and leaves (Blakeman and Brodie 1977; Andrews 1992). For example, pear and apple blossoms are protected against fire blight (caused by *Erwinia amylovora*) by the application of *E. herbicola* or *Pseudomonas fluorescens.* The data suggest that competitive or pre-emptive exclusion is a primary mechanism by which pathogen populations are inhibited, though antibiotic production by the antagonist may enhance control. On leaves, a number of different bacterial inoculants have been used in the control of frost injury caused by high populations of ice nucleation–active bacteria (Lindemann and Suslow 1987; Lindow et al. 1983; Lindow 1987). The lack of any significant correlation between antibiotic production and levels of frost control achieved by different bacterial biocontrol agents, plus the success of non-antibiotic-producing mutants relative to their parent strains in biocontrol (Lindow 1988), is consistent with the hypothesis that competitive interactions account for the inhibition of ice-nucleating bacterial populations. Inoculation of biocontrol agents prior to the arrival of ice-nucleating populations is critical to effective biocontrol in this system (Lindemann and Suslow 1987).

In the rhizosphere, competition for iron that is mediated through the production of siderophores by *Pseudomonas fluorescens* is a factor in the biological control of "minor" pathogens (Kloepper et al. 1980; Handelsman and Parke 1989). Addition of iron to the soil reduces the level of control achieved using *P. fluorescens* (Kloepper et al. 1980). Resource competition between nonpathogenic and pathogenic *Fusarium oxysporum* can provide biological control of fusarium wilt in a number of different crops (Lemanceau et al. 1993; Park et al. 1988; Elad and Baker 1985). Carbon (glucose) has been suggested to be limiting to the pathogen in the rhizosphere (Lemanceau et al. 1993).

## Developing Competition-based Biocontrol

Although competition has been implicated as a mechanism of biocontrol in a wide variety of systems, we lack a logical strategy for enhancing competition-based biocontrol. Most efforts have used a trial and error approach: "superior" competitors are sought based on lab, greenhouse, or (rarely) field assays that evaluate population outcomes or biocontrol efficacy among a random collection of potential antagonist strains. Not surprisingly, field evaluations of strains identified through such assays often show a tremendous amount of variability in biocontrol efficacy. Efforts to understand how the physical or the biological environment influences the outcome of the interaction have been limited, and have often focused primarily on the survival and establishment of the biocontrol population (Howie et al. 1987; Zhou and Reeleeder 1991; Mazzola and Cook 1991). While survival of the antagonist is obviously critical to the successful outcome of the interaction, it is probably insufficient for developing consistent and effective biocontrol.

Can we do a better job developing competition-based biocontrol systems? As a logical starting point, consider the two fundamental conditions that must be met for competitive interactions to occur in any environment (Arthur 1987). First, there must be some degree of niche overlap between the potential competitors. Second, there must be resource limitation. Resource(s) must be the critical factor limiting population size within the specified habitat as opposed to physical constraints such as temperature, relative humidity, or free moisture. These two conditions suggest areas of research that may help in enhancing biocontrol. The first condition focuses on the niche requirements of the microbial antagonist and pathogen populations of interest. Specifically, it prompts the need to characterize the diversity of resource requirements among individuals within a particular pathogen or pest population, and to determine the factors that can predict competitive exclusion by a particular antagonist. In contrast, the second condition for competition to occur, limited resources, focuses on habitat charactersitics. In particular, under what conditions are resources limiting to microbial populations on plant surfaces? These lines of investigation may provide a logical strategy for identifying the conditions under which competition may prove most successful in biocontrol, for developing logical strategies to select biocontrol antagonists, and for optimizing the ability of these antagonists to exclude pathogen populations.

## An Example: *Pseudmonas syringae* on Leaves

### *Niche Overlap*

Niche overlap is a prerequisite for effective competition-based biological control. Theory suggests that the intensity of competitive interactions between coexisting populations will be partly a function of the degree of niche overlap that exists between them (Arthur 1987; Christiansen and Loeschke 1990). Unfortunately, there is no simple method for measuring the niche overlap of organisms. However, microbes that are closely related should have similar resource needs (Christiansen and Loeschke 1990). Will microbes that are very closely related to the target population be most effective in competitive exclusion? To investigate this possibility in the competitive exclusion of ice nucleation–active strains of *P. syringae* from plants, non-ice nucleation–active strains were constructed from naturally occurring *P. syringae* isolates by deletion of the *ice* gene (Lindow 1986; Lindemann and Suslow 1987). In coinoculation experiments, competitive exclusion of parental strains by the isogenic (with the exception of the *ice* gene) ice bacteria was not consistently more effective than exclusion of non-related *P. syringae* strains (Lindow 1987; Lindemann and Suslow 1987; Kinkel and Lindow 1993). Despite the fact that taxonomic relatedness may reflect similarities in resource needs, in this system relatedness was not a prerequisite for effective competition-based biocontrol.

Focusing on the degree of niche overlap alone is insufficient for understanding the potential of antagonists in biocontrol for another important reason. In our example of an isogenic pair of organisms with presumably identical niche requirements and competitive abilities, the interaction between organisms is probably symmetric. Specifically, the influences of competition on each of the two populations should be equivalent. However, an asymmetric competitive interaction in which one population (the pathogen) is affected more strongly than the other (the biocontrol agent) is preferred in biocontrol. Thus, in addition to resource needs, the ability to efficiently acquire and use resources relative to competitors is an important characteristic.

Successful biocontrol will also depend upon the diversity of resource needs within the pathogen population. Pathogen populations have been shown to exhibit a tremendous amount of variability among individuals in a wide range of traits. For example, there is great diversity in nutrient utilization patterns among individuals in *P. syringae* populations (Morris and Rouse 1985). Even in cases where niche requirements between pathogens and antagonists may differ substantially, however, a single, shared, limiting resource may result in intense competition (e.g., siderophores; Leong and Expert 1989). The extent to which a single antagonist may effectively exclude a wide range of possible pathogen isolates will depend upon both the diversity in niche requirements among pathogen individuals and on the specific limiting resource(s) for those individuals.

To investigate the diversity and specificity of competitive interactions among *P. syringae* strains on leaves, a random collection of twenty-nine *P. syringae* strains, which had been isolated from healthy plant tissue, was inoculated onto leaves in pairwise combinations (Kinkel and Lindow 1993). Each strain was inoculated both as an antagonist (inoculated onto plants on day one) and as a challenge strain (inoculated seventy-two hours later). A total of 107 pairwise combinations were evaluated. Populations of both antagonist and challenge strains were quantified seventy-two hours after the challenge strain inoculation. Plants were maintained under moist conditions conducive to epiphytic bacterial growth throughout the experiment to maximize the probability that nutrients would be limiting and competitive interactions would occur.

The ability of individual strains to colonize leaves on which another strain had already established a population varied significantly among strains, and this invasion ability was dependent upon the identity of the preexisting strain. A strain that was good at successfully invading a leaf already occupied by one *P. syringae* strain was not necessarily good at invading leaves occupied by another strain. Similarly, a strain successful in competitively excluding one *P. syringae* strain from establishing a population on a leaf was not necessarily successful in excluding all possible *P. syringae* colonists. These data indicate a tremendous amount of both diversity and specificity in competitive interactions among *P. syringae* strains on leaves, and are consistent with nutrient data that suggest successful competitive exclusion of a large and diverse naturally occurring *P. syringae* population on leaves by a single bacterial antagonist may be unlikely. However, there were *P. syringae* strains that were more successful on average than other strains in successfully excluding later arriving strains. Among the most successful of these strains, competitive interactions were asymmetric: the excluded strain was strongly influenced by competition while there was little or no influence on the excluding strain. Strains that were less successful in excluding later arrivals either showed no evidence for competitive interactions (no influence of either population on the other, suggesting no niche overlap or no shared, limiting resources), or showed symmetric competitive interactions (suggesting that niche overlap and resource limitation were present and that the strains exhibited similar efficiencies in the acquisition and utilization of the limiting resource[s]).

In summary, niche overlap is certainly a prerequisite for the success of a microbial antagonist in competitively excluding a pathogen population. The data indicate, however, that the degree of niche overlap is not entirely predictive of success in competitive exclusion. In addition to shared resource needs, the abilities to efficiently acquire and utilize resources are critical to determining the outcome of the competitive interaction. Because these are not easily measured, consideration of alternative measures is important to predict competitive success. In particular, the extent to which either microbial growth rate or carrying capacity may be predictive of success in competitive exclusion needs to be investigated for microbes on plants (Grime 1979; Tilman 1982). In the experiments with *P. syringae* on leaves, population size was not generally predictive of the ability of a strain to exclude subsequent arrivals, but among those strains that did successfully exclude later arrivals, higher populations were correlated with more effective competitive exclusion (Kinkel and Lindow 1993). Thus, the best competitors tended to reach relatively higher populations than poorer competitors, but population size alone was not predictive of competitive ability. Finally, our data indicate that the outcome of competitive interactions will be a function of the characteristics of both the antagonist and the specific pathogen strain. Therefore, competition-based biocontrol strategies must consider the diversity in resource needs among individuals within the target population.

### *Resource Limitation*

The second condition required for competition to occur and for useful competitive biocontrol of pathogen populations, is that resources must limit population size.

How frequently and under what conditions are resources limiting to microbes on plant surfaces? This is difficult to address experimentally. Morris and Rouse (1985) suggest that, to demonstrate nutrient limitation, bacterial populations must increase following the application of nutrient to the leaves. As noted earlier, however, there are other possible explanations for an increase in epiphytic populations following nutrient applications. In cases where the limiting nutrient has been determined, conditions under which the nutrient is limiting are less difficult to evaluate. Unfortunately, this is not generally the case. As an alternative, one may determine the conditions under which there is evidence that competitive interactions have a significant influence on epiphytic populations (e.g., Rodger and Blakeman 1984). In these cases, the conclusion is that resources are limiting and that niche overlap occurs among coexisting populations.

To determine the physical conditions under which competition has the greatest influence on population dynamics of *P. syringae* strains on leaves, strain pairs were coinoculated onto potato plants and maintained under varying environmental conditions over five to ten day periods (Kinkel and Lindow 1993). Plants were cycled through twenty-four hour periods of either cool and moist or hot and dry conditions. Competitive interactions were quantified by comparing the mean population size for each bacterial strain alone on leaves with the mean bacterial population size for that strain on leaves to which a second strain had been coinoculated. Among all strain pair combinations (a total of five different strain pairs evaluated over 105 samples), mean bacterial population sizes were more likely to be significantly reduced by competition (e.g., populations were significantly smaller in the presence of a coexisting strain than when alone on leaves) following wet than dry incubation conditions ($\chi^2 = 4.75$, $p = 0.029$). In addition, competition (the presence or absence of a coinoculated bacterial strain) accounted for a significantly greater proportion of the sums of squares in bacterial population size among leaves following wet than dry incubation conditions (mean for wet conditions = 0.266, dry = 0.1474; *t*-test for unequal variance = 3.3974, $p = 0.0009$). Overall, the data indicate that competitive interactions are more likely to be effective as a biocontrol strategy for *P. syringae* under conditions conducive to bacterial growth on leaves. Perhaps more importantly, the data show that the significance of competitive interactions to leaf surface populations varies over time as a function of the physical environment. The varying importance of competition to epiphytic populations may be because under hot, dry conditions the physical environment limits microbial populations, or because the physical environment influences the availability of nutrients on plant surfaces and determines whether specific resources are limiting. Temporal variability in the significance of competitive interactions to the population dynamics of coexisting species may reduce the ultimate likelihood of success of competitive exclusion of a pathogen population by an antagonist (Chesson and Warner 1981; Grover 1988). Alternatively, it suggests there may be windows of opportunity where the physical environment is conducive to competitive exclusion, and that our biocontrol efforts should focus on those conditions. An understanding of when competition is most likely to be successful may help in forecasting biocontrol success or in deciding between biological and chemical control strategies. Such an approach may be more relevant to the phyllosphere, where physical conditions seem to be more variable over short periods of time than in the rhizosphere (Andrews 1992). However, understanding the physical conditions when competition is most successful is critical to enhancing biocontrol success in both the rhizosphere and the phyllosphere.

In addition to the temporal variability in competitive interactions among epiphytic microbes, the characteristic aggregation of nutrients and microbial populations on plant surfaces above- and below-ground (Tukey 1971; Bahme and Schroth 1987; Kinkel and Lindow 1990) also may influence the ability of competitive interactions to consistently exclude a particular target population (Slatkin 1974; Chesson 1985; Hanski 1983). Simulation models suggest that complete competitive exclusion of one population by another is unlikely in spatially or temporally patchy habitats. A better understanding of the role of aggregation in competition among microbial populations in space and in time is needed to determine when resource competition-based biocontrol may be most successful.

## Summary

Competitive interactions can provide the basis for biological control of plant disease, and are important to successful biocontrol even in cases where alternative mechanisms of interaction are critical. Future efforts to enhance competition-based biocontrol must focus on the development of a logical and consistent strategy for enhancing competitive exclusion. Specifically, an evaluation of the diversity and specificity of competitive abilities in pathogen populations and identification of factors (e.g., population growth rate or carrying capacity) that may be predictive of competitive ability should help in screening potential biocontrol agents and in determining the biological settings in which competition may be effective. In addition, critical investigation of the physical conditions in which competitive interactions have the greatest influence on pathogen population dynamics should permit the forecasting of biocontrol efficacy and inform decisions about the use of biocontrol. Currently, we lack information on whether a given set of biological or physical conditions may be generally conducive to competition or whether the conditions required for successful competitive exclusion on plant surfaces are specific to the organisms involved. Such information is critical to determining whether there is a basis for the development of a general model for competition-based biocontrol on plant surfaces, and should be sought in future work.

## References

Alabouvette, C., Y. Couteaudier, and P. Lemanceau. 1986. Nature of intrageneric competition between pathogenic and non-pathogenic *Fusarium* in a wilt-suppressive soil. In *Iron, siderophores, and plant diseases,* ed. T. R. Swinburne, 165–78. New York: Plenum.

Andrews, J. H. 1992. Biological control in the phyllosphere. *Annual Review of Phytopathology.* 30:603–35.

Arthur, W. 1987. *The niche in competition and evolution.* New York: Wiley.

Bahme, J. B., and M. N. Schroth. 1987. Spatial-temporal colonization patterns of a rhizobacterium on underground organs of potato. *Phytopathology* 77:1093–100.

Blakeman, J. P., and I.D.S. Brodie. 1977. Competition for nutrients between epiphytic microorganisms and germination of spores of plant pathogens on beetroot leaves. *Physiological Plant Pathology* 10:29–42.

Chesson, P. L. 1985. Coexistence of competitors in spatially and temporally varying environments: A look at the combined effects of different sorts of variability. *Theoretical Population Biology* 28:263–87.

Chesson, P. L., and R. R. Warner. 1981. Environmental variability promotes coexistence in lottery competitive systems. *American Naturalist* 117:923–43.

Christiansen, F. B., and V. Loeschcke. 1990. Evolution and competition. In *Population biology: Ecological and evolutionary viewpoints,* ed. K. Wohrman and S. K. Jain, 367–94. New York: Springer-Verlag.

Connor, E. F., and D. Simberloff. 1986. Competition, scientific method, and null models in ecology. *American Scientist* 74:153–62.

Elad, Y., and R. Baker. 1985. The role of competition for iron and carbon in suppression of chlamydospore germination of *Fusarium* spp. by *Pseudomonas* spp. *Phytopathology* 75: 1053–59.

Fokkema, N. J. 1971. The effect of pollen in the phyllosphere of rye on colonization by saprophytic fungi and on infection by *Helminthosporium sativum* and other leaf pathogens. *Netherlands Journal of Plant Pathology* 77, suppl. 1:1–60.

Fokkema, N. J., J. G. Den Houter, Y. J. C. Kosterman, and A. L. Nelis. 1979. Manipulation of yeasts on field-grown wheat leaves and their antagonistic effect on *Cochliobolus sativus* and *Septoria nodorum. Transactions of the British Mycological Society* 72:19–29.

Fokkema, N. J., I. Riphagen, R. J. Poot, and C. De Jong. 1983. Aphid honeydew, a potential stimulant of *Cochliobolus sativus* and *Septoria nodorum* and the competitive role of saprophytic mycoflora. *Transactions of the British Mycological Society* 81:355–63.

Grime, J. P. 1979. *Plant strategies and vegetation processes.* New York: Wiley.

Grover, J. P. 1988. Dynamics of competition in a variable environment: Experiments with two diatom species. *Ecology* 69:408–17.

Handelsman, J., and J. L. Parke. 1989. Mechanisms of biocontrol of soilborne plant pathogens. In *Plant-microbe interactions: Molecular and Genetic Perspectives,* Volume 3, ed. T. Kosuge and E. W. Nester, 27–61. New York: McGraw-Hill.

Hanski, I. 1983. Coexistence of competitors in patchy environment. *Ecology* 64:493–500.

Hartley, C. 1921. Damping-off in forest nurseries. *U.S. Department of Agriculture Bulletin* 934:1–99.

Howie, W. J., R. J. Cook, and D. M. Weller. 1987. Effects of soil matric potential and cell motility on wheat root colonization by fluorescent pseudomonads suppressive to take-all. *Phytopathology* 77:286–92.

Kinkel, L. L. and S. E. Lindow. 1990. Spatial distributions of *Pseudomonas syringae* strains on potato leaves. *Phytopathology* 80:1030.

______. 1993. Invasion and exclusion among *Pseudomonas syringae* strains on leaves. *Applied and Environmental Microbiology* 59:3447–54.

Kloepper, J. W., J. Leong, M. Teintze, and M. N. Schroth. 1980. *Pseudomonas* siderophores: A mechanism explaining disease suppressive soils. *Current Microbiology* 4:317–20.

Last, F. T., and R. C. Warren. 1972. Non-parasitic microbes colonizing green leaves: Their form and functions. *Endeavour* 31:143–50.

Leben, C. 1965. Epiphytic microorganisms in relation to plant disease. *Annual Review of Phytopathology* 3:209–30.

Lemanceau, P., P. Bakker, W. J. De Kogel, C. Alabouvette, and B. Schippers. 1993. Antagonistic effect on non-pathogenic *Fusarium oxysporum* Fo47 and Pseudobactin 358 upon pathogenic *Fusarium oxysporum* f. sp. *dianthi*. *Applied and Environmental Microbiology* 59: 74–82.

Leong, S. A., and D. Expert. 1989. Siderophores in plant-pathogen interactions. In *Plant-microbe interactions: Molecular and genetic perspectives,* Volume 3, ed. T. Kosuge and E. W. Nester, 63–83. New York: McGraw-Hill.

Lewin, R. 1983. Predators and hurricanes change ecology. *Science* 221:737–40.

______. 1983. Santa Rosalia was a goat. *Science* 221:636–39.

Lindemann, J. and T. Suslow. 1987. Competition between ice nucleation–active wild type and ice nucleation-deficient deletion mutant strains of *Pseudomonas syringae* and *P. fluorescens* Biovar I and biological control of frost injury on strawberry blossoms. *Phytopathology* 77:882–86.

Lindow, S. E. 1986. Construction of isogenic strains of *Pseudomonas syringae* for evaluation of specificity of competition on leaf surfaces. In *Current perspectives in microbial ecology,* ed. F. Megusar and M. Gantar, 508–15. Lubljana: Slovene Society for Microbiology.

______. 1987. Competitive exclusion of epiphytic bacteria by ice *Pseudomonas syringae* mutants. *Applied and Environmental Microbiology* 53:2520–27.

______. 1988. Lack of correlation of in vitro antibiosis with antagonism of ice nucleation active bacteria on leaf surfaces by non-ice nucleation–active bacteria. *Phytopathology* 78: 444–50.

Lindow, S. E., D. C. Arny, and C. D. Upper. 1983. Biological control of frost injury: An isolate of *Erwinia herbicola* antagonistic to ice nucleation active bacteria. *Phytopathology* 73:1097–102.

Mazzola, M., and R. J. Cook. 1991. Effects of fungal root pathogens on the population dynamics of biocontrol strains of fluorescent pseudomonads in the wheat rhizosphere. *Applied and Environmental Microbiology* 57:2171–78.

Morris, C. E., and D. I. Rouse. 1985. Role of nutrients in regulation epiphytic bacterial populations. In *Biological control on the phylloplane,* ed. C. Windels and S. Lindow, 63–82. St. Paul, Minn.: American Phytopathological Society.

Park, C. S., T. C. Paulitz, and R. Baker. 1988. Biocontrol of *Fusarium* wilt of cucumber resulting from interactions between *Pseudomonas putida* and non-pathogenic isolates of *Fusarium oxysporum*. *Phytopathology* 78:190–94.

Paulitz, T. C. 1990. Biochemical and ecological aspects of competition in biological control. In *New directions in biological control: Alternatives for suppressing agricultural pests and diseases,* ed. R. R. Baker and P. E. Dunn, 5:713–24. New York, Liss.

Peters, R. H. 1991. *A critique of ecology.* Cambridge: Cambridge University Press.

Rishbeth, J. 1957. *Fomes annosus* on stumps. *Transactions of the British Mycological Society* 40:167.

______. 1963. Stump protection against *Fomes annosus*. *Annals of Applied Biology.* 52:63–77.

Rodger, G., and J. P. Blakeman. 1984. Microbial colonization and uptake of 14C label on leaves of sycamore. *Transactions of the British Mycological Society* 82:45–51.

Roughgarden, J. 1979. *Theory of population genetics and evolutionary ecology: An introduction.* New York: Macmillan.

Slatkin, M. 1974. Competition and regional coexistence. *Ecology* 55:128–34.

Tilman, D. 1982. *Resource competition and community structure.* Princeton, N.J.: Princeton University Press.

Tukey, H. B. Jr. 1971. Leaching of substances from plants. In *Ecology of leaf surface microorganisms,* ed. T. F. Preece and C. H. Dickinson, 67–80. London: Academic Press.

Weller, D. M. 1988. Biological control of soilborne plant pathogens in the rhizosphere with bacteria. *Annual Review of Phytopathology.* 26:379–407.

Wilson, M. and S. E. Lindow. 1993. Interactions between the biological control agent *Pseudomonas fluorescens* A506 and *Erwinia amylovora* in pear blossoms. *Phytopathology* 83:117–23.

Zhou, T., and R. D. Reeleder. 1991. Colonization of bean flowers by *Epicoccum purpurascens. Phytopathology* 81:774–78.

# 8

# Ecology of Rearing: Quality, Regulation, and Mass Rearing

*Linda A. Gilkeson*

A 1992 list of biological control agents sold commercially in North America listed 105 species of arthropods to control arthropods and 27 species of arthropods to control weeds (Anon 1992). This is more than double the number of arthropods listed in 1989 (Bezak 1989) and shows the rapid growth in the commercial biological control industry. Although several established insectaries in North America have been producing large quantities of biological control arthropods for more than a decade, it is only within the past four years that industrial-scale production facilities have been built in North America. These produce phytoseiid mites and *Trichogramma* spp. parasitoids for use over very large areas. For example, the phytoseiid mite *Phytoseiulus persimilis* is used to control spider mites in strawberries in California (Grossman 1989), and it may be feasible to release them from the air over corn acreage (Pickett et al. 1987). With the increasing scale of inundative biological control programs, greater numbers of arthropod biological control agents are reared in semi-industrial conditions that differ in many respects from the natural conditions of the intended release site. At this time, the majority of mass-reared species are used in greenhouses, in an artificial environment similar to mass-rearing conditions. They are mostly used on annual crops grown in a protected climate that changes relatively little, with the exception of day length, throughout the season. Crop plants are generally uniform, and predators are distributed manually by greenhouse staff. With greater application outdoors, however, questions arise about the behavior, genetics, health, and overall quality and suitability for release of the arthropods reared in mass-production facilities. This chapter describes some of the goals and constraints of commercial mass production of arthropods, how those goals may conflict with requirements of biological control programs, and quality issues as they apply in mass production of arthropod natural enemies.

## Mass-Production Goals

To be viable, commercial mass rearing depends on an economical, predictable, and uniform supply of arthropods. While a researcher may consider only the intrinsic characteristics of a species in determining which species are promising for biological control, once that species is taken up for commercial rearing, the economics, efficiency, and ease of production ultimately determine whether the species becomes more widely available. Unfortunately, researchers may not understand the practical and financial constraints on commercial insectary managers in attempting to produce a profitable product. To maximize the investment in buildings and equipment, arthropods are reared at the highest possible population densities, sometimes on artificial diets or on alternate prey. Usually, the production cultures, which yield the arthropods to be sold, are inoculated at the start of the rearing cycle from parent cultures held separately or taken from a previous rearing cycle. In either case, tens or hundreds of generations of the parent cultures may cycle under controlled environmental conditions, with no exposure to seasonal climatic changes. Most species are also exposed to artificially high host populations during the rearing cycle, and thus may not be forced to employ long-distance searching or host recognition behaviors necessary to be successful in the field. After the production cycle, mass-reared arthropods experience handling, packaging, and shipping. This can directly affect their survival and may also have unforseen, and possibly subtle, effects on their viability and behavior after release. To what extent these rearing conditions influence the success of the biocontrols once they are released in the field is generally unknown.

## Rearing at High Population Densities

Rearing under conditions that are very productive per unit area nearly always requires the organisms to feed and reproduce for all or part of the rearing cycle under conditions of high population density. Through the use of greenhouses, screens, containers, or cages, mass-reared arthropods are usually prevented from escaping until they are harvested at the peak of the production cycle. Animals held under crowded conditions may be stressed from continual association, irritation, and a high rate of encounter with others of the same species. Through effects on feeding and other behaviors, longevity may decrease and other stress-related effects may be noticeable. Under these conditions, density-dependent effects become increasingly important as they affect both the behavior and the health of the arthropods. Some of these effects are listed in the following sections.

### *Oviposition Behavior*

Mutual interference between adults leads to a density-dependent decrease in egg production in anthocorids (Evans 1976). In some parasitoids, such as *Encarsia for-*

*mosa,* males are produced as a result of superparasitism, which occurs when population density is high (Stenseth 1985). Although *E. formosa* usually disperse when all suitable hosts on a leaf are parasitized (van Lenteren et al. 1980), in mass rearing the females are prevented from dispersal, which can result in a higher incidence of hyperparasitism.

### *Feeding Behavior*

The irritation of frequent encounters with other individuals reduces the time spent feeding and may stimulate repeated attempts to disperse. Also, some predators, such as anthocorids (Parker 1981) and lacewings (*Chrysopa* spp.), are notable for their cannibalistic behavior (Nordland and Morrison 1992), especially in early instars under crowded conditions.

### *Searching Behavior*

Because the prey species are also reared under high-density conditions, predators and parasitoids may not be as likely to use long-range searching behavior to locate prey or oviposition sites as they would be in the wild. Under mass-rearing conditions, less fit individuals, with impaired searching abilities or physical deformities, are at less of a disadvantage and would be more likely to reproduce. For example, prolonged rearing of *Trichogramma* parasitoids under a system that only requires them upon eclosion to walk to host eggs to oviposit may result in lines with a higher than normal proportion of individuals with wing deformities. Recent draft quality control guidelines require producers to ensure that females must fly to oviposition sites and set limits on the percentage of females with deformed wings acceptable in samples (Bigler et al. 1991).

### *Health*

Probably the most noticeable health effect in high-density conditions is the increased likelihood of disease outbreaks (from viruses, fungi, protozoa, rickettsia, and bacteria) (reviewed in Soares 1992). For example, protozoan infections in *A. cucumeris* and *A. barkeri* rearing have plagued several industrial production facilities (Gilkeson 1992), and fungal diseases that attack aphids (Hagen and van den Bosch 1968) and spider mites (van der Geest 1985) are not infrequent disasters that befall commercial production. Although lethal epizootics cause losses, sometimes severe, in commercial rearing, they are usually obvious to the insectary manager and are not likely to be exported from the insectary because the arthropods do not live long enough to be shipped. Sublethal or chronic infections are harder to diagnose, however, and are very likely to be exported to release sites. This is because actual mortality may be low at the insectary and, although productivity may be depressed, there may be several possible causes to investigate. Unless there has been a history of a disease problem, the insectary manager may search initially for an environmental factor that affected production rather than for an infection that may only be visible with the aid of electron microscopy. Compounding the problem is the fact that major insectaries frequently purchase stock from each other, often to compensate for unexpected increases in demand or to recolonize cultures depopulated by disease or system failures. This means that pathogens may be transmitted in large numbers from one insectary to another. Unfortunately, little is known about the invertebrate pathology that affects biological control organisms.

Another aspect of health is the increasing likelihood that the high concentration of a particular species in the mass rearing units will attract native predatory and parasitic species. When these infiltrate the rearing system they are often extremely difficult to eliminate, whether they are attacking the host species or the biological control agent. Mass rearing of the mealybug predator *Cryptolaemus montrouzieri* was severely affected for several years in California due to parasitism of the host mealybugs by a native parasitoid (Hale, personal communication). Native hyperparasitoids attack the aphid parasitoid *Aphidius matricariae* during spring and summer (Gilkeson 1990) and due to their small size can frequently infiltrate rearing units. The native parasitoid *Aphanogmus fulmeki* can infest commercial cultures of the aphid predator *Aphidoletes aphidimyza* (Gilkeson et al. 1993). Screens and sanitation practices can prevent such attacks, but these measures increase rearing expenses and are usually not instituted until a problem is known to exist.

## Rearing Under Artificial Environmental Conditions

In the interests of predictability and uniform production, mass-reared cultures are usually kept at a constant temperature or under a simple diurnal variation. Artificial lighting may be used continuously or as a supplement to natural lighting. Long-term laboratory rearing of arthropods that results in measurable genetic changes in strains is well documented (reviewed in Mangan 1992). Attributes that may change as a result of inadvertent selection under artificial conditions or random genetic drift include sex ratio, fecundity, searching ability, host preferences, temperature tolerance, and life span. Whether these strain changes are detrimental to the population when it is released under natural conditions is an important question. The accumulation of deleterious genes may also lead to a reduction in vigor, fecundity, and viability of eggs. Examples of possibly deleterious strain changes in insect rearing are numerous, but some traits that make a strain easier to rear may have no effect on (and may even improve) their fitness once they are released (Mangan 1992).

Prolonged rearing under a uniform climate may have two main effects. One possible effect is inadvertent selection of a population that is adapted to the mass-production temperature, lighting, and humidity regime. In uniform conditions there is no selective advantage for those genotypes able to withstand variable weather, or that are cold or heat tolerant.

A second possible effect is lack of selection for appropriate diapause or aestivation response as would occur every year under natural conditions for temperate zone species. After continuous rearing under non-diapause-inducing conditions, the frequency of diapause-related genes may change in the population. Inadvertent genetic selection of nondiapausing lines has occurred relatively frequently (Hoy 1977). In some cases, this may be because development rate and tendency to diapause may be linked, resulting in decreased incidence of diapause for lines selected for a higher rate of development (Henrich and Denlinger 1982). For example, a drift to a lower incidence of diapause without intentional selection pressure occurred in laboratory-reared strains of *A. aphidimyza* (Gilkeson and Hill 1986). Inappropriate diapause or aestivation response has been suggested as the reason for failure of some outdoor releases of biological control agents. Poor host-parasite synchronization because of diapause can cause failure of a biocontrol program (Schlinger 1960).

More research is required to define methods of avoiding genetic drift and inadvertent selection for strains adapted to artificial conditions. Starting with a representative founder population is generally recommended, but the ideal size of a founder population is debatable and there is little agreement in the literature on what constitutes a sufficient size or collection system to ensure an optimum gene pool. Regardless of the genetic complement of the original population, commercial arthropod producers occasionally suffer population crashes that reduce the rearing colony to low numbers. The degree or quality of genetic drift away from the original population as a result of such genetic bottlenecks may depend on why the crash occurred (malfunction of environmental controls, disease, shortage of hosts, and so on).

Although the suggestion is often made that the genetic complement of mass-reared populations should be periodically refreshed by the addition of wild-type or naturalized populations from release sites, this is feasible for commercial producers only if the arthropods are not too rare or difficult to find and collect. For example, when greenhouse crop plants are being removed, it is occasionally possible to collect large numbers of *Orius* spp. adults or *Encarsia* pupae. An overriding concern, however, is the risk of importing other species, especially parasites or diseases, with field-collected arthropods, which are a prime source of contaminants (Ravensberg 1992). Similar or related species can contaminate a culture and be very difficult to separate if their host range and rearing requirements are the same as those of the desired species. Importation of field-collected material may also have other effects, such as disrupting a stable breeding population of the aphid midge, *A. aphidimyza.* All offspring from one female midge are of the same sex (Sell 1976), therefore, sex ratio depends on the frequency in the population of two genotypes of females. Once a stable sex ratio is achieved, which can take about fifteen generations, it is not advisable to introduce wild-type individuals (Gilkeson 1986) because the sex ratio can become skewed to the point that the colony dies out. Taking into consideration these risks and the cost of labor to collect large numbers of arthropods, insectaries rarely find it practical to introduce field-collected populations into rearing cultures.

## Rearing on Artificial Diets and Alternate Hosts

The goal of some proponents of industrial-scale mass production is the wholly artificial diet, synthesized entirely from nonarthropod ingredients. In practice, very few predators or parasitoids have been reared for successive generations on artificial diets. A successful example, *Trichogramma* spp. parasitoids reared in the People's Republic of China on artificial eggs, relies on the inclusion of insect haemolymph in the mixture to stimulate the appropriate oviposition response and provide the correct balance of nutrients (Zhong 1987). What effect prolonged rearing on artificial diets may have on viability and behavior of biological control agents requires much more research than is usually possible through commercial companies.

Economical rearing depends on the use of alternate prey in some systems; for example, the phytoseiid mites *Amblyseius cucumeris* and *A. barkeri* and the laelapid mite, *Hypoaspis miles,* are reared worldwide on stored product mites, such as *Tyrophagus putrescentiae, Acaris farris,* or *A. siro* (Gilkeson 1992). Although these species may be within the normal host range for the laelapids, they would not be part of the normal diet of the phytoseiids, who are phyloplane species. *Amblyseius cucumeris* readily attacks western flower thrips, its intended prey, even after a lifetime of rearing on stored products mites. We do not know, however, whether that response is "normal" or whether after hundreds of generations of rearing on the alternate diet it will change.

There has been a great deal of research on host acceptance by hymenopterous parasitoids reared from different hosts. It is a common observation that parasitoids reared from one host species are better able to locate and more likely to use that host species (at least for one or two generations) rather than an alternate host. The parasitoid response to the prey may also change depending on the host plant the prey is reared upon. The implications for biological control are that parasitoid species reared on nontarget species because of ease of culture, availability of supply, and other reasons may not perform as expected once they are released (for detailed discussion see Bigler et al., Chapter 15 in this volume).

## Mass Extraction and Handling

Mass-reared arthropods are collected or harvested and spend at least some time in a package on the way to the release site. Some degree of stress may be inevitable during harvesting, especially where collection methods depend on behavior induced by high-density rearing conditions. For example, although phytoseiid mites are often reared from a small initial inoculation of predators into a good supply of host spider mites on plants, the numbers they reach before harvest are generally so high that the predators exhibit pronounced wandering and searching behavior (Gilkeson 1992). For these predators, some harvesting methods currently in use depend on the mites being prompted to disperse by lack of prey, low humidity, and overcrowding.

Although packages and shipping containers are designed to minimize obviously detrimental effects such as exposure to freezing or high temperatures, little controlled research has been done on developing arthropod containers that ensure a high level of survival and limit detrimental effects. Insectaries have largely been on their own in developing containers, often on a trial and error basis. Packaging is a critical component of a commercial product because customers unfamiliar with arthropod biology must rely on the package to provide clear directions and to prevent inadvertent errors in use that would reduce the effectiveness of the product. This is complicated because each species has different requirements for ventilation, humidity, temperature, and food supply. Some species must be kept at room temperature; others must be cooled to slow them down and to reduce their metabolic rate so that they will not deplete their body reserves or the food supply that may be shipped with them. Those shipped in the egg or pupal stage are generally the most robust. Many, however, are shipped as adults, which must be prevented from escaping by securely sealed containers and cool conditions. Ideally, after the arthropods leave the rearing facility, they are transported within twenty-four hours to release sites; too often, however, they are subjected to several days of shipment and handling, rarely under well-controlled conditions. The effects of package design and shipping conditions on the eventual success of the biological control organisms are generally unknown and likely to be variable.

## Quality Considerations

Although, as van Lenteren (1992) points out, it may not be necessary to maintain mass-produced populations identical to the field populations, we still need to define and ensure acceptable quality so that the beneficial species is in a condition to control the pest. Issues related to quality have already been mentioned. It is important to realize, however, that there are two different levels of quality: that of the packaged product, which must contain the right number, alive, of the correct species, and that of the arthropod itself, which includes its genetic, behavioral, and health characteristics. Many insectaries are still struggling with the first aspect of quality. It is not a trivial problem to devise techniques or automated equipment to ensure that the same number of arthropods are placed in each package. Arthropods are rarely counted individually; therefore, methods of estimating numbers by volume or by weight have been most frequently developed. The accuracy of these counting methods depends on the size of the arthropods, which can change with temperature and diet and with sex ratio in the culture (males may be smaller).

The optimum temperature for rearing is a simple example of the trade-offs commercial insectaries face on quality issues. Within the range of temperatures tolerated by a species, generally the higher the temperature, the faster they reproduce, therefore the less time it takes to produce a certain quantity. Higher-temperature rearing, therefore, provides the best return on the capital investment in rearing facilities. For many arthropods, however, rearing at the high end of the tolerable temperature range is stressful. It is common for insects to be larger when they are reared under cool conditions, which also adds several weeks to the life cycle. In many insects there is a positive correlation of size with fecundity; therefore, cooler rearing conditions may produce a higher-quality (more fecund) insect. But would such a difference in quality be sufficient to justify a higher cost of production?

At present, we know far too little about most species (even those that have been in commercial production for decades) to be sure what constitutes quality of the individual. Insectary managers exchange information at industry meetings to some degree, but concern over proprietary information limits the availability of documented results, and little is published. Certainly, progress is being made, especially for species such as *Trichogramma* that are now being reared on an extremely large scale. An International Organization for Biological Control working group on "quality control of mass-reared arthropods" has coordinated the development of quality control tests for ten commonly mass-reared species (Bigler et al. 1991), which will be tested over the next three years in eleven European countries.

## Conclusion

The mass production of biological controls has largely been an unregulated industry in the past. This is changing in response to the increasing demand by consumers for standardized assurances of quality and efficacy. To address these concerns, commercial biological control producers in North America have organized into a working group (the Association of Natural Biocontrol Producers) to develop label and quality guidelines and to protect industry interests. Regulators in the United States and Canada and internationally are developing guidelines for registration processes for biological controls. In Canada, regulatory guidelines are being developed by the Agriculture Canada Plant Industry Directorate, while in the United States, separate initiatives are under way in the Biological Control Institute and the Plant Protection and Plant Quarantine Programs, both agencies under the Animal and Plant Health Inspection Service (APHIS). The Food and Agriculture Organization (FAO) has also developed a Draft Code of Conduct to regulate importation and use of biological controls.

It is essential that government laboratories and university researchers become involved in research on problems relating to mass production of biological control agents. Biological control researchers, unfortunately, may feel that their job is done when the biology and host range of a promising species has been determined, laboratory rearing methods have been established, and experimental release trials have given satisfactory results. The next stage of research, involving development of industrial production, quality control tests and health assessments, harvesting, and packaging and storage methods remains to be done. Given the substantial costs involved and the slim margin of profit in the biological control industry, this work is not likely to be conducted by private companies unless it is supported by, or done in partnership with, university or government researchers.

## References


Anonymous. 1992. Directory of producers of natural enemies of common pests. *IPM Practitioner* 14(3):8–18.

Bezak, L. G. 1989. *Suppliers of beneficial organisms in North America.* Sacramento: California Department of Food and Agriculture.

Bigler, F., F. Cerutti, and J. Laing. 1991. First draft of criteria for quality control (product control) of *Trichogramma.* In *Proceedings 5th workshop IOBC working group: Quality control of mass reared arthropods,* ed. F. Bigler, 200–1. Wageningen, The Netherlands: IOBC.

Evans, H. F. 1976. Mutual interference between predatory anthocorids. *Ecological Entomology* 1:283–86.

Gilkeson, L. A. 1986. Genetic selection, evaluation and management of nondiapause *Aphidoletes aphidimyza* (Rondani) (Diptera: Cecidomyiidae) for control of greenhouse aphids in winter. Ph.D. thesis, McGill University, Montreal.

______. 1990. Biological control of aphids in greenhouse sweet peppers and tomatoes. *SROP/WPRS Bulletin* 13(5):64–70.

______. 1992. Mass-rearing of phytoseiid mites for testing and commercial application. In *Advances in insect rearing for research and pest management,* ed. T. E. Anderson and N. C. Leppla, 489–506. Boulder, Colo.: Westview.

Gilkeson, L. A., and S. B. Hill. 1986. Genetic selection and evaluation of nondiapause lines of predatory midge, *Aphidoletes aphidimyza* (Rondani) (Diptera: Cecidomyiidae). *Canadian Entomologist* 118:869–79.

Gilkeson, L. A., J. P. McLean, and P. Dessart. 1993. *Aphanogmus fulmeki* Ashmead (Hymenoptera: Ceraphronidae), a parasitoid of *Aphidoletes aphidimyza* Rondani (Diptera: Cecidomyiidae). *Canadian Entomologist* 125:161–62.

Grossman, J. 1989. Update: Strawberry IPM features biological and mechanical control. *IPM Practitioner* 11(5):1–4.

Hagan, K. S., and R. van den Bosch. 1968. Impact of pathogens, parasites, and predators on aphids. *Annual Review of Entomology* 13:325–84.

Henrich, V. C., and D. L. Denlinger. 1982. Selection for late pupariation affects diapause incidence and duration in the flesh fly, *Sarcophaga bullata. Physiological Entomology* 6:407–11.

Hoy, M. A. 1977. Rapid response to selection for a nondiapausing gypsy moth. *Science* 16:1462–63.

Mangan, R. L. 1992. Evaluating the role of genetic change in insect colonies maintained for pest management. In *Advances in insect rearing for research and pest management,* ed. T. E. Anderson and N. C. Leppla, 269–88. Boulder, Colo.: Westview.

Nordland, D. A., and R. K. Morrison. 1992. Mass rearing of *Chrysoperla* species. In *Advances in insect rearing for research and pest management,* ed. T. E. Anderson and N. C. Leppla, 427–39. Boulder, Colo.: Westview.

Parker, N. J. B. 1981. A method for mass-rearing the aphid predator *Anthocoris nemorum. Annals of Applied Biology* 99:217–23.

Pickett, C. H., F. E. Gilstrap, R. K. Morrison, and L. F. Bouse. 1987. Release of predatory mites (Acari: Phytoseiidae) by aircraft for the biological control of spider mites (Acari: Tetranychidae) infesting corn. *Journal of Economic Entomology* 80:906–10.

Ravensberg, W. J. 1992. Production and utilization of natural enemies in western European glasshouse crops. In *Advances in insect rearing for research and pest management,* ed. T. E. Anderson and N. C. Leppla, 465–87. Boulder, Colo.: Westview.

Schlinger, E. I. 1960. Diapause and secondary parasites nullify the effectivness of rose-aphid parasites in Riverside, California, 1957–1958. *Journal of Economic Entomology* 53:151–54.

Sell, P. 1976. Monogenie bei *Aphidoletes aphidimyza* (Rond.) (Diptera: Cecidomyiidae). *Zeitschrift für Angewante Entomologie* 82:58–61.

Soares, G. G. 1992. Problems with entomopathogens in insect rearing. In *Advances in insect rearing for research and pest management,* ed. T. E. Anderson and N. C. Leppla, 289–322. Boulder, Colo.: Westview.

Stenseth, C. 1985. Whitefly and its parasite, *Encarsia formosa.* In *Biological pest control: The glasshouse experience,* ed. N. W. Hussey and N. E. A. Scopes, 30–33. Ithaca, N.Y.: Cornell University Press.

van der Geest, L.P.S. 1985. Pathogens of spider mites. In *Spider mites: Their biology, natural enemies and control, vol. 1B,* ed. W. Helle and M. W. Sabelis, 247–56. Amsterdam: Elsevier.

van Lenteren, J. C. 1992. Quality control of natural enemies: Hope or illusion? In *Proceedings 5th workshop IOBC working group: Quality control of mass reared arthropods,* ed. F. Bigler, 1–14. Wageningen, The Netherlands: IOBC.

van Lenteren, J. C., H. W. Nell, and L. A. Sevenster-van der Lelie. 1980. The parasite-host relationship between *Encarsia formosa* (Hymenoptera: Aphelinidae) and *Trialeurodes vaporariorum* (Homoptera: Aleyrodidae): IV. Oviposition behaviour of the parasite, with aspects of host selection, host discrimination and host feeding. *Zeitschrift für Angewante Entomologie* 89:442–54.

Zhong, L. 1987. A preliminary study on culturing *Trichogramma ostriniae* in vitro. *Chinese Journal of Biological Control* 3: 112–13.

# 9

# Selection Pressures and the Coevolution of Host-Pathogen Systems

*Kurt J. Leonard*

## Introduction

In classical biocontrol of weeds, the biocontrol agent is introduced at a few release points to spread naturally over the range of the target weed (Templeton 1982). Biotrophic pathogens often are selected as classical biocontrol agents because they tend to be highly host specific and not damaging to crop or desirable native plant species. Host specificity, however, does not stop at the species level. During coevolution, plants accumulate genes for resistance to their pathogens, which in turn accumulate an array of virulence genes to overcome the effects of the resistance genes. This chapter explores in theoretical terms the evolutionary processes that may lead to accumulating resistance and virulence genes, and it examines how these processes may determine the long-term success of the use of biotrophic pathogens in classical biocontrol of weeds. Specifically, the chapter examines the theoretical possibilities that weed and pathogen populations may achieve equilibrium levels of resistance and virulence and defines the conditions that may determine the stability of such an equilibrium.

With diseases caused by biotrophic pathogens, the genetic variation typically follows a gene-for-gene pattern (Day 1974; Thompson and Burdon 1992) (Fig. 9.1). Each gene for resistance in the host is matched by a gene for virulence that allows the pathogen to overcome that resistance. For example, Clarke et al. (1990) found twenty-four factors for powdery mildew resistance in groundsel, *Senecio vulgaris,* and a corresponding number of virulence factors in the powdery mildew fungus. Obviously, any attempt to control groundsel with powdery mildew would have to take this variation into account.

Typically, resistance genes are most common and most diverse in environments that favor disease development. This has been well documented for resistance to crown rust in wild oats (Wahl 1970; Burdon 1987). As Burdon (1987) pointed out, this leads us to believe that the diversity is maintained in balanced rather than

| Pathogen Genotype | Host Genotype | |
|---|---|---|
| | rr | R- |
| A- | Susceptible | Resistant |
| aa | Susceptible | Susceptible |

Figure 9.1. Pattern of disease reactions in a typical gene-for-gene plant host–pathogen system. *Notes:* In this example the allele for resistance (R) in the host is dominant, and the allele for virulence (a) in the pathogen is recessive. Selection by avirulent pathogen populations favors resistance in the host population, and selection by resistant host plants favors virulence in the pathogen population. For balanced polymorphism to occur, selection must favor susceptibility either when there is no pathogen or when all pathogens are virulent. Similarly, for balanced polymorphism, selection on susceptible host plants must favor avirulent pathogen phenotypes.

transient polymorphisms. In balanced polymorphisms, two or more phenotypes are maintained more or less at equilibrium frequencies as a result of balancing forces of natural selection. Transient polymorphisms, on the other hand, endure only as long as it takes for one allele or morph to be replaced by a superior allele or morph.

If the polymorphisms are balanced, what forces of selection interact to determine equilibria at intermediate gene frequencies? We can explore this question with a model of host-pathogen coevolution, which I describe here.

In the gene-for-gene relationship illustrated in Fig. 9.1, it is obvious that selection imposed by avirulent pathogen populations will favor resistance, which protects the host from disease. However, if the polymorphism is balanced, there must be circumstances in which selection would favor the gene for susceptibility. Similarly, it is easy to see that a resistant host will cause selection for virulence in the pathogen population. But balanced polymorphism also requires that there be selection for avirulence in some circumstances. In the model, these balancing forces of selection are included as fitness costs of unnecessary or ineffective resistance genes and fitness costs of unnecessary virulence genes (Leonard 1977).

Two versions of the host-pathogen coevolution model are presented here. Each version has a single internal equilibrium point, the position of which can be calculated readily. However, mathematical analysis of the stability of equilibria in the model proved very difficult and gave inconclusive results (Fleming 1980; Leonard and Czochor 1980; Sedcole 1978). Therefore, I chose numerical analysis to reveal the behavior of the model and determine the stability of polymorphisms for resistance and virulence in the model (Leonard 1994). This involved thousands of simulation runs with 752 combinations of key parameters in the model and a variety of different starting frequencies of resistance and virulence.

## Description of the Model

The model is for plant species with discrete generations (Leonard and Czochor 1980). This is typical for annual plants that have a distinct growing season, so that the seedlings initiate growth at the same time each year and the adult plants mature and set seed more or less in synchrony. There is a delayed feedback between phenotype frequencies in the host and pathogen populations, because the pathogen does not kill the plants outright. A single plant may support hundreds, perhaps even thousands, of individual infections and still survive to produce some seeds at maturity. Host genotypes with severely impaired fitness produce fewer, and possibly weaker, offspring in the next generation.

For simplicity, selection coefficients are treated as constants in the model. The model considers only one locus for resistance and one locus for virulence. Each locus is assumed to have just two alleles: an allele for resistance and one for susceptibility in the host, and an allele for virulence and one for avirulence in the pathogen. In the model, fitness values for host plants are represented as deviations from 1 (Fig. 9.2). The parameter $s$ takes into account the sum of the effects of environment, host population density, and other factors that determine the severity of the disease in the area in which the host grows. This parameter $s$ is multiplied by the fitness of the pathogen phenotype in combination with each host phenotype. The parameter $c$ represents the cost of the resistance when the resistance is either unnecessary or totally ineffective, as it would be if the entire pathogen population were made up of the virulent phenotype. Indirect evidence suggests that the value of $c$ must be low (i.e., less than 0.05). Plant breeders generally do not detect any measurable cost of resistance in terms of yield loss when resistance genes are added to cultivars by backcrossing.

Fitness of pathogen phenotypes is expressed in terms of deviations from the fitness of the avirulent phenotype on the susceptible host (Fig. 9.3). The parameter $t$ represents the effectiveness of the resistance. When the resistance totally suppresses the ability of the avirulent pathogen to infect and reproduce on the resistant host, the value of $t$ is 1. For partial resistance, the value of $t$ is less than 1, possibly as low as 0.5. The parameter $k$ is the cost of virulence. Two versions of the model were considered with regard to the cost of virulence in the pathogen (Leonard 1994). In the hard selection version of the model, the virulent phenotype of the pathogen cannot reproduce as rapidly as the avirulent phenotype can when it is on the suitable susceptible host (Fig. 9.3). This can be described as hard selection in which fitness is a measure of inherent rates of reproduction. In hard selection, fitness differences do not depend upon actual competition between phenotypes for nutrients (Wallace 1975). In the competition (or soft selection) version of the model, the virulent phenotype of the

| Pathogen Phenotype | Host Phenotype | |
|---|---|---|
| | Susceptible | Resistant |
| Avirulent | $1 - s(W_{AS})$ | $1 - c - s(W_{AR})$ |
| Virulent | $1 - s(W_{VS})$ | $1 - c - s(W_{VR})$ |

Figure 9.2. Fitness of susceptible and resistant host phenotypes exposed to avirulent and virulent pathogen phenotypes in a gene-for-gene model of host-pathogen coevolution. *Notes:* The model is for a single locus for resistance in the host and a single locus for virulence in the pathogen. $W_{AS}$ and $W_{AR}$ represent fitness of the avirulent pathogen phenotype on the susceptible and resistant hosts, respectively. $W_{VS}$ and $W_{VR}$ represent fitness of the virulent pathogen phenotype on the susceptible and resistant hosts, respectively. The parameter $s$ indicates severity of the disease; $c$ is the cost of resistence when it is unnecessary or ineffective, as it is when the plant is attacked by the virulent pathogen phenotype.

Hard Selection Model

| Pathogen Phenotype | Host Phenotype | |
|---|---|---|
| | Susceptible | Resistant |
| Avirulent | 1 | 1 - t |
| Virulent | 1 - k | 1 - k |

Competition Model

| Pathogen Phenotype | Host Phenotype | |
|---|---|---|
| | Susceptible | Resistant |
| Avirulent | 1 | 1 - t |
| Virulent | 1 - k | 1 |

Figure 9.3. Fitness of avirulent and virulent pathogen phenotypes attacking susceptible and resistant host plants in two versions of a gene-for-gene model of host-pathogen coevolution. *Notes:* In the model, $t$ represents the effectiveness of the resistance and $k$ is the cost of virulence. In the hard selection version of the model, virulence is associated with a decrease in the intrinsic rate of reproduction by the pathogen, so there is a cost of virulence even on the resistant hosts. In the competition version of the model, there is no cost of virulence on the resistant host where the virulent phenotype does not compete with the avirulent phenotype, which is not supported by the resistant host.

pathogen can reproduce as rapidly as the avirulent phenotype, but only if it does not have to compete directly with it (Fig. 9.3).

### Equilibrium Conditions

Positions of the equilibrium points in the model are shown in Fig. 9.4. For oat stem rust, Leonard (1969) measured values of $t$ approximately equal to 1 and values of $k$ in the range of 0.05 to 0.40. For those values, the resistant host phenotypes should occur at a frequency of approximately 5 to 40 percent in wild host populations if they are at equilibrium. For the competition version of the model, the equilibrium frequencies of resistant phenotype would be slightly less than with the hard selection version. Equilibrium frequencies of virulent phenotypes of the pathogen depend upon the value of the parameter $s$, which indicates the suitability of the environment for disease development. Virulence should occur at higher frequencies in areas where the environment favors severe epidemics. If the resistance genes that correspond to these virulence genes have low fitness costs, the equilibrium frequencies of virulence genes will be high, often in the range of 90 to 99 percent for the hard selection version of the model or 70 to 90 percent for the competition version, depending on the magnitude of $k$, the cost of virulence.

### Results of Simulations

Fig. 9.5 shows two simulation runs with the hard selection version of the model with $s = 0.8$, $c = 0.03$, $t = 1.0$, and $k = 0.3$. With these parameter values, the equilibrium frequency of virulence is 0.96, and the equilibrium frequency of resistance is 0.16. The simulation shown in Fig. 9.5a was run for 165 host generations (165

| Phenotype | Equilibrium Frequencies | |
|---|---|---|
| | Hard Selection | Competition |
| Resistant | k/t | k/(t + k) |
| Virulent | (st - c)/st | (st - c)/s(t + k) |

Figure 9.4. Frequencies of resistant and virulent phenotypes at equilibrium in a gene-for-gene-for-gene coevolution model for plants and pathogens. *Notes:* In the model, $c$ = cost of resistance, $k$ = cost of unnecessary virulence, $t$ = effectiveness of resistance, and $s$ = severity of disease in terms of decreased fitness of a susceptible host phenotype.

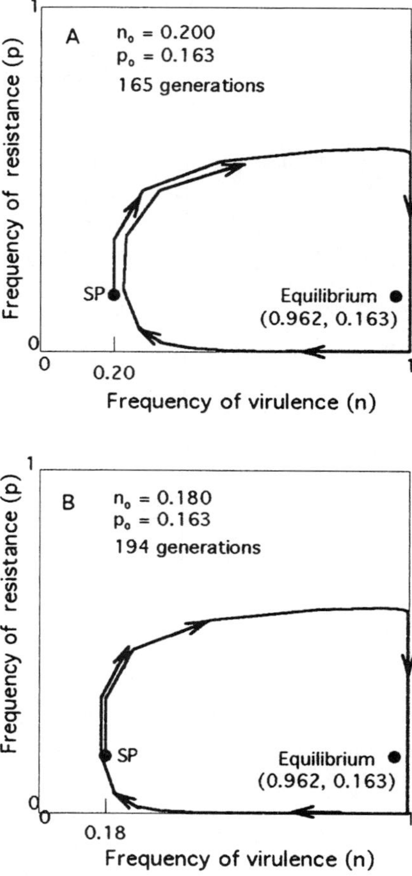


Figure 9.5. Phase planes from simulations with a host-pathogen coevolution model in which $p$ is the frequency of a dominant allele for resistance in the host and $n$ is the frequency of virulence in a haploid pathogen. *Notes:* Simulations were run for the hard selection version of the model with high disease severity ($s$ = 0.8), complete resistance ($t$ = 1.0), low fitness cost of resistance ($c$ = 0.03), and moderately high fitness cost of virulence ($k$ = 0.3). With these parameter values, the equilibrium frequencies of alleles for resistance and virulence are $p_{eq}$ = 0.163 and $n_{eq}$ = 0.962. (a) Initial frequencies (SP) started at $p_0$ = 0.163 for resistance and $n_0$ = 0.200 for virulence. The simulation was run for 165 generations. Notice the inward spiral of the allele frequencies. (b) Parameter values are the same as in (a), but initial allele frequencies (SP) are $p_0$ = 0.163 for resistance and $n_0$ = 0.180 for virulence. The simulation was run for 194 generations. The outward spiral of allele frequencies indicates an unstable limit cycle that passes to the left of the equilibrium point at approximately $n$ = 0.19. Allele frequencies starting anywhere inside the limit cycle spiral inward, and allele frequencies starting outside the limit cycle spiral outward until the alleles for virulence and susceptibility become fixed in the pathogen and host populations.

years). The gene for resistance started at its equilibrium frequency of 0.16, and the initial frequency of the virulence gene was 0.20. From this starting point, the gene frequencies cycled in toward the equilibrium point as though the equilibrium were stable. However, when the initial frequency of virulence started further from the equilibrium point at 0.18, the gene frequencies spiraled outward, eventually ending up in the lower right corner of the phase plane where the frequency of virulence equals 1.0 and the frequency of resistance is 0.0 (Fig. 9.5b).

Fig. 9.5 illustrates an unstable limit cycle that passes to the left of the equilibrium point at approximately $n = 0.19$. If the gene frequencies start within the limit cycle, they spiral toward the equilibrium point. If they start outside the limit cycle, they spiral outward to fixation of virulence and loss of resistance.

For both the hard selection version and the competition version of the model, there is an unstable limit cycle for virtually all combinations of parameters that might be considered (Leonard 1994). The position of the limit cycle, however, varies drastically depending on the values of the parameters that are chosen. If the limit cycle is very large, it will follow close to the boundaries of the phase plane. In that case, new genes for resistance and virulence introduced into the host and pathogen populations by mutation or migration would always start within the limit cycle, and the gene frequencies would always spiral toward equilibrium. One measure of the size of the limit cycle is the value of *n,* the frequency of virulence, in the limit cycle when resistance is at its equilibrium frequency. This measure indicates how far the limit cycle extends to the left of the equilibrium point.

Figs. 9.6 and 9.7 summarize the results of thousands of simulations run to locate the unstable limit cycles for 752 combinations of parameter values. Combinations included *s* from 0.2 to 0.8, *t* from 0.5 to 1.0, *c* from 0.00 to 0.30, and *k* from 0.01 to 0.40. The size of the limit cycle for each combination of parameter values is indicated by the negative logarithm of the frequency of virulence ($-\log n$) at the leftmost point of the limit cycle. This value is plotted on the y axis. For large values of $-\log n$, the equilibrium can be regarded as evolutionarily stable. That is, even for very rare genes in a population, the force of selection will be toward an internal equilibrium point rather than toward extinction. For example, if the unstable limit cycle encloses all values of $n > 10^{-10}$ ($-\log n > 10$), any new genes for virulence introduced into the pathogen population by mutation or migration would appear at frequencies within the limit cycle, thus leading to an inward spiral toward the equilibrium point. On the other hand, values of $-\log n < 3$ in Figs. 9.7 and 9.8 indicate unstable situations in which polymorphisms would not be maintained for virulence genes starting at frequencies below $10^{-3}$. Mutant genes for virulence almost certainly would enter the pathogen populations at frequencies less than $10^{-3}$.

Fig. 9.6 shows that for the hard selection version of the model, the equilibria are evolutionally unstable for all combinations of parameter values except those with high values of *c,* the cost of resistance. With the hard selection version of the model, the equilibria become less evolutionarily stable with increasing values of *s,* the severity of disease. These features of the hard selection version of the model

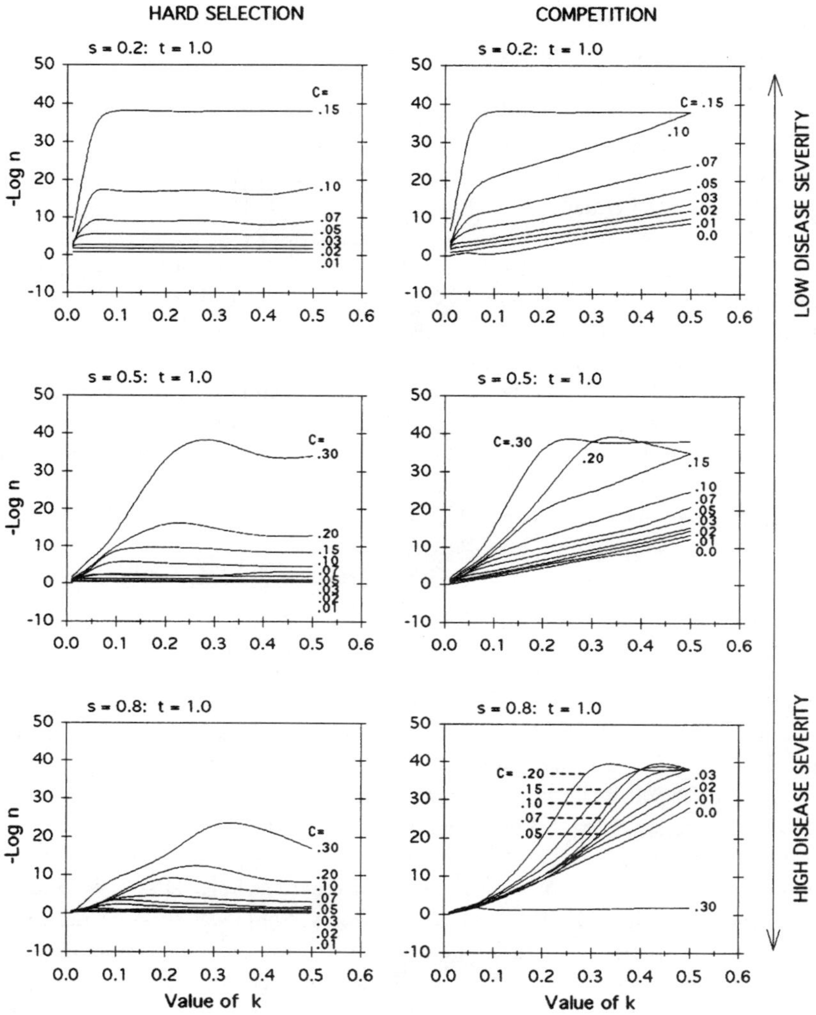


Figure 9.6. Effect of parameter values on stability of polymorphisms in a host-pathogen coevolution model: *s* represents disease severity, *t* effectiveness of resistance (completely effective at $t = 1$), *c* cost of resistance, and *k* cost of virulence. *Notes:* Simulations were run for the hard selection version (graphs on the left) and the competition version (graphs on the right) of the model. The position of the unstable limit cycle for each combination of parameter values is indicated by $-\log n$, where *n* is the frequency of virulence at its lowest point on the limit cycle. Combinations of parameters represented by points on the curves with $-\log n > 10$ indicate stable equilibria, because new genes for virulence would enter the system at $n > 10^{-10}$, causing gene frequencies to spiral toward the equilibrium point. Notice that equilibria are more stable for the competition version of the model. Also notice that stability increases with increasing disease severity in the competition version but decreases in the hard selection version of the model. *Source:* From Leonard (1994) *Phytopathology* 84: 70–77.

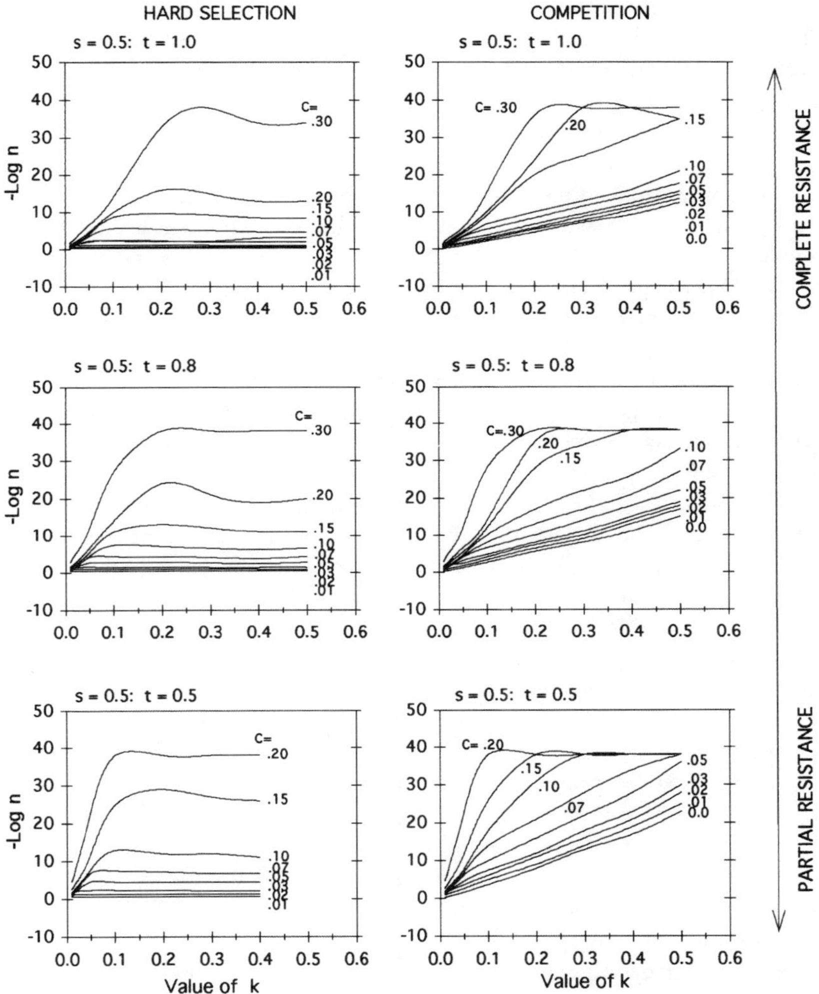


Figure 9.7. Effect of partial resistance on stability of equilibria in the hard selection version (graphs on the left) and the competition version (graphs on the right) of the host-pathogen coevolution model. See Figure 9.6 for details of parameters. *Notes:* The position of the unstable limit cycle for each combination of parameter values is indicated by – log *n,* where *n* is the frequency of virulence at its lowest point on the limit cycle. Notice that equilibria are more stable for partial resistance ($t < 1.0$) than for complete resistance ($t = 1.0$). This effect is more pronounced with the competition version of the model ($a = k$) than with the hard selection version. *Source:* From Leonard (1994) *Phytopathology* 84: 70–77.

make it unacceptable as an explanation for balanced polymorphisms of resistance and virulence in natural host-pathogen systems of plant disease. As I stated earlier, observations of natural host-pathogen systems indicate that they contain resistance genes with low fitness costs and that their diversity is greater rather than lower in areas of greater disease severity.

The behavior of the competition version of the model differs from that of the hard selection version. In the competition version, stable polymorphisms are possible for resistance genes with low fitness costs if the cost of unnecessary virulence, *k*, is moderately high. In fact, with the competition version, stable polymorphisms may occur even when the resistance genes have no fitness cost ($c = 0.0$). Furthermore, the evolutionary stability of equilibria increases with increasing disease severity in the competition version of the model (Fig. 9.6).

Fig. 9.7 shows that with both versions of the model, equilibria are more evolutionarily stable for genes for partial resistance than for complete resistance. In all cases, however, the polymorphisms are evolutionarily unstable if the cost of virulence is very low (i.e., < 0.02).

Another aspect of stability of the host and pathogen polymorphisms in coevolving populations is the rate at which the resistance and virulence gene frequency oscillations damp toward the equilibrium values. This can be thought of as the strength of attraction toward an internal equilibrium point when initial gene frequencies are within the unstable limit cycle. Figs. 9.8 and 9.9 illustrate the greater attraction toward the equilibrium point in the competition than in the hard selection version of the model. Changes in mean fitness of the host population are plotted through one cycle of gene frequency oscillations around the equilibrium point. For both simulations the parameter values and starting gene frequencies are as illustrated in Fig. 9.5a. With the hard selection version of the model (Fig. 9.8) starting at $n_0 = 0.2$ and $p_0 = 0.163$ (equilibrium value), it took 160 generations to complete one circuit of gene frequencies around the equilibrium point. On the other hand, with the competition version of the model starting at $n_0 = 0.2$ and $p_0 = 0.123$ (equilibrium value), it required only 25 generations to complete the cycle. Thus, the hard selection simulation completed just one cycle in 160 host generations and moved the gene frequencies only slightly closer to the equilibrium point. The competition simulation, however, completed more than six cycles in 160 generations and moved the gene frequencies much closer to the equilibrium point.

Changes in host fitness in response to changing frequencies of resistance and virulence genes shown in Figs. 9.8 and 9.9 reflect the changes in levels of biocontrol that might be expected in a coevolving host-pathogen system. In the hard selection version of the model (Fig. 9.8), changes in host fitness are marked by a distinct peak and a deep valley. The peak in host fitness occurs when the frequency of resistance is high and the frequency of virulence is still low. The valley occurs shortly before the peak when both resistance and virulence are at low frequencies. In the competition version of the model (Fig. 9.9), a peak in host fitness also occurs when the frequency of resistance is high and the frequency of virulence is low, and a valley occurs when

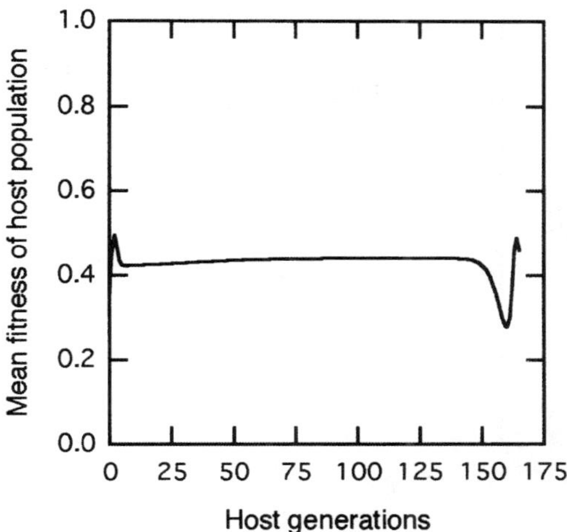


Figure 9.8. Changes in mean fitness of the host population through one cycle of gene frequency oscillations in the hard selection version of a host-pathogen coevolution model. *Notes:* Parameter values and initial frequencies of resistance in the host and virulence in the pathogen are as in Figure 9.5.a. As indicated in Figure 9.5.a, a single cycle around the equilibrium point required 160 host generations.

both resistance and virulence are at low frequencies. In addition, in the competition version of the model, there is a second, broader peak of host fitness through the part of the cycle when the frequency of virulence is high and that of resistance is low.

The reasons for the patterns of host fitness changes in simulations with the hard selection and competition versions of the model can be seen in Table 9.1. Fitness values in Table 9.1 are relative to a fitness of 1.0 for the susceptible host phenotype when the pathogen is not present. If all the host plants were susceptible in an environment where $s = 0.8$ and if the pathogen population were made up exclusively of avirulent phenotypes, disease would reduce host fitness to 0.2 in both the hard selection and the competition versions of the model. The greater host fitness of 0.44 for a completely susceptible host population attacked by a completely virulent pathogen population is due to the cost of virulence ($k = 0.3$ in this example) with both versions of the model. In the presence of the pathogen, the greatest possible host fitness (0.97) would occur if all the host plants were resistant and none of the pathogen phenotypes could overcome that resistance. This, of course, is the worst case scenario for biocontrol of a weed with a pathogen.

The main difference between the hard selection and the competition versions of the model in terms of host fitness is seen in the case of a completely resistant host population attacked by a completely virulent pathogen population. In the

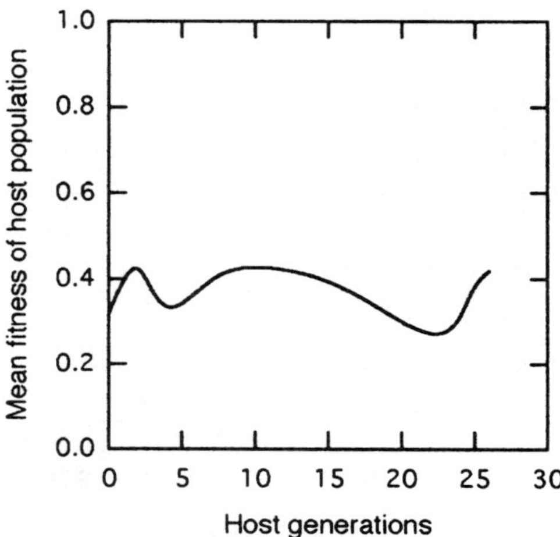


Figure 9.9. Changes in mean fitness of the host population through one cycle of gene frequency oscillations in the competition version of a host-pathogen coevolution model. *Notes:* Parameter values are as in Figure 9.8 (see also Figure 9.5.a). As in Figure 9.8, the frequency of the resistance gene started at its equilibrium frequency (0.123 in the case of the competition version of the model) and the frequency of the virulence gene started at 0.2. A single cycle around the equilibrium point took 25 generations.

Table 9.1. Relative fitness values for susceptible or resistant host plants attacked by avirulent or virulent pathogen populations compared with relative fitness of mixed host populations in equilibrium in a hypothetical host-pathogen system[a].

| Host Type | Pathogen type | Host Fitness | |
|---|---|---|---|
| | | Hard Selection Model | Competition Model |
| Susceptible | No pathogen | 1.00 | 1.00 |
| Susceptible | Avirulent | 0.20 | 0.20 |
| Resistant | Avirulent | 0.97 | 0.97 |
| Resistant | Virulent | 0.41 | 0.17 |
| Susceptible | Virulent | 0.44 | 0.44 |
| Equilibrium mixture | Equilibrium mixture | 0.43 | 0.38 |

[a] Fitness values for host and pathogen phenotypes are as shown in Figs. 9.2 and 9.3 with $s = 0.8$, $t = 1.0$, $c = 0.03$, and $k = 0.3$. Equilibrium frequencies of resistance and virulence are as shown in Fig. 9.4.

competition version of the model, this combination yields the minimum host fitness. In this situation, the host bears the cost of resistance but the pathogen suffers no cost of virulence because there is no competition from the avirulent pathogen phenotype.

Neither host fitness nor pathogen fitness is maximized at equilibrium. The reason for this is that maximum fitness in the host is attained by minimizing fitness of the pathogen, and conversely, host fitness is minimized as a consequence when pathogen fitness is maximized. The host population has maximum fitness when all the pathogen population is avirulent and the host is resistant (Table 9.1). Maximum pathogen fitness occurs when all the host plants are susceptible and the pathogen is avirulent; in the case of the competition version of the model, it may also occur when the hosts are all resistant and the pathogen population is completely virulent (Fig. 9.3). At equilibrium, both host and pathogen fitness are balanced somewhere between the minimum and maximum levels that they can reach.

## Conclusions

Both the hard selection and competition versions of the host-pathogen coevolution model indicate that at equilibrium, resistance should occur at relatively low frequencies in wild plant populations while the frequency of virulence should be high. Wild host-pathogen systems that have been studied thoroughly exhibit these characteristics. Clarke et al. (1990) found that resistance genes in groundsel occur in relatively low frequencies over a wide geographical area, although individual resistance genes can occur at high frequency in small plots within the area. In contrast, virulence occurs at high frequency, and most mildew isolates have complex genotypes with many virulence genes. This pattern of moderate to low frequencies of resistance genes and high frequencies of virulence genes has been demonstrated also for wild oats, *Avena* spp., and the crown rust fungus, *Puccinia coronata,* in both Israel (Wahl 1970) and Australia (Burdon 1987). Wahl (1970) and Burdon (1987) showed that wild oats in areas with environments unfavorable to crown rust have less resistance to the disease. Burdon (1987) showed that both frequency and diversity of virulence in *P. coronata* populations is reduced in environments unfavorable to the disease.

Both the hard selection and competition versions of the model have unstable limit cycles. The position of the limit cycle varies depending upon the values assigned to the key parameters in the model. Because gene frequencies inside the unstable limit cycle spiral in toward the equilibrium point, we can use the size of the limit cycle as a measure of the relative evolutionary stability of the equilibrium for each set of parameter values. With both versions of the model, there is a tendency for the equilibria to be more evolutionarily stable for genes that have high fitness costs associated with them, but this is less pronounced with the competition version than with the hard selection version. In general, equilibria are more evolutionarily stable in the competition version, particularly for resistance genes with low associated fitness costs. Also, in the competition version, the evolutionary stability of equilibria increased with increasing disease pressure, whereas the reverse was true with the hard selection version. Thus, the evidence indicates that the competition version of the model is a more realistic representation of natural host-pathogen systems than the hard selection version is. This suggests that the cost of unnecessary virulence is manifested only under conditions of direct competition between virulent and avirulent pathogen phenotypes on susceptible plants. It also suggests that the cost of unnecessary virulence should be both density dependent and frequency dependent. Further modifications of the coevolution model should take this into account.

The second measure of stability of equilibria is in the strength of attraction to the equilibrium point. By this measure, the competition version of the coevolution model also displays greater stability of polymorphisms than does the hard selection version. Gene frequency oscillations cycle faster and move toward equilibrium more rapidly in the competition version of the model than in the hard selection version.

If the competition version of the model reasonably represents the process of host-pathogen coevolution in diseases of wild plants, we can draw some conclusions relative to biocontrol of weeds. In their natural habitats, the weeds and their coevolved pathogens are likely to exist in more or less stable equilibria that maintain a diversity of genes for resistance in the weed and virulence in the pathogen. The equilibria can be disrupted by moving a small part of the weed's diversity to a new isolated habitat. Introducing the pathogen into the new habitat may eventually reestablish an equilibrium, but that is likely to take hundreds of years. Host resistance is not likely to be a great impediment to effective biocontrol of weeds by biotrophic pathogens, if one starts with a mixture of pathogen isolates that represents an adequate sample of the pathogen population in its native habitat. The best place to find an abundance of highly virulent pathogen isolates would be in areas where the disease is severe, preferably in the center of origin of the weed and its pathogen.

In a system of genetically diverse host and pathogen populations in an environment consistently favorable for disease development, fluctuations in mean host fitness will occur in response to oscillations in gene frequencies, but the simulations indicate that these fluctuations should be relatively mild. This, of course, depends on the presence of virulence in the pathogen population to match all resistance genes in the host population. It also depends upon a highly mobile pathogen that responds quickly to the selection pressures imposed by the resistant host. These traits are shared by biotrophic pathogens such as the rust and powdery mildew fungi, which infect foliar parts of plants and produce wind-borne spores.

It is significant also that the simulations show that in unstable situations, the gene frequencies disappeared into the lower right corner of the phase plane. The lower right corner represents the situation in which virulence is fixed in the pathogen population and susceptibility is fixed in the host population. Thus, in the absence of a stable equilibrium between the weed and its pathogen, it is more likely that the weed will lose its resistance than that the pathogen will lose its virulence. While this combination of susceptible host and virulent pathogen genotypes does not minimize host fitness in either the hard selection or the competition version of the model, it does reduce host fitness to a level very near that which would be reached at the internal equilibrium point. For successful biocontrol of a weed with a biotrophic pathogen, it is not necessary that resistance be completely absent from the host population, nor is it even necessary that an equilibrium be reached between resistance and virulence. What is absolutely essential is that the pathogen must have virulence genes to match any resistance genes present in the host population.

## References

Burdon, J. J. 1987. *Diseases and plant population biology.* Cambridge: Cambridge University Press.

Clarke, D. D., F. S. Campbell, and J. R. Bevan. 1990. Genetic interactions between *Senecio vulgaris* and the powdery mildew fungus *Erysiphe fischeri.* In *Pests, pathogens and plant communities,* ed. J. J. Burdon and S. R. Leathers, 189–217. Oxford: Blackwell.

Day, P. R. 1974. *Genetics of host-parasite interaction.* San Francisco: Freeman.

Fleming, R. A. 1980. Selection pressures and plant pathogens: Robustness of the model. *Phytopathology* 70:175–78.

Leonard, K. J. 1969. Selection in heterogeneous populations of *Puccinia graminis* f.sp. *avenae. Phytopathology* 59:1851–57.

———. 1977. Selection pressures and plant pathogens. *Annals of the New York Academy of Sciences* 287:207–22.

———. 1985. Population genetics of gene-for-gene interactions between plant host resistance and pathogen virulence. In *Proceedings 15th International Congress of Genetics, New Delhi, India. Vol. 4. Applied Genetics,* ed. V. L. Chopra, B. C. Joshi, R. P. Sharma, and H. C. Bansal, 131–48. New Delhi: Oxford & IBH.

———. 1994. Stability of equilibria in a gene-for-gene coevolution model of host-parasite interactions. *Phytopathology* 84:70–77.

Leonard, K. J., and R. J. Czochor. 1980. Theory of genetic interactions among populations of plants and their pathogens. *Annual Review of Phytopathology* 18:237–58.

Sedcole, J. R. 1978. Selection pressures and plant pathogens: Stability of equilibria. *Phytopathology* 70:175–78.

Templeton, G. 1982. Status of weed control with plant pathogens. In *Biological control of weeds with plant pathogens,* ed. R. Charudattan and H. L. Walker, 29–44. New York: Wiley.

Thompson, J. N., and J. J. Burdon. 1992. Gene-for-gene coevolution between plants and parasites. *Nature* 360:121–25.

Wahl, I. 1970. Prevalence and geographic distribution of resistance to crown rust in *Avena sterilis. Phytopathology* 60:746–49.

Wallace, B. 1975. Hard and soft selection revisited. *Evolution* 29:465–73.

# 10

# Deleterious Rhizobacteria and Weed Biocontrol

*Ann C. Kennedy*

Public concern over surface water, groundwater, and food contamination has resulted in stricter pesticide regulations, fewer approved pesticides reaching the market, and the withdrawal of previously registered pesticides. As a result, growers are being forced to reduce the use of synthetic chemical pesticides and to rely on cultural and biological methods to control pests. Biological control of weeds is based on the premise that biotic factors have a significant influence on the competitive abilities of plant species. Phytotoxic effects of microorganisms are often plant species and cultivar specific. Microorganisms that selectively suppress weed species may alter competition among plants. These plant pathogens potentially may be used to regulate the growth of unwanted plant species growing simultaneously with more desirable plants (Templeton 1982). This would be true especially if competitive weed growth coincided with environmental factors conducive to microbial growth and weed-suppressive activity. Biological control offers alternative means of suppressing weed growth and establishment in agricultural and range systems. Before biological control methods can be fully integrated into weed management systems, however, we need a greater understanding of the processes of host recognition, specificity, and colonization. Awareness of the ecological constraints of the pathogen and the weed is critical to successful biological weed control.

Most of the research on microbial control of weeds has concentrated on fungal plant pathogens for broadleaf weed control. Most notable is the use of rusts (*Puccinia jaceae* Otth.) for the control of diffuse knapweed (*Centuria diffusa* Lam.) (Mortensen 1986; Watson and Clement 1986), and skeleton weed (*Chondrilla juncea*) control with the use of *Puccinia chondrillina* (Cullen et al. 1973). Mycoherbicides, such as the fungal pathogens of weeds sold under the trade names of DeVine and Collego, are commercially available. DeVine is being used to control stranglervine (*Morrenia odorata*) in citrus (Ridings 1986), and Collego (*Colletotrichum gloeosporioides* f. sp. *aeschynomene*) (Daniel et al. 1973) is used for the control of northern jointvetch (*Aeschynomene virginica*) (Templeton 1982) in rice and soybean. BioMal has just been registered for the control of round-leaf mallow (*Malva pusilla*) (Makowski, Chapter 5 in this volume). The success of these commercial mycoherbicides indicates the potential for the use of microorganisms in modern weed control technology.

Bacterial plant pathogens can have a role in weed control, although their potential contribution often is overlooked. Bacteria can exert a subtle yet profound effect on plant growth. Over the past ten years, research has been conducted on the negative effects of rhizosphere bacteria on plant health (Woltz 1978; Suslow and Schroth 1982; Elliott and Lynch 1985; Schippers et al. 1987). Many of these bacteria are saprophytic on plant roots but do not parasitize the plant (Salt 1979). These rhizobacteria can negatively affect plant growth by reducing the supply of water, nutrients, or plant growth substances to the root, or by limiting root growth by phytotoxic substances (Salt 1979; Alstrom and Burns 1989; Bolton and Elliott 1989). Rhizobacteria that produce phytotoxic compounds can suppress plant growth by reducing germination or altering plant development (Suslow and Schroth 1982; Alstrom 1987; Schippers et al. 1987). These bacteria can exist in intimate association with the root, occupying sites in the root mucigel, epidermis, cortex, or intercellular spaces. In such close contact with the plant cell, the bacterial metabolites can profoundly affect plant growth. Because these bacteria act at the critical stage of early shoot and root growth, they may have greater impact on subsequent plant growth than if the plants were exposed to the bacteria at later stages in their development.

Because of their negative effect on plant growth, these organisms have been termed deleterious rhizobacteria (DRB) (Suslow and Schroth 1982). DRB had originally been considered "minor" pathogens because they were primarily found colonizing juvenile tissue such as root hairs, tips, and cortical cells and did not cause major disease symptoms (Salt 1979). In actuality, DRB are not minor and can dramatically alter plant growth. Their presence is often difficult to demonstrate, because there are no visible signs of pathogenicity or any distinct symptoms; consequently, these organisms often are overlooked. Often the impact of DRB on the host is dependent upon the environment, climate, soil type, and host vigor. Cropping practices also may greatly affect the occurrence of these organisms and their effect on plant growth (Schippers et al. 1987).

Deleterious rhizobacteria have reduced crop growth and yield of potatoes (*Solanum tuberosum* L.) (Bakker and Schippers 1987), sugar beets (*Beta vulgaris* L.) (Suslow and Schroth 1982), and winter wheat (*Triticum aestivum* L.) (Fredrickson and Elliott 1985). A negative growth effect has been seen for citrus (Gardner et al. 1984), and bean (*Phaseolus vulgaris* L.) (Alstrom 1987). These reductions in plant growth and subsequent yield may be as significant as that from traditional pathogens. Many DRB are effective colonizers of roots and residues (Suslow and Schroth 1982; Fredrickson and Elliott 1985; Stroo et al. 1988). In investigations of the early spring growth of winter wheat (Fredrickson and Elliott 1985), stunted wheat plants were heavily colonized by bacteria that produced plant inhibitory compounds. These compounds specifically inhibited winter wheat and did not injure other small grains or legumes (Bolton and Elliott 1989). Host specificity of these compounds produced by DRB has also been seen in laboratory studies with a number of other plant species (Woltz 1978; Cherrington and Elliott 1987; Astrom 1991).

A wide variety of DRB were found to inhibit velvetleaf (*Abutilon theophrasti*), morning glory (*Ipomoea* sp.), cocklebur (*Xanthium canadense* Mill.), pigweed (*Amaranthus* sp.), lambsquarter (*Chenopodium* spp.), smartweed (*Polygonum* sp.), and jimson weed (*Datura stramonium*) (Kremer 1986; Kremer et al. 1990). The potential of DRB for biological weed control has been reported for broadleaf weeds (Kremer 1986; Kremer et al. 1990) and the grass weeds downy brome (Cherrington and Elliott 1987; Kennedy et al. 1991) and jointed goatgrass (Kennedy et al. 1992b). DRB were selective in their inhibition to growth of several crop species and the grass weed downy brome (Cherrington and Elliott 1987). Further investigation illustrated that these DRB could be used for control of downy brome in the field (Kennedy et al. 1991).

## Deleterious Rhizobacteria for Downy Brome Control

### *The Bacteria*

Deleterious rhizobacteria that specifically inhibit various grass weeds, but do not affect the crop, have been isolated from soil (Kennedy et al. 1991, 1992b) and inhibit plant growth by the production of plant-suppressive compounds (Tranel et al. 1993). These bacteria are excellent biological control agents because they are aggressive colonizers of the roots and residue, often representing up to 95 percent of the total pseudomonads on the plant root (Stroo et al. 1988; Kennedy et al. 1992a). The bacteria can function as a direct delivery system for the natural plant-suppressive compounds they produce. They tend to be fairly tolerant of low soil moisture, although they do not survive well under hot, dry conditions. They survive well at low temperatures with an optimum temperature often below 15°C. The greatest amount of suppression occurs at low temperatures and in moist soils (Johnson et al. 1993). These DRB are most prevalent in the soil in late fall and early spring at the times of germination and seedling growth of grass weeds.

### *The Weed*

Downy brome (*Bromus tectorum* L.), commonly called cheatgrass, was chosen as a biological control target because it is an especially troublesome weed. It is an invading species from Eurasia that germinates in fall or spring, over wide ranges of temperature and moisture (Hitchcock and Chase 1971; Mack and Pyke 1983). It is often considered a forage species providing early spring grazing in range systems (Klemmedson and Smith 1964); however, its short growth period, fluctuating forage production, and high fire hazard make it less desirable than other species. Cultivated fields, overgrazed rangelands, abandoned farms, and road cuts provide an optimal environment for the proliferation of downy brome. Its present distribution includes Canada, Mexico, and the United States except for a small portion in the southeast (Mack 1981). The downy brome plant exhibits tremendous phenotypic plasticity, with plants flowering and producing seed over a wide range of environmental conditions (Stewart and Hull 1949; Hulbert 1955). It is a prolific seed producer, yielding up to 450 kg of seed per hectare, with a kilogram of seed containing 330,000 propagules (Hull and Pechanec 1947). Seeds germinate and emerge at onset of late summer and early fall rains (Jurhen et al. 1965; Wicks et al. 1971; Thill et al. 1984), with germination often continuing through winter and into spring (Mack and Pyke 1983). Fall-germinated plants can overwinter in a semidormant state (Klemmedson and Smith 1964), and continue to add leaves and expand their root system if soil temperatures remain above 2 to 3°C (Harris 1967; Uresk et al. 1979). This growth characteristic allows downy brome to be highly competitive in grasslands (Harris 1967) and cultivated fields (Rydrych 1974; Thill et al. 1984; Stahlman 1986). Downy brome poses a serious problem to small-grain production in the western United States and Canada (Rydrych and Muzik 1968; Rydrych 1974; Thill et al. 1984; Stahlman 1986), causing growers to lose $300 million annually due to crop yield loss with an added $35 million to $70 million annually for chemical control (Donald and Ogg 1991).

Downy brome control in alfalfa (*Medicago sativa* L.) (Kapusta and Strieker 1975), perennial ryegrass (*Lolium perenne* L.) seed production (Lee 1965), and rangeland (Eckert et al. 1974; Morrow et al. 1977) with herbicides has had limited success. Similarities in downy brome and winter wheat life cycles prevent consistent chemical control in winter wheat croplands (Morrow and Stahlman 1984). Downy brome and winter wheat both emerge in the fall and overwinter as seedlings. The herbicides metribuzin, trifluralin, and diclofop are registered for the preemergent control of downy brome. After the crop emerges, however, chemical control options are limited (Peeper 1984; Stahlman 1986). Control has been best accomplished by an integrated approach of crop rotation (Wicks 1984), fall and spring tillage, herbicides to kill emerging seedlings (Wicks et al. 1971; Massee 1976), and delayed planting of the crop (Fenster and Wicks 1978). With lack of adequate selective weed control and the erosive practices of tillage for control, downy brome is an excellent model for biological weed control efforts.

### *The Biological Control Pair*

These DRB and the grass weed downy brome make a good match as a biological control system because the bacteria suppress weed growth at a point in the weed life cycle that reduces its competitiveness with the crop. In field studies conducted in eastern Washington (Kennedy et al. 1991), bacteria were applied to wheat fields infested with natural populations of downy brome. Reduction in downy brome growth varied and depended on the bacterial strain. One strain of inhibitory bacteria, *Pseudomonas fluorescens* strain D7, reduced plant populations and aboveground growth of downy brome 31 percent and 53 percent, respectively (Fig. 10.1). In the same experiment, seed production of downy brome was reduced 64 percent. Winter wheat yields were increased by 35 percent with the application of the bacteria and subsequent suppression of downy brome growth (Fig. 10.2). This increase in yield is similar to the yield increase expected from the elimination of a moderate infestation of downy brome. Another bacteria, *Pseudomonas syringae* strain 2V19, suppressed downy brome growth by 25 percent in the same field study, with a 27 percent increase in winter wheat yield.

The plant-suppressive compounds produced by the bacteria reduced germination, overwinter vigor, and root growth, thereby reducing downy brome's competitive ability and giving wheat the competitive edge. The toxin also reduced tiller number and the total number of seeds, resulting in fewer plants with fewer tillers and fewer seeds. Thus, in theory, we have a good match for a biological control pair. Even though the bacteria did not control the downy brome 100 percent, they suppressed downy brome growth so that winter wheat could outcompete the weed.

Application of DRB and the resultant suppression of downy brome root growth may allow the crop to outcompete weeds. If this suppression occurs during the seedling stage, the crop may gain a competitive growth advantage. This increase in crop competitive ability may further suppress weed growth at later growth stages. Early seed colonization appears to be critical to the success of the bacteria, with greater populations of the introduced DRB found in the spermosphere of downy brome than of winter wheat (Doty 1992). The ecology of these DRB and their interaction with the germinating seed greatly affects the outcome of this weed-microbe interaction and needs further study. Seed and root exudate quality or quantity (Van Vuurde and Schippers 1980) and microfloral competition also play a part in bacterial survival (Suslow and Schroth 1982). The production of plant-suppressive compounds and subsequent weed suppression will change with colonization of the root by the bacteria (Bolton and Elliott 1989).

Root colonization and plant suppression by these bacteria may be sensitive to many soil factors, including soil water content (Fredrickson and Elliott 1985; Howie et al. 1987; Schippers et al. 1987; Liddell and Park 1989) and soil temperature (Loper et al. 1985; Kenerley and Jeger 1990). In subsequent field studies with *Pseudomonas fluorescens* strain D7, time of application was critical to suppression of downy brome. Greatest inhibition of downy brome occurred when the bacterium was applied just prior to a significant rain event, indicating the importance of bacterial survival on biological control efficacy (Kennedy and Young, unpublished data).

Combinations of biological control agents and reduced rates of chemicals may improve control and broaden the spectra of control (Charudattan and DeLoach 1988). Integration of stresses has been demonstrated for velvetleaf control using the fungal pathogen *Fusarium oxysporum* and chemical herbicides (Kremer and Schulte 1989) or the fungal pathogen and a seed-feeding insect (Kremer and Spencer 1989).

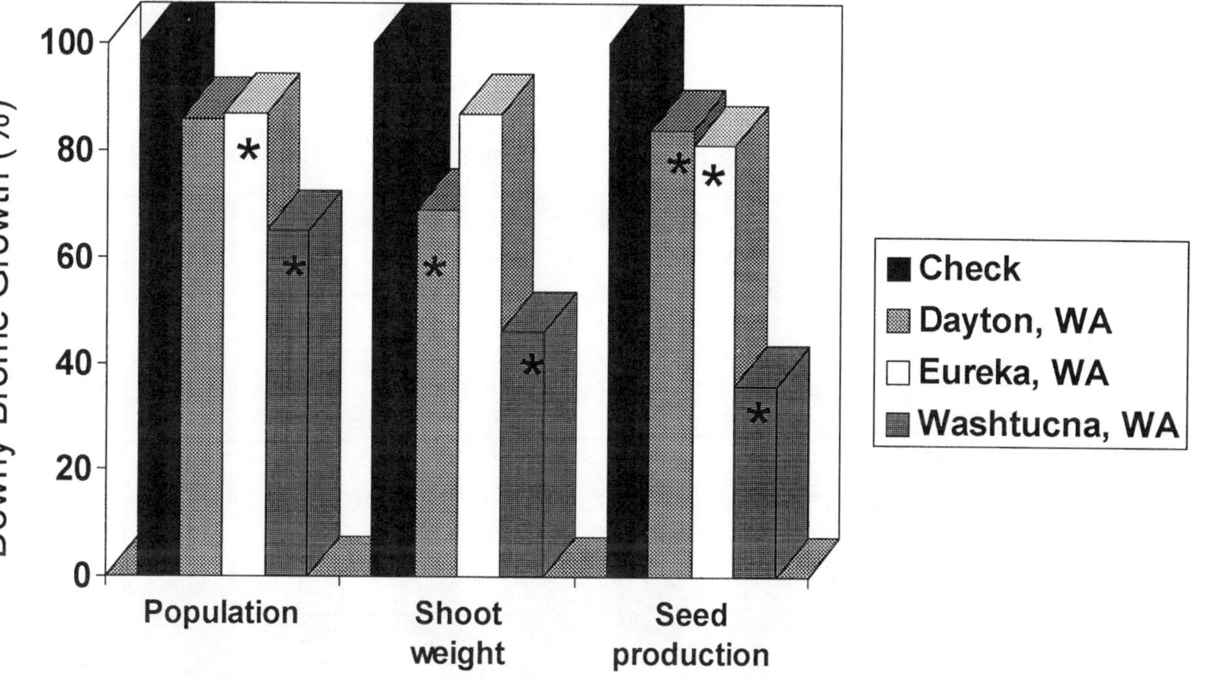


Figure 10.1. Downy brome growth and seed production from fields inoculated with rhizobacteria and planted to winter wheat at three locations in eastern Washington. *Note:* Asterisks indicate significant differences at $p \leq 0.05$. *Source:* Kennedy et al. 1991.

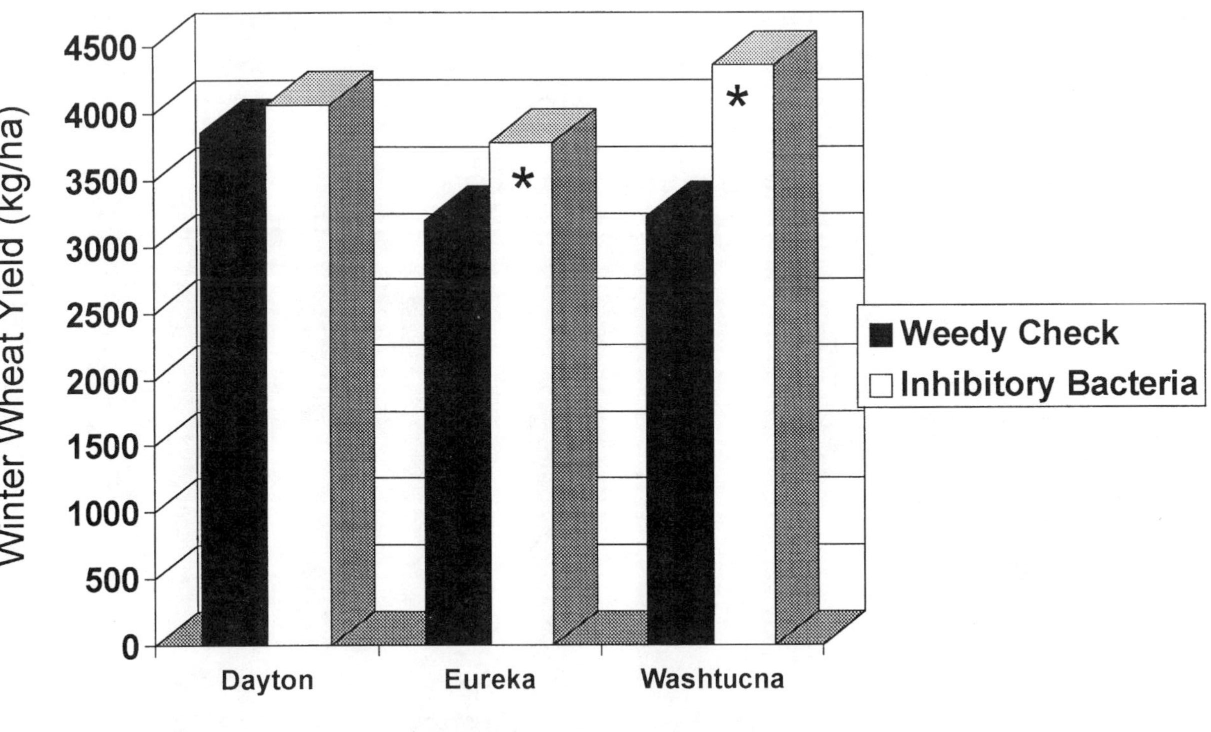


Figure 10.2. Winter wheat population and yield from fields inoculated with rhizobacteria and planted to winter wheat at three locations in eastern Washington. *Note:* Asterisks indicate significant differences at $p \leq 0.05$. *Source:* Kennedy et al. 1991.

The rhizobacteria can be compatible with several herbicides for grass control. Compatibility would be advantageous because it would allow for the bacteria to be used to supplement herbicides with poor activity on downy brome. Thus, a combination of the biological control agent and reduced rates of herbicides may reduce significantly the infestations of downy brome, and the biological control agent may act synergistically with herbicides to increase activity, selectivity, or spectrum of control.

## Future Need and Considerations

There are many challenges ahead for the use of deleterious rhizobacteria for grass weed control. Future research endeavors will need to include the following areas.

### *Biological Control Agents*

Investigations should include a wide range of microorganisms, both fungi and bacteria, to develop a variety of candidates and hosts for biological control of broadleaf and grass weeds. With the development of each new system, new concepts and problems will arise and deepen our understanding as they increase the acceptance of microorganisms for weed control.

### *Plant Suppressive Compounds*

Knowledge of the structure and mode of action of each plant-suppressive compound is imperative to determine the mode of action and enhance the activity or specificity of these compounds.

### *Molecular Investigations*

Molecular investigations of DRB will uncover the genetic basis of virulence and host specificity and broaden our appreciation of the range of diversity in pathogens. Genetic manipulation of the microbes can increase metabolite production, regulate host specificity or host range, and regulate production of the plant-suppressive compounds. This regulation will allow us to control the expression of plant-suppressive genes by specific promoters or under specific environmental conditions.

### *Ecology*

A deeper understanding of the ecology of the introduced bacteria is imperative to increase survival and efficacy, and thus improve the effectiveness of biological control agents. Survival of the bacteria is critical to the success of the agent. Moreover, the environmental constraints to survival of the microorganism and toxin production need to be determined. Microbial detection levels at present are inadequate, and enumeration methods need further study. The environmental effects on bacterial survival that need to be considered include physical constraints, competition with other microbes, and colonization of the seed and root. The ecology of the weed and the weed-crop competition is crucial to the success of these biological control agents and must also be considered. The colonization of the seed and root by the bacteria is also an ecological consideration that involves the weed as well. Of importance too is the weed-herbicide interaction and the physiological responses of the weed when stress, either chemical or environmental, is imposed.

### *Fermentation Technology*

The production of these DRB in large-scale batches has been explored in only a few cases. A greater understanding of the commercial inoculum production is needed for a practical, economical production of these bacteria. The shelf life of the organism after it is produced also demands consideration.

### *Delivery Systems*

A greater understanding of the ecological constraints of each delivery system will foster development of systems or formulations that will reduce the impact of the biotic and abiotic forces and increase the viability, efficacy, and ease of application. Many of the problems in the survival of the bacterium, such as moisture, temperature, and radiation sensitivities, may be reduced through formulation technology.

### *Microbe-Herbicide Interactions*

The compatibility and synergistic effects of microbe-herbicide interactions will assist in the integration of biological control agents with synthetic chemical herbicides and plant growth regulators. This information may differ with each system, but it is critical for user recommendations. The concept of multiple stresses as a means of plant suppression needs further study, whether it be bacteria and herbicide or bacteria and other growth regulators.

### *User Education*

Producers need to shift their thinking to accept some weed populations as economically acceptable. The concept of a weed-free field as a necessity for economical crop protection is no longer valid. Growers need to be made aware of what constitutes a suitable economic suppression versus 100 percent weed control. A greater appreciation of the influence that biotics can have on plant growth will be beneficial. Educating producers and research weed scientists about plant-microbe and plant-plant interactions should result in the abandonment of the "spray and pray" mentality and a shift to an ecological paradigm. The public also needs a greater familiarity with the possible limitations of biocontrol methods, because these often are slower and results are less visible than with herbicides. The environmental constraints that affect biological control agents need to be recognized.

### *Integration with Current Practices*

For any biological control system to be successful, it will need to be used with present management practices or be able to be easily implemented so that it becomes a standard practice. The feasibility of integration with other management practices and the ease of use requires consideration.

### *Regulatory Issues*

Regulation and registration of natural microorganisms is an issue that needs to be further defined to facilitate the timely release of these products to market.

## The Outlook

The research on DRB has demonstrated the potential for use of naturally occurring plant-suppressive bacteria as a novel, nonchemical approach for controlling grass weeds in cereal crops. Deleterious rhizobacteria as microbial weed control agents have a place in modern weed management and should significantly reduce costs, the need for tillage, and chemical pesticide use. These reductions will in turn reduce erosion and water pollution. The research on DRB opens the door for control of other weeds in field and horticulture crops. The "spray and pray" approach to weed control is a thing of the past, and with greater awareness of nonpoint source pollution, the race for novel methods of weed control continues to intensify. Biological control offers an alternative by using natural pathogen-plant interactions and plant-plant competition. Biological systems, however, need special consideration to enhance their activity to achieve a reduction in weed growth. It is important to realize that the goal with biological, especially augmentative, control is not 100 percent control, but rather suppression of weed growth. Suppression of weeds by multiple stresses may be more successful than relying on other means of control. Biological control agents will not replace herbicides; rather, these agents can be used in concert with reduced levels of synthetic chemicals for economical suppression of weeds. Biological control demands a greater understanding of the ecology of the system, its participants, and the specificity of their interactions.

A much larger outcome of this research will be the expansion of knowledge on microbial processes such as survival in soil, specificity, secondary metabolite production, and rhizosphere colonization for purposes of using microorganisms as a mechanism for delivery of any desired compound to the root surface. These concepts will assist scientists in understanding how the soil microbial component can be managed to reverse soil degradation and aid in maintaining a healthy and productive soil. The knowledge we gain from this line of research will not only further environmentally sound methods of weed control, but also enhance the use of microorganisms in other areas of agriculture and the environment for as many applications as we care to imagine and discover.

## Summary

Microorganisms can significantly influence the distribution, abundance, and competition among plant species, and thus they can be used in biological weed control programs. Deleterious rhizobacteria (DRB) are soil bacteria that colonize the rhizosphere and suppress plant growth by the production of phytotoxic substances. These substances can impair seed germination and delay plant development in specific plant species and cultivars. These DRB have the potential to be used to regulate the growth of weeds. Biological control of weeds using DRB while reducing weed pressures can reduce costs, the need for tillage, and synthetic pesticide use. We have isolated soil bacteria that are selective in their suppression of various weed species and are effective in reducing weeds when they are applied to the field. These bacteria are excellent biological control agents because they are aggressive colonizers of the roots and residue, thereby functioning as a direct delivery system for the natural, plant-suppressive compounds they produce. A better understanding of the ecology of these bacteria and the weeds they suppress is needed for the successful use of this technology.

## References

Alstrom, S. 1987. Factors associated with detrimental effects of rhizobacteria on plant growth. *Plant and Soil* 102:3–9.

Alstrom, S., and R. G. Burns. 1989. Cyanide production by rhizobacteria as a possible mechanism of plant growth inhibition. *Biology and Fertility of Soils* 7:232–38.

Astrom, B. 1991. Intra- and interspecific variations in plant response to inoculations with deleterious rhizosphere pseudomonads. *Journal of Phytopathology* 131:184–92.

Bakker, A. W., and B. Schippers. 1987. Microbial cyanide production in the rhizosphere in relation to potato yield reduction and *Pseudomonas* spp.–mediated plant growth reduction. *Soil Biology and Biochemistry* 19:452–58.

Bolton, H., and L. F. Elliott. 1989. Toxin production by a rhizobacterial sp. that inhibits wheat root growth. *Plant and Soil* 114:269–78.

Charudattan, R., and C. J. DeLoach Jr. 1988. Management of pathogens and insects for weed control in agroecosystems. In *Weed management in agroecosystems: Ecological approaches,* ed. M. A. Altieri and M. Liebman, 245–64. Boca Raton, Fla.: CRC Press.

Cherrington, C. A., and L. F. Elliott. 1987. Incidence of inhibitory pseudomonads in the Pacific Northwest. *Plant and Soil* 101:159–65.

Cullen, J. M., P. F. Kable, and M. Catt. 1973. Epidemic spread of a rust imported for biological control. *Nature* 244:462–64.

Daniel, J. T., G. E. Templeton, R. J. Smith Jr., and W. T. Fox. 1973. Biological control of northern jointvetch in rice with an endemic fungal disease. *Weed Science* 21:303–7.

Donald, W. W., and A. G. Ogg Jr. 1991. Biology and control of jointed goatgrass (*Aegilops cylindrica*): A review. *Weed Technology* 5:3–17.

Doty, J. A. 1992. Spermosphere colonization by a plant-suppressive rhizobacterium. Ph.D. thesis, Washington State University, Pullman.

Eckert, R. E. Jr., J. E. Asher, M. D. Christensen, and R. A. Evans. 1974. Evaluation of the atrazine-fallow technique for weed control and seedling establishment. *Journal of Range Management* 27:288–92.

Elliott, L. F., and J. M. Lynch. 1985. Plant growth–inhibitory pseudomonads colonizing winter wheat (*Triticum aestivum* L.) roots. *Plant and Soil* 84:57–65.

Fenster, C. R., and G. A. Wicks. 1978. Know and control downy brome. *Nebraska Cooperative Extension Service Bulletin* G–78–422.

Fredrickson, J. K., and L. F. Elliott. 1985. Colonization of winter wheat roots by inhibitory rhizobacteria. *Soil Science Society of America Journal* 49:1172–77.

Gardner, J. M., J. L. Chandler, and A. W. Feldman. 1984. Growth promotion and inhibition by antibiotic-producing fluorescent pseudomonads on citrus roots. *Plant and Soil* 77: 103–13.

Harris, G. A. 1967. Some competitive relationships between *Agropyron spicatum* and *Bromus tectorum*. *Ecological Monographs* 37:89–111.

Hitchcock, A. S., and A. Chase. 1971. *Manual of the grasses of the United States*. 2d ed. New York: Dover.

Howie, W. J., R. J. Cook, and D. M. Weller. 1987. Effects of soil matric potential and cell motility on wheat root colonization by fluorescent pseudomonads suppressive to take-all. *Phytopathology* 77:286–92.

Hulbert, L. C. 1955. Ecological studies of *Bromus tectorum* and other annual bromegrasses. *Ecological Monographs* 25:181–213.

Hull, A. C. Jr., and J. F. Pechanec. 1947. Cheatgrass: A challenge to range research. *Journal of Forestry* 45:555–64.

Johnson, B. N., A. C. Kennedy, and A. G. Ogg Jr. 1993. Suppression of downy brome growth by a rhizobacterium in controlled environments. *Soil Science Society of America Journal* 57:73–77.

Jurhen, M. F., W. Went, and E. Phillips. 1965. Ecology of desert plants IV. Combined field and laboratory work on germination of annuals in the Joshua Tree National Monument, California. *Ecology* 37:318–30.

Kapusta, G., and C. F. Strieker. 1975. Selective control of downy brome in alfalfa. *Weed Science* 22:202–6.

Kenerley, C. M., and M. J. Jeger. 1990. Root colonization by *Phymatotrichum omnivorum* and symptom expression of *Phymatotrichum* root rot in cotton in relation to planting date, soil temperature and soil water potential. *Plant Pathology* 39:489–500.

Kennedy, A. C., H. Bolton Jr., H. F. Stroo, and L. F. Elliott. 1992a. The competitive abilities of the Tn5 Tox–mutants of a rhizobacterium inhibitory to wheat growth. *Plant and Soil* 44:143–53.

Kennedy, A. C., L. F. Elliott, F. L. Young, and C. L. Douglas. 1991. Rhizobacteria suppressive to the weed downy brome. *Soil Science Society of America Journal* 55:722–27.

Kennedy, A. C., A. G. Ogg Jr., and F. L. Young. 1992b. Biocontrol of jointed goatgrass. Patent number 07/597,150. November 17, 1992.

Klemmedson, J. O., and J. G. Smith. 1964. Cheatgrass (*Bromus tectorum* L.). *Botanical Review* 30:226–62.

Kremer, R. J. 1986. Antimicrobial activity of velvetleaf (*Abutilon theophrasti*) seeds. *Weed Science* 34:617–22.

Kremer, R. J., M.F.T. Begonia, L. Stanley, and E. T. Lanham. 1990. Characterization of rhizobacteria associated with weed seedlings. *Applied and Environmental Microbiology* 56: 1649–55.

Kremer, R. J., and L. K. Schulte. 1989. Influence of chemical treatment and *Fusarium oxysporum* on velvetleaf. *Weed Technology* 3:369–74.

Kremer, R. J., and N. R. Spencer. 1989. Impact of seed-feeding insects and microorganisms on velvetleaf (*Abutilon theophrasti*) seed viability. *Weed Science* 37:211–16.

Lee, W. O. 1965. Selective control of downy brome and rattail fescue in irrigated perennial grass seed fields of central Oregon. *Weeds* 13:205–8.

Liddell, C. M., and J. L. Parke. 1989. Enhanced colonization of pea taproots by a fluorescent pseudomonad biocontrol agent by water infiltration into soil. *Phytopathology* 79:1327–32.

Loper, J. E., A. Haack, and M. N. Schroth. 1985. Population dynamics of soil pseudomonads in the rhizosphere of potato (*Solanum tuberosum* L.). *Applied and Environmental Microbiology* 49:416–22.

Mack, R. N. 1981. Invasion of *Bromus tectorum* L. into western North America: An ecological chronicle. *Agroecosystems* 7:145–65.

Mack, R. N., and D. A. Pyke. 1983. The demography of *Bromus tectorum*: Variation in time and space. *Journal of Ecology* 71:69–93.

Massee, T. W. 1976. Downy brome (*Bromus tectorum*) control in dryland wheat with stubble-mulch fallow and seedling management. *Agronomy Journal* 68:952–55.

Morrow, L. A., and P. W. Stahlman. 1984. The history and distribution of downy brome (*Bromus tectorum*) in North America. *Weed Science* 32:2–6.

Morrow, L. A., C. R. Fenster, and M. K. McCarty. 1977. Control of downy brome on Nebraska rangeland. *Journal of Range Management* 30:293–96.

Mortensen, K. 1986. Biological control of weeds with plant pathogens. *Canadian Journal of Plant Pathology* 8:229–31.

Peeper, T. F. 1984. Chemical and biological control of downy brome (*Bromus tectorum*) in wheat and alfalfa in North America. *Weed Science* 32:18–25.

Ridings, W. H. 1986. Biological control of stranglervine (*Morrenia odorata* Lindl.) in citrus—A researcher's view. *Weed Science Supplement* 34:Supp. 1.

Rydrych, D. J. 1974. Competition between winter wheat and downy brome. *Weed Science* 22:211–14.

Rydrych, D. J., and T. J. Muzik. 1968. Downy brome competition and control in dryland wheat. *Agronomy Journal* 60:279–80.

Salt, G. A. 1979. The increasing interest in "minor pathogens". In *Soil borne plant pathogens,* ed. B. Schipper and W. Gams, 289–313. New York: Academic.

Schippers, B., A. W. Bakker, and P. A. Bakker. 1987. Interactions of deleterious and beneficial rhizosphere microorganisms and the effect of cropping practices. *Annual Review of Phytopathology* 25:339–58.

Stahlman, P. 1986. Competition and control of annual bromes in winter wheat. AES Bulletin 89.

Stewart, G., and A. C. Hull. 1949. Cheatgrass (*Bromus tectorum* L.): An ecologic intruder in southern Idaho. *Ecology* 30:58–74.

Stroo, H. F., L. F. Elliott, and R. I. Papendick. 1988. Growth, survival and toxin production of root-inhibitory pseudomonads on crop residues. *Soil Biology and Biochemistry* 20: 201–7.

Suslow, T. V., and M. N. Schroth. 1982. Role of deleterious rhizobacteria as minor pathogens in reducing crop growth. *Phytopathology* 72:111–15.

Templeton, G. E. 1982. Status of weed control with plant pathogens. In *Biological control of weeds with plant pathogens,* ed. R. Charudattan and H. L. Walker, 29–44. New York: Wiley.

Thill, D. C., K. G. Beck, and R. H. Callihan. 1984. The biology of downy brome (*Bromus tectorum* L.). *Weed Science* 32:7–12.

Tranel, P. J., D. R. Gealy, and A. C. Kennedy. 1993. Inhibition of downy brome (*Bromus tectorum* L.) root growth by a phytotoxin from *Pseudomonas fluorescens* strain D7. *Weed Technology* 7:134–39.

Uresk, D. W., J. F. Kline, and W. H. Rickard. 1979. Growth rates of a cheatgrass community and some associated factors. *Journal of Range Mangement* 32:168–70.

Van Vuurde, J.W.L., and B. Schippers. 1980. Bacterial colonization of seminal wheat roots. *Soil Biology and Biochemistry* 12:559–65.

Watson, A. K., and M. Clement. 1986. Evaluation of rust fungi as biological control agents of weedy *Centaurea* in North America. *Weed Science* 34:7–10.

Wicks, G. A. 1984. Integrated systems for control and management of downy brome (*Bromus tectorum*) in cropland. *Weed Science* 32 (Supplement 1):26–31.

Wicks, G. A., O. C. Burnside, and C. R. Fenster. 1971. Influence of soil type and depth of planting on downy brome seed. *Weed Science* 19:82–86.

Woltz, S. S. 1978. Nonparasitic plant pathogens. *Annual Review of Phytopathology* 16: 403–30.

# 11

## Spring-Seeded Smother Plants for Weed Control in Corn and Other Annual Crops

*Robert L. De Haan, Donald L. Wyse, Nancy J. Ehlke, Bruce D. Maxwell, and Daniel H. Putnam*

Corn producers in developed countries rely heavily on herbicides and mechanical tillage to control weeds. Both herbicides and tillage, however, can cause undesirable environmental effects. Herbicides have been detected in surface water and groundwater at concentrations that exceed the lifetime health advisory limit for drinking water as established by the Environmental Protection Agency (Goolsby et al. 1991; Klaseus et al. 1988). Mechanical tillage breaks down soil structure and decreases the amount of plant residue on the soil surface, which increases the risk of soil erosion (Laflen et al. 1980; Langdale et al. 1991). Soil erosion causes on-site and off-site environmental degradation and poses a threat to long-term crop production (Larson et al. 1983).

Cover crops or smother crops have been proposed as an alternative weed control method in corn (Liebman 1988; Palada et al. 1983; Worsham 1991). Cover crops that suppress weeds could reduce herbicide and tillage inputs (Palada et al. 1983), increase soil water infiltration (Bruce et al. 1992), contribute nitrogen to the main crop (Corak et al. 1991; Decker et al. 1994; Maskina et al. 1993), and reduce economic risk (Hanson et al. 1993).

Research investigating biological weed control through plant interference has focused on the use of winter annual or perennial species sown in the fall and suppressed or killed with a herbicide in the spring (Curran et al. 1994; Eadie et al. 1992; Eberlein et al. 1992; Echtenkamp and Moomaw 1989; Enache and Ilnicki 1990; Fischer and Burrill 1993; Grubinger and Minotti 1990; Hoffman et al. 1993; Johnson et al. 1993; Kumwenda et al. 1993; Mohler 1991; Teasdale 1993; Tollenaar et al. 1993). Species that have been evaluated as cover crops include alfalfa (*Medicago sativa* L.), Austrian winter pea (*Pisum sativum* L. subsp. *sativum* var. *arvense* (L.) Poir), barley (*Hordeum vulgare* L.), chewings fescue (*Festuca rubra* L.), crimson clover (*Trifolium incornatum* L.), hairy vetch (*Vicia villosa* Roth), ladino clover (*Tri-*

*folium repens* L.), subterranean clover (*Trifolium subterraneum* L.), white clover (*Trifolium repens* L.), winter rye (*Secale cereale* L.), and wheat (*Triticum aestivum* L.). In the central United States, hairy vetch, ladino clover, subterranean clover, and white clover show promise, but in the Upper Midwest none of these species is consistently winter hardy. In addition, soil water resources can be depleted by spring growth of these species, and all of them except subterranean clover require chemical or mechanical suppression to limit their competitive effects on corn.

These results indicate that there is a need to identify or develop adapted cover crop species for use in the Upper Midwest (Karlen 1990; Lal et al. 1991; Power and Biederbeck 1991). Adapted cover crops could be developed by incorporating greater winter hardiness into existing perennial or winter annual cover crops (Karlen 1990), but this will be difficult, as winter hardy germplasm is not available for most cover crop species (Palada et al. 1983; Robinson 1956).

The development of short-lived spring-seeded cover crops is an alternative method of generating cover crops adapted for use in the Upper Midwest (De Haan et al. 1994; Power and Biederbeck 1991). Spring-seeded cover crops or smother plants would not need to be winter hardy, and may be easier to develop than winter hardy perennial or winter annual cover crops.

Spring-seeded smother plants are defined as specialized cover crops that have been selected for their ability to suppress weeds without affecting main crop development or yield. Spring-seeded smother plants could be planted simultaneously with the main crop, or shortly after it.

The feasibility of short-lived spring-seeded smother plants is supported by plant competition research. Studies indicate that weed interference with corn for the first two to eight weeks after corn emergence may not be detrimental to crop yield (Bunting and Ludwig 1964; Hall et al. 1992; Neito et al. 1968; Thomas and Allison 1975; Zimdahl 1988). It is possible that a smother plant could interfere with corn for a similar period of time without reducing crop yield. Other researchers have reported that weeds that do not emerge until three to six weeks after corn do not reduce corn yield (Hall et al. 1992; Knake and Slife 1965; Thomas and Allison 1975). Annual weeds suppressed for four weeks by the smother plant, therefore, would not compete effectively with corn. Ideally, the smother plant would suppress weeds early in the growing season and then senesce, minimizing its competitive effect on the main crop.

Successful smother plants need to be selective; they should suppress weeds, but not the main crop. In many cases an important source of selectivity may be based on the seed size of the main crop versus that of weeds (Mohler 1994). Most crop seeds are much larger than weed seeds, and therefore crop plants reach a height of 4 to 5 cm soon after emergence. Plants from small seeded weeds typically reach a height of 4 to 5 cm a week or more later than the crop. A smother plant that is taller than the weeds but shorter than the crop during this time period (perhaps due to an intermediate seed size) would compete for light more effectively with the weeds than with the crop.

Precision fertilizer placement may be another way to develop selectivity. If plant nutrients were more readily available to the main crop than to the weeds or the smother plant, the crop would have a competitive advantage. Crops grown with this advantage might be more tolerant of smother plant interference than crops grown under conditions where plant nutrients were uniformly available.

A model spring-seeded smother plant, or ideotype, for use in corn production systems has been proposed (De Haan et al. 1994). Ideotypes are hypothetical plants described in terms of traits that are thought to enhance performance potential (Donald 1968; Rasmussen 1987). The spring-seeded smother plant ideotype has the following characterics: seedling emergence four to five days after planting under cool conditions; a horizontal leaf angle; a mature leaf size of 2 cm by 3 cm; a rooting depth of 25 cm; a maximum height of 10 cm; a life cycle of five weeks or less; and seed production potential of at least 500 kg $ha^{-1}$.

Spring-seeded smother plants corresponding closely to the ideotype described here could be developed from numerous plant families and genera. In this chapter we describe our efforts to develop and evaluate a *Brassica* species smother plant for the Upper Midwest.

## Development of a *Brassica* Species Smother Plant

Plants in the genus *Brassica* appear to be well suited for smother crop use. Unselected *Brassica* species possess many of the characteristics of the proposed smother plant ideotype, and the genus *Brassica* contains species with diverse growth habits and life histories. In addition, related *Brassica* species cross readily, giving the breeder access to extensive genetic diversity.

The objectives of this research program were to develop a prototype spring-seeded *Brassica* species smother plant using classical plant breeding methods and to evaluate the effect of the new *Brassica* species smother plant on corn under field conditions.

### *Plant Breeding and Selection*

A rapid cycling, short-statured *Brassica campestris* line from the Crucifer Genetics Cooperative (CrGC 1-21) was chosen as a parental line for the breeding program. It flowered less than twenty-two days after planting and had a maximum height of 25 cm. It was phenotypically a rapid cycling dwarf and was homozygous for a partially dominant dwarf gene (dwf2/dwf2) that causes extreme compaction of the internodes. Six locally adapted *Brassica campestris, B. juncea,* and *B. napus* stocks were obtained from commercial sources. These *Brassica* spp. are diploid, insect pollinated, and strongly outbreeding with self-incompatibility controlled by a multiple allelic series of genes (Williams 1985). The six locally adapted lines were reciprocally crossed with CrGC 1-21. F1 seed from the crosses was planted, and days to emergence, days from planting to flowering, and height twenty-eight days after planting were recorded. Upon flowering, plants from each cross with similar height and days from emergence to first flower were put in a bee cage to produce F2 seed. An F1 dwarf *Brassica* population (1-21 x 101a in Table 11.1) was selected for field evaluation because it was phenotypically most similar to the spring-seeded smother plant ideotype described previously (De Haan et al. 1994). F2 seed from this population was planted in the greenhouse and did not segregate noticeably for plant height or maturity. F3 seed was produced in the greenhouse; F4 and F5 seed was produced in the field at St. Paul during the summer of 1990; and F6 seed was produced in Arizona during the winter of 1991. The 1990 field evaluation utilized F3 seed, and the 1991 field trials utilized F5 and F6 seed.

### *Field Evaluation*

The dwarf *Brassica* smother plant was evaluated in the field at the University of Minnesota St. Paul Experiment Station during 1990 and at the St. Paul and Waseca experiment stations in 1991 (De Haan 1992). Hybrid field corn (Garst 8851 in 1990 and Pioneer 3772 in 1991) was planted in 76 cm wide rows at 65,000 seeds ha$^{-1}$ in late May or early June on conventionally tilled sites. Each site was divided into weedy and weed-free plots. The most abundant weeds were yellow foxtail (*Setaria glauca*

Table 11.1. Time of emergence in days after planting (dap), days to first flower, height twenty-eight days after planting, and number of plants for selected parent and $F_1$ *Brassica* populations grown in the greenhouse in 1990

| *Brassica* populations | Time to emergence (day) | Time to flowering (dap) | Height at 28 dap (cm) | Population size |
|---|---|---|---|---|
| Parents: | | | | |
| CrGC 1-21[b] | 3 | 19 | 10 | --- |
| 101[c] | 4 | C[1] | 6 | --- |
| E1 Progenye: | | | | |
| 1-21 x 101a | 3 | 20 | 13 | 56 |
| 1-21 x 101b | 3 | 35 | 5 | 5 |
| 1-21 x 101c | 3 | 40-50 | 5 | 20 |
| 101a x 1-21 | 3 | 27 | 20 | 3 |
| 101b x 1-21 | 3 | C[1] | 5 | 150 |

[1]Cold treatment required to initiate flowering.

*Note:* The female parent for each cross is listed first. CrGC 1-21 is a rapid cycling dwarf from the Crucifer Genetics Cooperative, and 101 is Chinese cabbage "Chinese mastus." Letters following the crosses indicate distinct subpopulations.

(L.) Beauv.), green foxtail (*Setaria viridis* (L.) Beauv.), common lambsquarters (*Chenopodium album* L.), and redroot pigweed (*Amaranthus retroflexus* L.).

The dwarf *Brassica* smother plant was seeded over the corn rows at rates of 0, 530, 1060, 2120, or 4240 seeds $m^{-2}$. Seeding was done within twenty-four hours of corn planting using a three row pattern. One row of dwarf *Brassica* smother plants was seeded directly over the corn row, and the other rows were seeded 10 cm to the left and right of the corn row.

Data were analyzed using analysis of variance and linear regression procedures. Nonlinear regression models were fit to the data to generate predictive equations for smother plant dry weight and corn grain yield response to smother plant seeding rate. The model used for smother plant dry weight was (Regression 1):

$$Y = (a \times d) / (b + d),$$

where $Y$ is actual smother plant dry weight, $a$ and $b$ are estimated parameters, and $d$ is the smother plant seeding rate in seeds $m^{-2}$. The response of corn grain yield to smother plant seeding rate was estimated using the hyperbolic model of Cousens (1991) (Regression 2):

$$Y = Y_{sf}[1 - (I \times d) / (1 + I \times d / c)],$$

where $Y$ is actual corn grain yield, $d$ is smother plant seeding rate in seeds $m^{-2}$, and $Y_{sf}$, $I$, and $c$ are estimated parameters denoting corn yield when grown without a smother plant, corn grain yield loss per unit smother plant seeding density as $d$ approaches 0, and maximum yield loss as $d$ approaches infinity, respectively.

## Discussion

F1 progeny from a single cross between CrGC 1-21 and locally adapted *Brassica* species generally segregated into two or three maturity groups or subpopulations that flowered at distinctly different times (Table 11.1). The earliest maturing subpopulation of the CrGC 1-21 by Chinese cabbage cross was chosen for further evaluation. In the greenhouse this subpopulation produced plants characterized by seedling emergence three days after planting; a generally horizontal leaf angle; a mature leaf size of approximately 2 by 3 cm; a maximum height of about 30 cm; and a life cycle of five to six weeks. Except for maximum height, these characteristics are similar to those described for the smother plant ideotype.

When they were evaluated in the field, dwarf *Brassica* smother plant growth patterns were similar across locations and weed treatments. The smother plant grew rapidly during the first three weeks after emergence, and ground cover increased (Fig. 11.1). Three weeks after emergence the smother plants started to flower, and five weeks after emergence they began to senesce. Smother plant ground cover was greatest three to four weeks after emergence. When smother plant seeding rates of

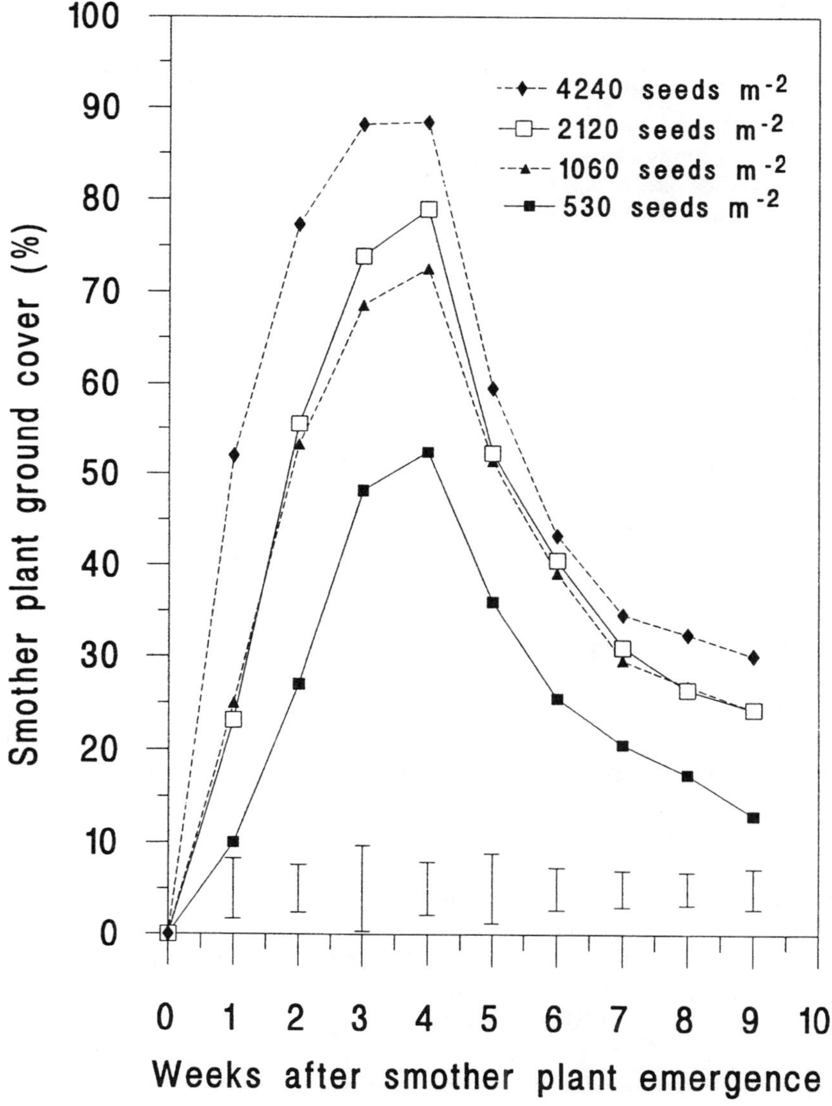


Figure 11.1. Soil surface covered by the smother plant as affected by time after smother plant emergence, averaged over weed treatments at St. Paul in 1990 and 1991 and Waseca in 1991. *Note:* Bars indicate standard errors for comparison of means ($n$ = 22) at each time interval.

530 to 4,240 seeds $m^{-2}$ were used, maximum ground cover of a 25 cm wide band over the corn row ranged from 52 percent to 88 percent, respectively.

Smother plant dry weight four weeks after emergence was different between locations, as determined by a reduced versus full model F test (Neter et al. 1983). Smother plant dry weight increased asymptotically with seeding rate, so intraspecific competition may have occurred at the higher seeding rates (Fig. 11.2). Dwarf *Brassica* seeding rates of 530 to 4,240 seeds $m^{-2}$ produced a mean biomass over locations, years, and weed treatments of 107 to 210 g $m^{-2}$, respectively. The relationship between smother plant height and corn height was similar across locations, years, and weed treatments. The highest smother plant seeding rate, 4,240 seeds $m^{-2}$, produced the tallest smother plants and the shortest corn (data not shown). At this seeding rate, smother plant height approached corn height two to three weeks after emergence, but never exceeded corn height (Fig. 11.3). The smother plants reached an average maximum height of 33 cm four weeks after emergence, substantially taller than the 10 cm maximum height for the smother plant ideotype.

The response of corn height seven weeks after smother plant emergence, a measure of corn vegetative development, to varying smother plant seeding densities was significantly different between the St. Paul 1991 site and the response at St. Paul in 1990 and at Waseca in 1991 (Fig. 11.4). At St. Paul in 1991 smother plant seeding rate did not affect corn height seven weeks after smother plant emergence. However, in the other two environments corn height seven weeks after smother plant emergence decreased with increasing smother plant seeding rate. At St. Paul in 1990 and Waseca in 1991, smother plant seeding rates of 530 to 4,240 seeds $m^{-2}$ reduced corn height seven weeks after smother plant emergence by 3 to 16 cm (2 to 10 percent) and 7 to 37 cm (4 to 19 percent), respectively, as compared to corn grown without a smother plant. Weed populations were not heavy enough to reduce corn height at seven weeks, and therefore smother plant presence in the weedy subplots did not increase corn height at seven weeks compared to weedy subplots without smother plants.

Previous research has indicated that competition between corn and an intercrop or cover crop is usually competition for nitrogen or water (Kurtz et al. 1952; Neito and Staniforth 1961; Staniforth 1961). Corn grown at St. Paul in 1990 did not show any visible symptoms of nitrogen deficiency, but corn grown at Waseca in 1991 did appear to be deficient in nitrogen. Competition for water at either Waseca in 1991 or St. Paul in 1990 was unlikely to be responsible for the reduction in corn height. Normal rainfall at St. Paul and Waseca from June 1 to July 31 is 20.2 cm and 21.6 cm, respectively. Rainfall during June and July totaled 41.0 cm at St. Paul in 1990 and 24.6 cm at Waseca in 1991. The St. Paul 1991 location, at which corn height was not reduced by smother plant interference, received only 18.6 cm of rainfall during June and July. It is likely that corn height reduction at the Waseca 1991 location was influenced by competition for nitrogen, and nitrogen competition may also have occurred at St. Paul in 1990, although it was not visually apparent. End of season corn height did not differ among treatments at any experimental site (data not shown).

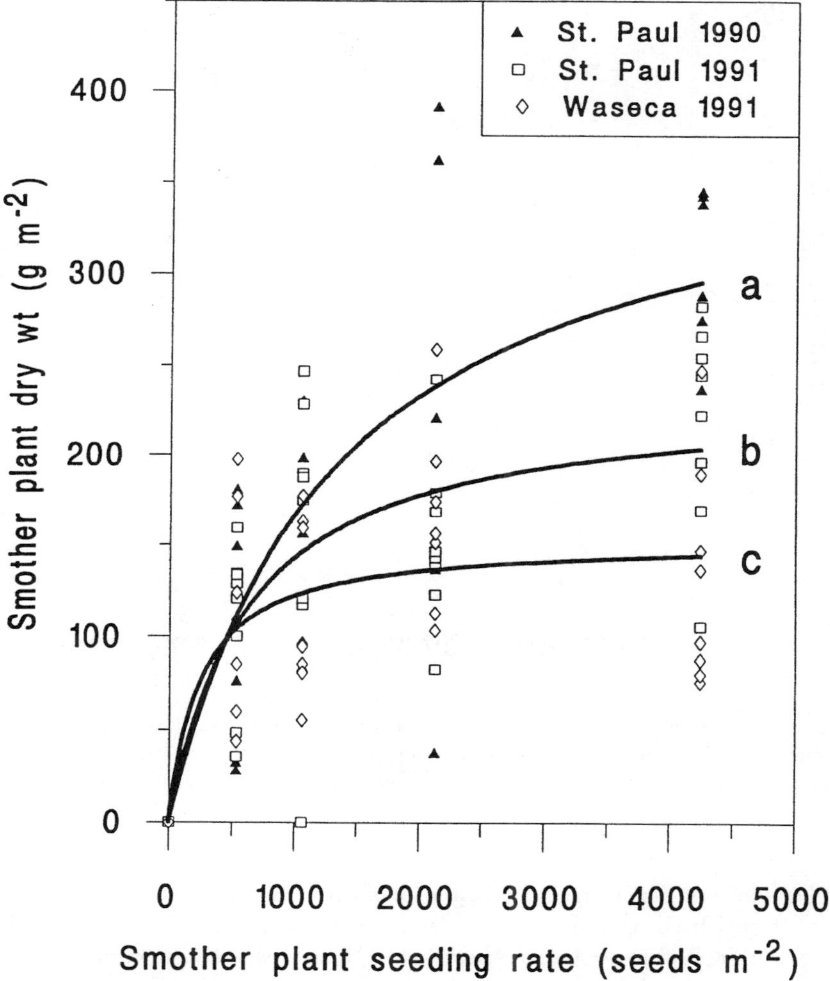


Figure 11.2. Smother plant dry weight four weeks after emergence as affected by smother plant seeding rate, and fitted lines (Regression 1). *Notes:* Data are from weedy and weed-free treatments. Estimates of *a* and *b* with standard errors ( ) are 387 (73) and 1325 (633), 233 (28) and 631 (268), 152 (19) and 245 (179), for St. Paul 1990 (a), St. Paul 1991 (b), and Waseca 1991 (c), respectively.

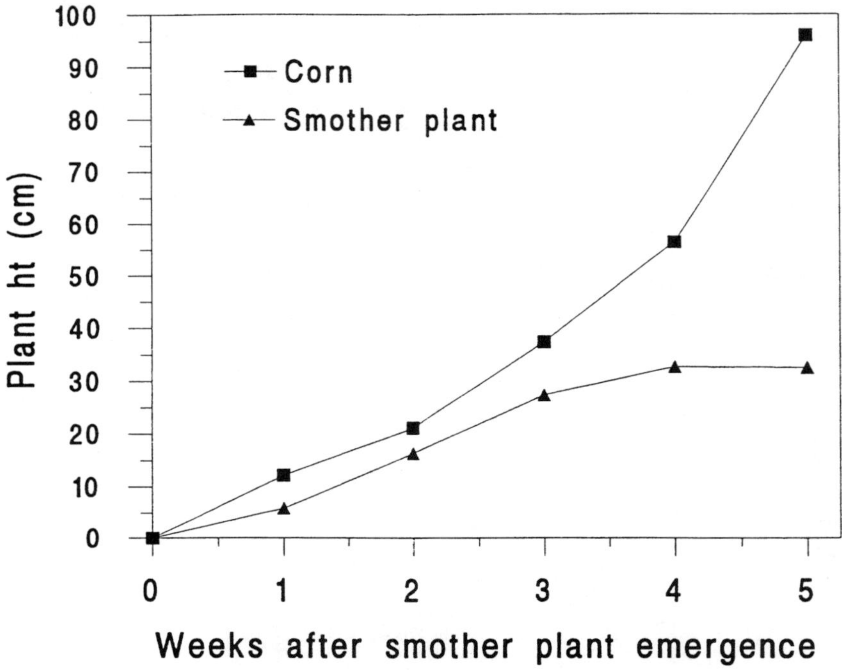


Figure 11.3. Height of corn grown with a smother plant seeded at 4,240 seeds m$^{-2}$, and height of smother plant seeded at 4,240 seeds m$^{-2}$, as influenced by time after smother plant emergence, averaged across locations, years, and weed treatments ($n$ = 22).

Days to corn silk emergence, an estimate of corn reproductive development, was not affected by increasing smother plant seeding rates at St. Paul in 1991 (Fig. 11.5). At St. Paul in 1990 and Waseca in 1991, however, days to corn silk emergence increased with increasing smother plant seeding rates. The slopes of the linear regression lines for the St. Paul 1990 and Waseca 1991 data were not different ($p$ = 0.05). Smother plant seeding rates of 530 to 4,240 seeds m$^{-2}$ delayed silk emergence by an average of 0.7 to 1.8 days and 1.5 to 4.9 days for the St. Paul 1990 and Waseca 1991 locations, respectively, as compared to subplots without smother plants. At Waseca in 1991, weed presence delayed silk emergence by 1.5 days when weedy and weed-free subplots without smother plants were compared, but at the other two experimental sites weed presence did not affect days to silk emergence.

Corn grain yield was not reduced by smother plant presence at St. Paul in 1990, but was reduced by smother plant presence at Waseca in 1991 (Fig. 11.6), where smother plant seeding rates of 530 to 4240 seeds m$^{-2}$ reduced mean grain yield by

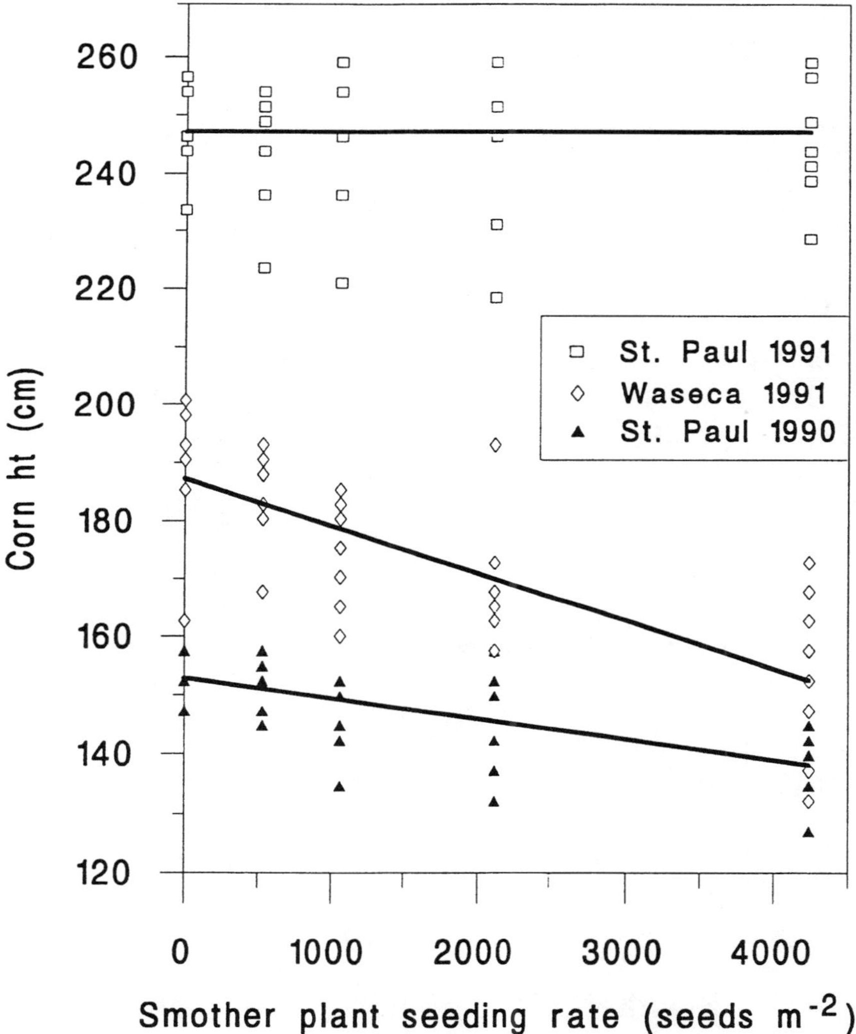


Figure 11.4. Corn height seven weeks after smother plant emergence as affected by smother plant seeding rate and location. *Notes:* Data from both weedy and weed-free treatments are shown. Regressions: St. Paul 1991, mean = 247; Waseca 1991, $y = 187 - .0082X$, $r^2 = .56$, $p < .01$; St. Paul 1990, $y = 152.9 - .0035X$, $r^2 = .38$, $p < .01$.

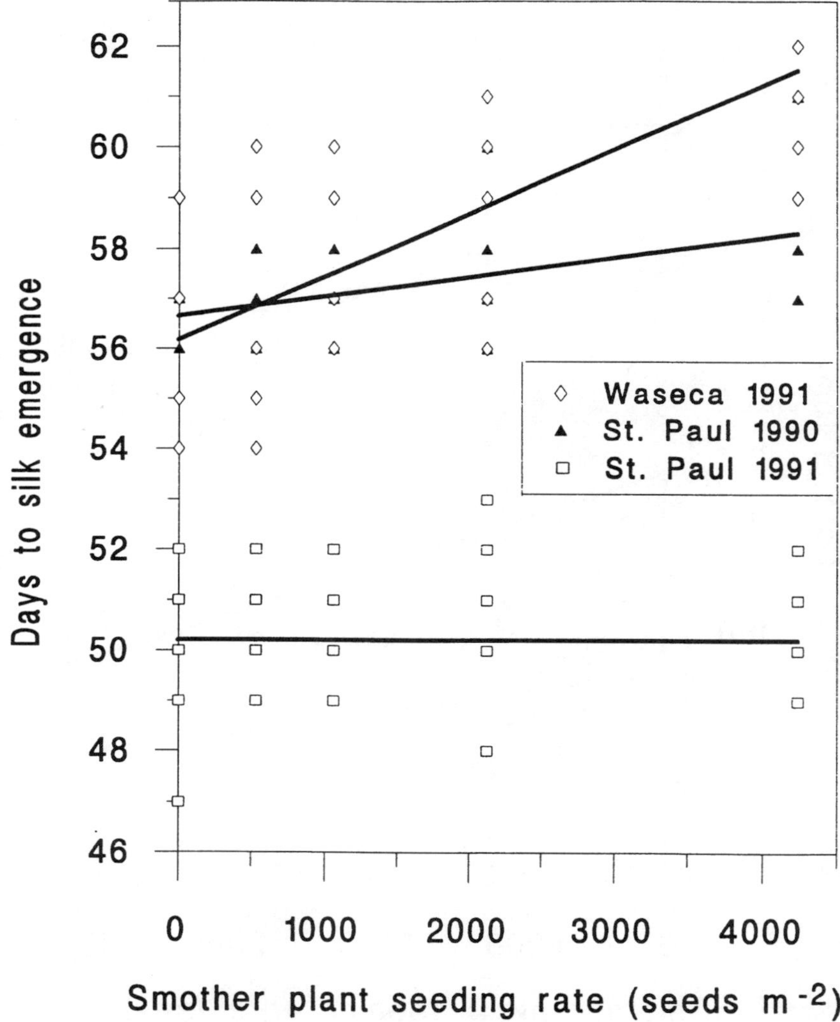


Figure 11.5. Time to corn silk emergence as affected by smother plant seeding rate and location. *Notes:* Data from both weedy and weed-free treatments shown. Regressions: Waseca 1991, $y = 56.2 + 0.00103X$, $r^2 = .42$, $p < .01$; St. Paul 1990, $y = 56.6 + .00039X$, $r^2 = .25$, $p < .01$; St. Paul 1991, mean = 50.2.

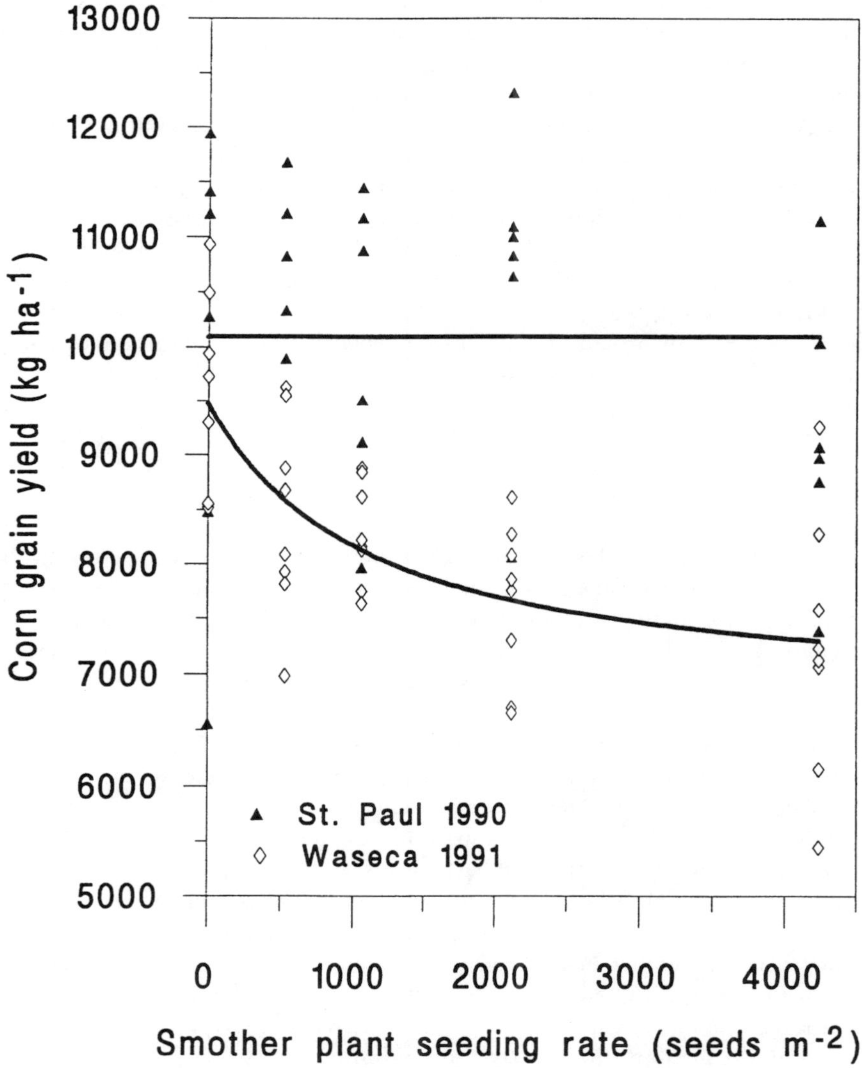


Figure 11.6. Corn grain yield as affected by smother plant seeding rate and location. *Notes:* Data from both weedy and weed-free treatments are shown with fitted lines (Regression 2). Estimates of $Y_{sf}$, $I$, and $c$ with standard errors ( ) are 9,475 (298), 0.00026 (0.00015), and 0.292 (0.068), for Waseca 1991; St. Paul, mean = 10,090.

1058 to 2239 kg ha$^{-1}$ (11 to 24 percent) as compared to subplots without smother plants. Weed infestations were not heavy enough at any of the experimental sites to reduce corn yield, and therefore smother plant presence in the weedy subplots did not increase corn yield compared to weedy subplots without smother plants.

Dry weight of annual grass and broadleaf weeds harvested eleven weeks after smother plant emergence generally decreased with increasing smother plant seeding rate, but the data were too variable to obtain a reliable estimate of weed suppression.

Smother plant evaluation under conditions of more intense weed competition than observed in this study will be required to fully assess the technology. In the absence of the smother plant, corn in weedy plots developed as rapidly and yielded as well as corn grown under weed-free conditions. In this study, therefore, even an ideal smother plant would not have had a beneficial effect. This suggests that a threshold level of weed competition may be necessary for smother plant use to be advantageous. Smother plant research employing several levels of weed competition could be used to investigate this possibility.

## Conclusions

The dwarf *Brassica* smother plant evaluated in this research was competitive with the main crop in two of the three testing environments. Smother plants that reduce main crop yield by 10 percent or more in some years are not likely to be adopted by producers. Researchers involved in smother plant development may be able to make the technology more economically feasible by employing a combination of the following approaches.

First, researchers need to develop a better understanding of the factors involved in determining whether the crop and smother plant will compete for the same resources. Corn development and yield results indicate that the dwarf *Brassica* smother plant may have a minimal impact on corn under some (favorable) environmental conditions, but that it could have a negative effect if the availability of nitrogen or other nutrients is limited. Additional research investigating the effects of the availability of nitrogen, water, and other nutrients on the competitive interactions between corn, smother plants, and weeds could facilitate the development of smother plant lines with greater selectivity than observed with the dwarf *Brassica.*

It is possible that manipulation of the smother plant genotype alone will not be successful. Main crop cultivars may also need to be selected for performance in a smother plant system before smother plant technology becomes feasible. Current crop cultivars have been selected for performance in high nutrient status monoculture cropping systems, and may not be adapted for use with smother plants. Interdisciplinary research involving plant breeders, weed scientists, soil scientists, and perhaps others is needed to investigate this question.

Some method of controlling smother plant height, life cycle length, or both would be a useful way of regulating its interference with crop growth. It is possible that a gene or genes could be altered or inserted into the smother plant so that the application of a common, otherwise nontoxic substance would cause the smother plant to senesce. Another possibility would be the incorporation of a lethal gene into the smother plant so that it dies after four to five weeks of growth, unless a specific management tool is utilized. Producers would then have a way to control the growth of the smother plant and obtain maximum benefit from its use.

The *Brassica* smother plant, along with other cover crops that have been evaluated, does not have any direct economic value of its own. Consequently, its use can cause little, if any, main crop yield loss if it is to be an economically viable weed control method. Development of smother plants that suppress weeds for the first four to six weeks of the growing season and then can be harvested and marketed may be an important element in the feasibility of weed control methods utilizing plant interference.

The development of smother plants designed for specific crop production conditions may be necessary to facilitate smother plant use. Currently, corn and soybeans can be treated with one, or a combination, of many available herbicides. Producers select herbicide treatments based on their production practices, the weed species present in their fields, and the cost of the treatment. Smother plant weed control technology is still in its infancy. It is likely that a range of smother plants, each adapted to a particular crop species, soil type, and weed spectrum, will need to be developed before smother plants can be a reliable, widely adopted weed suppression tool.

## Summary

Herbicides and tillage are the predominant means of weed suppression in industrialized countries, but use of these methods often results in environmental degradation. Biological weed management through plant interference may be a viable alternative. Winter annual and perennial cover crops or smother plants have been evaluated for their ability to suppress weeds, but the species investigated were not winter hardy in the Upper Midwest United States, and most were competitive with the main crop. Development of effective spring-seeded annual smother plants would provide producers with a nonchemical weed control option, and smother plant use could also reduce soil erosion. Research on a *Brassica* species smother plant for weed control in corn (*Zea mays* L.) is described as an example of the development and use of spring-seeded annual smother plants. Crosses were made between rapid cycling, short-statured *Brassica campestris* entries and locally adapted *Brassica* species. A population derived from a dwarf *Brassica* (*Brassica campestris* "CrGC 1-21") by Chinese cabbage (*Brassica campestris* ssp. *pekinensis* "Chinese mastsus") cross was chosen for field evaluation. Field grown dwarf *Brassica* smother plants flowered three weeks after emergence, had a maximum height of 33 cm, and began to senesce five weeks after emergence. Field studies were conducted to determine the effect of the dwarf *Brassica* smother plant on corn. Corn silk emergence date at St. Paul in 1991 was not affected by the *Brassica* smother plant, but at St.

Paul in 1990 and Waseca in 1991 the smother plant delayed silk emergence by 0.7 to 1.8 days, and by 1.5 to 4.9 days, respectively, depending on seeding rate. At St. Paul in 1990 corn grain yield was not affected by smother plant presence, but at Waseca in 1991 the smother plant reduced corn grain yield by 11 to 24 percent. These results indicate that it may be possible to develop spring-seeded smother plants that reduce corn yield by less than 10 percent. Development of smother plants may be facilitated by research into the mechanisms of smother plant interference with the main crop and with weeds, by genetic manipulation and selection of smother plant cultivars, and by the development of smother plants that produce a marketable product. Selection of main crop genotypes tolerant of smother plants may also be important.

## References

Bruce, R. R., G. W. Langdale, L. T. West, and W. P. Miller. 1992. Soil surface modification by biomass inputs affecting rainfall infiltration. *Soil Science Society of America Journal* 56:1614–20.

Bunting, E. S., and J. W. Ludwig. 1964. Plant competition and weed control in maize. *Proceedings 7th British Weed Control Conference* 1:385–88.

Corak, S. J., W. W. Frye, and M. S. Smith. 1991. Legume mulch and nitrogen fertilizer effects on soil water and corn production. *Soil Science Society of America Journal* 55: 1395–1400.

Cousins, R. 1991. Aspects of the design and interpretation of competition (interference) experiments. *Weed Technology* 5:664–73.

Curran, W. S., L. D. Hoffman, and E. L. Werner. 1994. The influence of a hairy vetch (*Vicia villosa*) cover crop on weed control and corn (*Zea mays*) growth and yield. *Weed Technology* 8:777–84.

Decker, A. M., A. J. Clark, J. J. Meisinger, F. R. Mulford, and M. S. McIntosh. 1994. Legume cover crop contributions to no-tillage corn production. *Agronomy Journal* 86: 126–35.

De Haan, R. L. 1992. Simulation, development, and evaluation of spring seeded smother plants for weed control in corn (*Zea mays*). M.S. thesis, University of Minnesota, St. Paul.

De Haan, R. L., D. L. Wyse, N. J. Ehlke, B. D. Maxwell, and D. H. Putnam. 1994. Simulation of spring-seeded smother plants for weed control in corn. *Weed Science* 42:35–43.

Donald, C. M. 1968. The breeding of crop ideotypes. *Euphytica* 17:385–403.

Eadie, A. G., C. J. Swanton, J. E. Shaw, and G. W. Anderson. 1992. Integration of cereal cover crops in ridge-tillage corn (*Zea mays*) production. *Weed Technology* 6:553–60.

Eberlein, C. V., C. C. Sheaffer, and V. F. Oliveira. 1992. Corn growth and yield in an alfalfa living mulch system. *Journal of Production Agriculture* 5:332–39.

Echtenkamp, G. W., and R. S. Moomaw. 1989. No-till corn production in a living mulch system. *Weed Technology* 3:261–66.

Enache, A. J., and R. D. Ilnicki. 1990. Weed control by subterranean clover (*Trifolium subterraneum*) used as a living mulch. *Weed Technology* 4:534–38.

Fischer, A., and L. Burrill. 1993. Managing interference in a sweet corn–white clover living mulch system. *American Journal of Alternative Agriculture* 8:51–56.

Goolsby, D. A., R. C. Coupe, and D. J. Markovchick. 1991. Distribution of selected herbicides and nitrate in the Mississippi River and its major tributaries, April through June 1991. U.S. Geological Survey. Water-Resources Investigations Report 91–4163.

Grubinger, V. P., and P. L. Minotti. 1990. Managing white clover living mulch for sweet corn with partial rototilling. *American Journal of Alternative Agriculture* 5:4–12.

Hall, M. R., C. J. Swanton, and G. W. Anderson. 1992. The critical period of weed control in grain corn (*Zea mays*). *Weed Science* 40:441–47.

Hanson, J. C., E. Lichtenberg, A. M. Decker, and A. J. Clark. 1993. Profitability of no-tillage corn following a hairy vetch cover crop. *Journal of Production Agriculture* 6:432–37.

Hoffman, M. L., E. E. Regnier, and J. Cardina. 1993. Weed and corn (*Zea mays*) responses to a hairy vetch (*Vicia villosa*) cover crop. *Weed Technology* 7:594–99.

Johnson, G. A., M. S. Defelice, and Z. R. Helsel. 1993. Cover crop management and weed control in corn (*Zea mays*). *Weed Technology* 7:425–30.

Karlen, D. L. 1990. Conservation tillage research needs. *Journal of Soil and Water Conservation* 45:365–69.

Klaseus, T. G., G. C. Buzicky, and E. C. Schneider. 1988. Pesticides and groundwater: Surveys of selected Minnesota wells. Report to Minnesota Legislative Commission on Minnesota Resources. St. Paul, Minn.: Legislative Commission on Minnesota Resources.

Knake, E. L., and F. W. Slife. 1965. Giant foxtail seeded at various times in corn and soybeans. *Weeds* 13:331–34.

Kumwenda, J.D.T., D. E. Radcliffe, W. L. Hargrove, and D. C. Bridges. 1993. Reseeding of crimson clover and corn grain yield in a living mulch system. *Soil Science Society of America Journal* 57:517–23.

Kurtz, T., S. W. Melsted, and R. H. Bray. 1952. The importance of nitrogen and water in reducing competition between intercrops and corn. *Agronomy Journal* 44:13–17.

Laflen, J. M., W. C. Moldenbauer, and T. S. Colvin. 1980. Conservation tillage and soil erosion on continuously row-cropped land. In *Crop production with conservation in the 80's,* 121–33. St. Joseph, Mich.: American Society of Agricultural Engineering.

Lal, R., E. Regnier, D. J. Eckert, W. M. Edwards, and R. Hammond. 1991. Expectations of cover crops for sustainable agriculture. In *Cover crops for clean water,* ed. W. L. Hargrove, 1–11. Ankeny, Ia.: Soil and Water Conservation Society.

Langdale, G. W., R. L. Blevins, D. L. Karlen, D. K. McCool, M. A. Nearing, E. L. Skidmore, A. W. Thomas, D. D. Tyler, and J. R. Williams. 1991. Cover crop effects on soil erosion by wind and water. In *Cover crops for clean water,* ed. W. L. Hargrove, 15–22. Ankeny, Ia.: Soil and Water Conservation Society.

Larson, W. E., F. J. Pierce, and R. H. Dowdy. 1983. The threat of soil erosion to long-term crop production. *Science* 219:458–65.

Liebman, M. 1988. Ecological suppression of weeds in intercropping systems: a review. In *Weed management in agroecosystems: Ecological approaches,* ed. M. Altieri and M. Liebman, 197–212. Boca Raton, Fla.: CRC Press.

Maskina, M. S., J. F. Power, J. W. Doran, and W. W. Wilhelm. 1993. Residual effects of no-till crop residues on corn yield and nitrogen uptake. *Soil Science Society of America Journal* 57:1555–60.

Mohler, C. L. 1991. Effects of tillage and mulch on weed biomass and sweet corn yield. *Weed Technology* 5:545–52.

______. 1994. Ecological basis for the cultural control of annual weeds. *Agronomy Abstracts 1994.* Annual Meetings 89.

Neito, J., and D. W. Staniforth. 1961. Corn-foxtail competition under various production conditions. *Agronomy Journal* 53: 1–5.

Neito, J. H., M. A. Brando, and J. T. Gonzales. 1968. Critical periods of the crop growth cycle for competition from weeds. *Pest Article News Summary* 14:159–66.

Neter, J., W. Wasserman, and M. H. Kutner. 1983. *Applied linear regression models.* Homewood, Ill.: Richard D. Irwin.

Palada, M. C., S. Ganser, R. Hofstetter, B. Volak, and M. Culik. 1983. Association of interseeded legume cover crops and annual row crops in year-round cropping systems. In *Environmentally sound agriculture,* ed. W. Lockeretz, 193–213. New York: Praeger.

Power, J. F., and V. O. Biederbeck. 1991. Role of cover crops in integrated crop production systems. In *Cover crops for clean water,* ed W. L. Hargrove, 167–74. Ankeny, Ia.: Soil and Water Conservation Society.

Rasmussen, D. C. 1987. An evaluation of ideotype breeding. *Crop Science* 27:1140–46.

Robinson, R. G. 1956. Winter annual field crops Minnesota. *Farm and Home Science* 14(1):16.

Staniforth, D. W. 1961. Responses of corn hybrids to yellow foxtail competition. *Weeds* 9: 132–36.

Teasdale, J. R. 1993. Reduced-herbicide weed management systems for no-tillage corn (*Zea mays*) in a hairy vetch (*Vicia villosa*) cover crop. *Weed Technology* 7:879–83.

Thomas, P.E.L., and J.C.S. Allison. 1975. Competition between maize and *Rottboellia exaltata* Linn. *Journal of Agricultural Science* 84:305–12.

Tollenaar, M., M. Mihajlovic, and T. J. Vyn. 1993. Corn growth following cover crops: Influence of cereal cultivar, cereal removal, and nitrogen rate. *Agronomy Journal* 85:251–55.

Williams, P. H. 1985. *Crucifer genetics cooperative resource book.* Madison: Crucifer Genetics Cooperative, Department of Plant Pathology, University of Wisconsin.

Worsham, D. A. 1991. Role of cover crops in weed management and water quality. In *Cover crops for clean water,* ed. W. L. Hargrove, 141–45. Ankeny, Ia.: Soil and Water Conservation Society.

Zimdahl, R. L. 1988. The concept and application of the critical weed-free period. In *Weed management in agroecosystems: Ecological approaches,* ed. M. A. Altieri and M. Liebman, 145–55. Boca Raton, Fla.: CRC Press.

# 12

# Issues in the Use of Microsporidia for Biological Control of European Corn Borer

*Timothy J. Kurtti and Ulrike G. Munderloh*

## Introduction

Microsporidia are important natural regulators of many insect populations and as such are suitable microbial agents for the biological control of pest insects. Their effects on pest insects, however, are often sublethal and less obvious than those induced by other entomopathogens. Instead, microsporidia exert their adverse effects by reducing longevity and fecundity. They generally do not cause epizootics, when large numbers of insects die in a short time. Generally, epizootics occur when host density is high and after the economic threshold is exceeded. To overcome this drawback, researchers have used inundative inoculation of spores into the habitat of pest insects. This method has yielded some success when spores were used alone or were integrated with other chemical or biological control strategies. Nevertheless, microsporidia are not considered pathogenic enough to be used as biological insecticides. Rather, their biological control potential should be exploited in programs aimed at long-term, not short-term, control of pest insects.

The characteristics of microsporidia important to biological control include the production of large numbers of infective units (spores) per host insect that can persist in the environment, a high degree of infectivity, efficient transmission mechanisms, and moderate pathogenicity. Some pests are susceptible to more than one microsporidian species that can be categorized on the basis of speed of action and impact on the host: those that cause chronic debilitating diseases, and others that are quite virulent and can kill the host quickly, even within twenty-four hours. The majority, however, including the genus *Nosema,* require time to multiply and develop to debilitate the host. In the species that cause chronic diseases, multiplication of the parasite within the tissues of the host insect is extensive. Meanwhile, the insect continues to eat and grow, and the microsporidia are often transmitted transovarially by the female to her progeny. In more virulent species, e.g., those of the genus *Vairimorpha,* this does not play a role.

Among the best-characterized microsporidia are those that infect the European corn borer, *Ostrinia nubilalis.* Hence we will focus our analysis on *Nosema pyrausta* and *Vairimorpha necatrix,* as well as the recently described *Nosema furnacalis* (Wen and Sun 1988). Maddox et al. (1992) have outlined criteria to guide the identification and selection of an appropriate entomopathogen for biological control purposes. These include evidence that the pathogen causes periodic epizootics that regulate the host population density, is moderately virulent and has efficient transmission mechanisms, and has minimal impact on other biological control agents of the pest. We will examine these criteria in light of the need to recognize or develop more effective strains of microsporidia for corn borer control. We will also review recent advances in the identification of microsporidia that may facilitate the detection of introduced microsporidia and their delineation from indigenous species. Populations of many microbial pathogens are clonal in structure. Nevertheless, geographical clones that differ in phenotypic features, such as virulence, can be isolated. Finally, we will evaluate what is known about microsporidian strains that offer different characteristics important to their use in biological control and the prospects for their manipulation via strain selection or genetic manipulation.

## Pathology of Microsporidiosis in European Corn Borer

The impact of microsporidia on European corn borer varies with the species, environmental conditions, and possibly strain. *Nosema pyrausta,* which causes a debilitating, multisystemic disease, replicates mainly in the malpighian tubules, silk glands, and gonads, though infections of the fat body, muscles, and hypodermis have been reported (Lewis et al. 1983). Examples of some of the sublethal effects of *N. pyrausta* include slower larval development (Solter et al. 1990) and reduced adult longevity and fecundity (Zimmack and Brindley 1957; Kramer 1959b; Van Denburgh and Burbutis 1962; Lewis and Lynch 1970; Windels et al. 1976; Siegel et al. 1986).

There have been conflicting reports on the impact *N. pyrausta* exerts on fecundity, such as a reduction in the number of egg masses laid per female or the number of eggs per egg mass. However, Siegel et al. (1986) observed that egg masses from naturally infected feral females were larger than those of uninfected females, possibly due to altered oviposition behavior. Mortality resulting from *per os* infection is generally modest and depends on larval age and the number of spores ingested. *Nosema pyrausta* causes mortality in all life stages, but its greatest effect is on larvae infected transovarially (Siegel et al. 1986). Mortality in transovarially infected larvae can be 30 to 80 percent higher than that for uninfected larvae (Kramer 1959b; Siegel et al. 1986; Maddox 1986). Kramer (1959b) observed that substantially fewer transovarially infected larvae reached adulthood (14 percent as compared to 75 percent for uninfected control). This mortality was greatest in larvae hatching from the last egg masses laid by an infected female (Siegel et al. 1986). Some colonies of *O. nubilalis* infected with *N. pyrausta* died out after three generations as a result of poor larval survival (Solter et al. 1991) and only healthy or lightly infected insects developed into adults. In contrast, Lewis and Lynch (1970) observed little mortality in a colony with an infection rate of 100 percent, which they maintained for twelve generations. In the laboratory, the impact of *N. pyrausta* on *O. nubilalis* may be lower than under field conditions where other stress factors contribute to its detrimental impact on feral insects (Lewis et al. 1983). All stages are more susceptible to stress in the form of crowding and temperature extremes. Both cold and hot weather have been cited as stress factors (Kramer 1959c; Siegel et al. 1986). Overwintering infected larvae experience considerably higher mortality rates than uninfected ones (Kramer 1959c; Andreadis 1984; Siegel et al. 1986). Under stress, laboratory-reared corn borers respond to microsporidia infections as do field populations. Solter et al. (1990) observed no detrimental effects of *N. pyrausta* in transovarially infected larvae reared at 30°C, but those reared at 24°C took longer than uninfected insects to develop into pupae. When the density (in both first and third instars) of infected *O. nubilalis* was raised from one to five larvae per rearing cup (Siegel et al. 1986), the resultant crowding led only to a 20 to 30 percent increase in mortality overall, compared to an 83 percent higher larval mortality in caterpillars from late egg masses laid by infected females. These important observations may affect the design of future field trials because they indicate that the route of infection may significantly affect pathogenicity of *N. pyrausta,* which is further modulated by environmental factors.

There is much less information on the impact of *V. necatrix* or *N. furnacalis* on *O. nubilalis.* Unlike *N. pyrausta, V. necatrix* may elicit an acute reaction in corn borers, and those that survive longer usually die during pupation. While its host range is wider (Maddox and Sprenkel 1978), its tissue specificity is narrow as *V. necatrix* replicates primarily in the fat body (Lewis et al. 1983). Thus, *V. necatrix* is transmitted horizontally and not vertically. In laboratory tests, oral *V. necatrix* infections can result in high mortality within a few days due to the destruction of midgut cells during spore germination. The polar filaments of the hatching spores produce lesions that allow secondary, opportunistic agents—bacteria, for example—to invade the hemocoel of the insect, which then dies of septicemia (Lewis et al. 1982). Field tests indicate *V. necatrix*–infected corn borers are also more sensitive to cold stress (Lewis et al. 1983). *Nosema furnacalis,* originally thought to be identical to *N. pyrausta,* was isolated from Asian corn borers (*Ostrinia furnacalis*) and differs from *N. pyrausta* in host range, ultrastructure, and histopathology (Wen and Sun 1988). Unlike *N. pyrausta,* which is host specific, *N. furnacalis* can infect several species of Lepidoptera, but does infect similar tissues in the host insect (i.e., malpighian tubules, silk glands, and gonads, but not fat body). Its impact on the host insect has not been reported, but it appears to be more virulent and cause greater mortality in horizontally infected *O. nubilalis* than *N. pyrausta* (Kurtti unpublished).

### Epizootiology: *Nosema pyrausta*-European Corn Borer

The epizootics caused by many microsporidia of phytophagous Lepidoptera and Coleoptera resemble those induced by *N. pyrausta* (Maddox 1986). Because little is known about the epizootiology of *V. neacatrix* and *N. furnacalis* in European corn borer, we shall limit our discussion to *N. pyrausta.*

*Nosema pyrausta* is widespread among North American populations of European corn borer and plays a prominent role in its population cycles. The ecological model developed by Onstad and Maddox (1989) demonstrates that *N. pyrausta* can effectively control *O. nubilalis* infestations at levels 15 percent below the "carrying capacity of the maize environment" (about 22 larvae/plant). Two factors influence the prevalence of *N. pyrausta* during these cycles: larval population densities, which influence horizontal infection rates, and oviposition by infected females, which affects the dynamics of vertical (transovarial) transmission. In much of its range, European corn borer is bivoltine, and the prevalence and transmission of *N. pyrausta* differs in the two generations (Andreadis 1986). Because persistence of *N. pyrausta* outside the host is limited, vertical transmission is the primary route by which infection is passed from the first to the second generation. Furthermore, larvae feed on foliage no longer than two weeks before tunneling, and spores persist on foliage for only about about the same time (Lewis 1982). Thus, applications have to be carefully timed to coincide with the period during which larvae are present outside the shelter of the stalk. *Nosema pyrausta* overwinters in diapausing larvae, and transovarial transmission is also the mechanism whereby *N. pyrausta* persists from year to year. Infected first-generation larvae are most likely to have acquired the parasites transovarially, and microsporidian prevalence remains quite stable during this generation (Andreadis 1984, 1987; Siegel et al. 1987). Any increase in prevalence occurs mainly in larvae that infest the same plant, as first-generation larval migration from plant to plant is minimal (Andreadis 1986). Also, the critical population density necessary for effective horizontal transmission is usually not reached in the first generation (Andreadis 1986; Siegel et al. 1987). The annual buildup of *N. pyrausta* is seen mainly in second-generation larvae, which also disperse more actively. Horizontal transmission is facilitated by heavy infections of the malpighian tubules and shedding of spores with the frass (Siegel et al. 1986; Solter et al. 1990). An average of 48 million spores can be released into the frass of transovarially infected corn borers (Solter et al. 1990). Thus, as infected larvae tunnel within the stalk, they contaminate it with spores that infect other larvae *per os.* The highest transmission rate of *N. pyrausta* occurs when uninfected and infected *O. nubilalis* inhabit the same feeding cavity (Andreadis 1987). Pathogen dispersal in second-generation populations is also promoted by higher larval densities and longer larval developmental time (Andreadis 1986). The corn borer population cycles seen in infected populations of second-generation larvae are characteristic of those regulated by a parasite (Andreadis 1984; Onstad and Maddox 1989). Features of these cycles include a peak of *N. pyrausta* prevalence that occurs shortly after the corn borer population peaks, cycles of corn borer abundance showing a slow rise followed by decline, coupled with cycles of *N. pyrausta* where a rapid rise in prevalence is followed by a slow decline, and a further drop in the incidence of infection when borer density has fallen beneath a critical threshold. Crashes in European corn borer populations occur in populations that have been on the rise for several years to reach levels of five or more larvae per plant and when the prevalence of *N. pyrausta* approaches 100 percent (Maddox 1986).

The prevalence of *N. pyrausta* in naturally infected corn borer populations is usually not very high. Attempts to boost the pathogen population to epizootic levels with foliar applications of spores generally have met with modest success (Lewis and Lynch 1978; Lewis et al. 1983; Laing and Jaques 1984). In spite of these efforts, *N. pyrausta* prevalence increases mainly when infected *O. nubilalis* females oviposit in a cornfield (Maddox 1986). Maddox (1986) notes that *N. pyrausta* is best conserved from one growing season to the next by plowing cornstalks under only after the emergence of infected adult females. He has suggested that a small amount of maize could be planted earlier or later than the main planting to attract ovipositing infected females and create epizootic centers with high population densities.

## Interaction of Microsporidia with Other Biological Control Agents

Very few studies have been done on the interactions that occur between microsporidia and other natural pathogens of *O. nubilalis.* Past studies have focused mainly on the effect of *N. pyrausta* on microbials intended for use as insecticides. The joint use of *V. necatrix* and *N. pyrausta* resulted in higher larval and pupal mortality than did use of either microsporidium alone (Lewis et al. 1983). The negative effects were greatest for feral overwintering insects, in which fewer larvae survived overwintering and survivors had lower body weights and higher rates of infection than larvae infected with either microsporidium alone. However, spore production (the number of spores per mg of body weight) was lowest in insects challenged with both microsporidia (Lewis et al. 1983). Spore identity was not determined, so the degree to which each microsporidium contributed to larval and pupal mortality was not known. Four foliar applications of *V. necatrix* or *N. pyrausta* spores to sweet and field corn infested with corn borers endogenously infected with *N. pyrausta* did not reduce crop damage (Jaques et al. 1984). The interaction of *N. furnacalis* with *V. necatrix* or *N. pyrausta* has not been investigated.

The effectiveness of *N. pyrausta* combined with *Bacillus thuringiensis* has been evaluated. Lublinkhof et al. (1979) sprayed plants with *N. pyrausta* spores one day after infestation with egg masses and a day prior to the application of Thuricide. The two treatments had an additive effect in reducing the number of larvae per plant, with the greatest reduction occurring shortly after the application of endotoxin. Thuricide did not enhance infection by *N. pyrausta* spores, nor did it boost the prevalence of *N. pyrausta* (Lublinkhof and Lewis 1980). Likewise, the intensity of infection (spore yield per mg of body weight) and horizontal infection rates were similar to those obtained when *N. pyrausta* was used alone. In laboratory trials, larval mortality was higher but the intensity of infection with microsporidia was lower when *N. pyrausta* was used in combination with Thuricide than when it was used alone (Lublinkhof and Lewis 1980). These authors speculated that spore germination in the insect's gut augmented Thuricide activity and toxicity.

Single applications of microbial insecticides to *N. pyrausta*–infected populations of *O. nubilalis* did not eradicate *N. pyrausta.* However, the effect of prolonged exposure or repeated use of these agents on *N. pyrausta* persistence has not been evaluated. The impact of transgenic corn carrying genes for *B. thuringiensis* endotoxin on *N. pyrausta* effectiveness and persistence in feral *O. nubilalis* will also be an important parameter to evaluate.

Surprisingly, a virus has never been isolated from *O. nubilalis,* and for that reason attempts have been made to cross infect European corn borers with other insect viruses (Lewis et al. 1977). The nuclear polyhedrosis virus of *Autographa californica* (ACNPV) had low infectivity and biological activity for laboratory-reared corn borers (Lewis et al. 1977; Lewis and Johnson 1982). The interaction of this virus with *N. pyrausta* in feral populations of *O. nubilalis* was likewise found to be negligible (Lewis and Johnson 1982; Laing and Jaques 1984).

The effect of microsporidia on parasitoids that infect the European corn borer has not received much attention, but may be important. *Macrocentrus grandii,* a braconid wasp that parasitizes corn borer larvae, suffered reduced pupal viability and adult longevity when it ingested *N. pyrausta* spores present in infected larvae (Andreadis 1980). High levels of *N. pyrausta* infections in *O. nubilalis* (*i. e.,* 57 percent) may play a major role in limiting parasitoid populations (Andreadis 1980), and the disappearance of *Lydella thompsoni* in Nebraska may be an example (Hill et al. 1978). Similarly, *N. furnacalis* infects *Lydella thompsoni* (= *grisecens*) and *Macrocentrus linearis* (Wen and Sun 1988). There is a clear need to evaluate more carefully the ecological range of these microsporidia and their impact on parasitoid populations before large-scale introduction into European corn borer populations are attempted.

## Virulence and Transmission

Several definitions of virulence have been put forward; they range from simply the ability of an organism to cause disease or mortality to its ability to reduce the reproductive success of its host (Herre 1993). MacVean and Capinera (1991) have challenged the concept that virulence and transmission potential should be regarded as "mutually exclusive, species specific traits." Highly virulent microsporidia, such as *V. necatrix,* when used as a microbial insecticide, are not expected to recycle or produce progeny for subsequent rounds of horizontal infection. Pathogens used in this approach should possess greater virulence, speed of killing activity, and persistence on the substrate. Such a microsporidium is not likely to produce large numbers of spores in the host, which is expected to succumb early. Some researchers do not consider this biological control, even though pest populations are suppressed naturally. Microsporidia with low or moderate virulence are more likely to cause a slow, debilitating disease in the host and produce higher numbers of spores. Thus, these organisms offer greater potential for long-term biological control. Such species also appear to have greater potential for vertical transmission and the ability to persist in the pest population for generations. Onstad and Maddox (1989) developed an ecological model of disease prevalence to simulate the long-term population dynamics of European corn borer–*N. pyrausta* interactions. Persistence of *N. pyrausta* is predicted to be most sensitive to reductions in spore dissemination, which causes decreased contamination of the plant. Their model also predicted that when high virulence is combined with low spore dissemination the probability of persistence is low. Increasing the virulence of a pathogen is assumed to increase the chances of its extinction.

MacVean and Capinera (1991) point out that it is not always necessary to chose between short-term virulence and longer-term transmission and persistence potential. Recent studies indicate that virulence and transmission can be positively correlated. A pathogen of enhanced virulence will be able to sustain itself in the field, even in the absence of a target species, when it is not exclusively dependent on that single host species for survival, and when opportunities for transmission to and among alternate hosts are improved (Herre 1993). Therefore, one might predict that making a pathogen less dependent on a single host would allow it to become more virulent and avoid self-extinction at the same time. An example of such a situation is *V. necatrix,* which has a wide lepidopteran host range. In systems where hosts are limiting at certain phases, or there are restrictions in transmission cycles, as in the *N. pyrausta–O. nubilalis* complex, the parasites tend to develop benign associations with their hosts, and become more dependent on vertical transmission for persistence. Thus, a low transmission potential would not only reduce the spread of the parasite but also cause a decline in its virulence. It is important to realize that microsporidian virulence is not a static feature but varies considerably with environmental conditions. This is especially true when we consider the impact of *N. pyrausta* on overwintering larvae. Pathogenicity and spore production in microsporidia differ with the infective dose and the developmental age of the host and are therefore amenable to manipulation. Microsporidia can cause high mortality when they are fed at high dosage to young insects, especially those in the first instar. Conversely, spore production is favored if insects are exposed to the minimum infective dose for their age. This seemingly contradictory behavior of microsporidia, often taken as a drawback, may actually serve as a double-edged sword. MacVean and Capinera (1991) have pointed out that "the dichotomy between pathogenicity and transmission potential does not force a choice between different species of pathogens, but rather implies a choice of inoculum levels and timing of application of a single species, depending on management objectives."

## Characteristics of Microsporidia Important to Biological Control Applications

Even though over 100 genera of entomopathogenic microsporidia species have been described and named (Sprague et al. 1992), there is no consensus on what characteristics justify the naming of a new species. The ability to discriminate between microsporidian species for taxonomic purposes is of paramount importance to their use as biological control agents. Methods for the reliable identification of different species in mixed infections of indigenous and introduced microsporidia are not available, not to mention means of detecting possible recombinants. These essential tools will be needed to detect artificially introduced microsporidia in nontarget species, and to enable measurement of their persistence in the pest insects' habitat. Unfortunately, some controversy exists over the appropriate diagnostic criteria for recognizing microsporidian species. Proposed methods include light and electron microscopy and gel electrophoresis (Hazard et al. 1981). More recently, molecular and immunological techniques have been used as diagnostic aids as well.

### *Microscopy and Life Cycle*

Microscopy alone, even electron microscopy, is not considered useful for the identification of microsporidia at the species level. Phenotypic features such as spore size are somewhat plastic and influenced by host physiology or environmental conditions. Nonetheless, Giemsa-stained smears or sections of infected host tissues are important aids in determining parasite developmental stages, determining which tissues (cell types) are infected, and approximating the intensity of infections or cause of death (by septicemia or microsporidiosis). Smears or sections of the malpighian tubules or silk glands are most often used to examine for infections of *N. pyrausta* or *N. furnacalis.* The light microscopic appearance, in Giemsa-stained preparations, of various stages in the life cycle of *N. pyrausta* in *O. nubilalis* has been given by Hall (1952) and Kramer (1959a). Andreadis (1980) used light and electron microscopy to compare the development of *N. pyrausta* in both corn borers and its parasite, *M. grandii.* The development of *N. pyrausta* in both was identical, and mature spores produced in both hosts were identical in exospore structure and number of coils in the polar filaments. Wen and Sun (1988) also used light and electron microscopy to compare the development and histopathology of *N. furnacalis* with *N. pyrausta. Nosema pyrausta* spores had polar filaments with eleven to twelve coils that were 64.1 ± 9.7 μm long when extruded. The polar filament of *N. pyrausta* had nine to ten coils and was shorter (47.3 ± 7.9 μm). Interestingly, these authors could distinguish between the two species on the basis of appearance of infected silk glands: silk glands infected with *N. pyrausta* were mottled (piebald), while white glands infected with *N. furnacalis* were more evenly opaque white. Spore size and polar tube length vary considerably and should not be used as sole criteria for identification. Average spore sizes and polar filament length reported for *N. pyrausta* are 4.2 × 2.1 μm with a tube length of 110 μm (Kramer 1959a); 4.16 × 1.76 μm (Andreadis 1980); and 4.7 × 1.7 with a polar tube length of 64.1 μm (Wen and Sun 1988). The development of *V. necatrix* in corn borers is less well characterized, and parasites are primarily seen in the fat body. Lewis et al. (1982) used midgut and fat body tissues of *O. nubilalis* to identify larvae that were killed by septicemia (midgut disruption) or microsporidiosis (replicative stages in the fat body). Peritoneal sheath tissues of the larval gonads have also been found to become infected (Lewis et al. 1983). The developmental dimorphism (i.e., the formation of uninucleate octospores) that has been described in *V. necatrix* infections of *Helicoverpa* and *Spodoptera* is likely to occur in *O. nubilalis* as well (Pilley 1976; Moore and Brooks 1992).

The stimuli that induce spore germination and the spore type play important roles in host and, possibly, tissue specificity. Recently, important observations have been made on the sporogonic development of *Nosema* and *Vairimorpha.* The observation of empty spore cases within host cells, sometimes with polar filaments still attached, has been interpreted to indicate that spores that form early in pathogenesis germinate intracellularly and are involved in the spread of the microsporidia within the host (Vavra and Undeen 1970; Avery and Anthony 1983; Odindo and Jura 1992). Early spores of *Nosema bombycis* are thin-walled and have fewer polar filament turns than those that differentiate during later stages of pathogenesis in silkworm larvae (Iwano and Ishihara 1991b, 1991c). Such spores have been observed to germinate intracellularly in vitro (Iwano and Ishihara 1989). Extended passaging of *N. furnacalis* in *H. zea* cell cultures also favors the production of spores with thinner spore walls and fewer coils of the polar filament (Iwano and Kurtti 1995). These observations point to the existence of spore types, at least in the genus *Nosema,* with specific functions: those that are responsible for the spread of infection to different tissues within one host, and those involved in horizontal transmission from larva to larva. The genetic mechanisms that enable microsporidia to produce the appropriate spore type at the required time in sufficient quantity may be important determinants of virulence, tissue specificity, and transmissibility.

### *Immunological Tests*

Serological tests developed to detect and quantify microsporidia include the enzyme-linked immunosorbent assay (ELISA) (Greenstone 1983, 1986; Kawarabata and Hayasaka 1987; Ke et al. 1990; Oien and Ragsdale 1992), the latex agglutination test utilizing antibody-sensitized latex polystyrene beads (Hayasaka and Ayuzawa 1987; Baig et al. 1992), and an immunoblot assay (Irby et al. 1986). Alkali extracts of spore surface antigens provided the main antigen for the development of these tests, and most showed some degree of species cross reactivity (Greenstone 1986; Ke et al. 1990; Kawarabata and Hayasaka 1987). Interestingly, the ELISA developed by Ke et al. (1990), which used a monoclonal antibody to spore proteins of *N. bombycis,* cross-reacted with *N. pyrausta* and *N. furnacalis,* indicating that these two species are antigenically related to *N. bombycis.* In spite of some cross-reactivity, Irby et al. (1986) found that banding patterns in an immunoblot analysis of exospore proteins enabled them to distinguish most species. A species specific serological test that used proteins extracted from a suspension of germinated *N. furnacalis* spores to immunize mice and rabbits was developed by Oien and Ragsdale (1992).

The authors postulated that a sporoplasm antigen was responsible for the specificity, although the antigen was not critically identified. This polyclonal antibody did not cross-react with five different species of *Nosema,* including *N. pyrausta,* or three species of *Vairimorpha.* An immunoblot analysis indicated that the polyclonal antibody reacted with a single polypeptide. These results indicate that species specific serological tests can be developed for the microsporidia and that the strategy of extracting antigens from germinated spores should be explored in other species.

Another consideration with serological tests is the sensitivity of the assay and its ability to detect low-grade infections in the insect. The ELISA may offer advantages in ease of use over other methods, and could be employed in epizootiological studies and field release trials. Unfortunately, most sensitivity assays used purified spores and the results were related back to the insect. A protein A–linked latex antisera (PALLAS) assay for *N. bombycis* detected $1 \times 10^6$ spores, which is ten times more sensitive than the ordinary latex agglutination test (Baig et al. 1992). The ELISA developed by Ke et al. (1990) detected as few as 400 *N. bombycis* spores, while the one described by Kawarabata and Hayasaka (1987) detected the equivalent of 2,000 spores of that species, similar to the sensitivity of an ELISA for *Amblyospora* spores (Greenstone 1983). Using their polyclonal antibody to *N. furnacalis,* Oien and Ragsdale (1992) followed microsporidia growth in the insect as soon as two days postinfection and detected $1 \times 10^5$ spore-equivalents. Overall, the ELISAs described here appear to vary in sensitivity by as much as three orders of magnitude. Thus, they will have to be carefully standardized, and positive endpoint dilutions need to be established, before they are generally applied as a diagnostic or taxonomic tool.

### *Chromosomes and Nucleic Acids*

The genus *Nosema* and the binucleated stages of *Vairimorpha* are thought to be diploid (Canning 1988). The two nuclei divide in synchrony, and no evidence for meiosis or syngamy of putative sexual stages has been reported. Mitosis of the replicating stage is intranuclear, and chromosome separation is accompanied by nuclear elongation and constriction. Neither light nor electron microscopy has revealed much detail about the chromosomal organization in these genera. Genetic studies and specific identification based on cytogenetic features are difficult because of the small size of microsporidia and their intracellular inhabitation. Recent studies using pulsed field electrophoresis to analyze intact chromosomal DNA promise to overcome some of these difficulties. To avoid the breakage of shear-sensitive DNA, Munderloh et al. (1990) germinated spores in vitro and separated whole chromosomes of *N. pyrausta* and *N. furnacalis* from agarose-immobilized sporoplasms by transverse alternating field electrophoresis (TAFE). In both species, thirteen chromosomal DNA bands, which ranged in size from 440 to 1,390 kb pairs, were indentified. The patterns were similar yet distinct, most notably in the area between approximately 600 to 900 kb pairs, and the genome size was approximated at 10,240 to 10,560 kb pairs. The chromosomes of *V. necatrix* are larger, and there are apparently fewer than in *N. pyrausta* or *N. furnacalis* (Munderloh et al., unpublished observations). Malone and McIvor (1993) compared the chromosomes of two *Vavraia oncoperae* isolates, a *Vairimorpha* species, and *Nosema costelytrae.* The chromosome size distribution was somewhat broader than that reported by Munderloh et al. (1990), between 130 and 1,930 kb pairs. This *Vairimorpha* species also had fewer and larger chromosomes than the *Nosema* and *Vavraia* species (Malone and McIvor 1993). No evidence for the existence of plasmids in microsporidia has been reported.

Molecular techniques promise to yield useful tools for the identification of microsporidia species. The organization of the ribosomal RNA (rRNA) of several species examined so far indicates that microsporidia are primitive eukaryotic cells (Vossbrinck et al. 1987). Microsporidia possess 70S ribosomes (Ishihara and Hayashi 1968) with 16S and 23S-like subunits that resemble those of prokaryotes (Curgy et al. 1980). In particular, the large ribosomal subunit does not have a 5.8S rRNA, and the sequence of the 16S rRNA of the small subunit of *V. necatrix* was found to be unlike the 18S rRNA of other eukaryotes (Vossbrinck and Woese 1986). The structure of the 5S rRNA of *N. bombycis,* however, is typical of eukaryotic cells (Kawakami et al. 1992).

Base sequence analysis of rDNA has yielded confirmatory data on the species status of several microspora (Vossbrinck et al. 1993) and has been suggested as an aid in the identification of microsporidia. Because rRNA genes exist in high copy number per genome, they are particularly good targets for the development of DNA probes. Used in conjunction with the polymerase chain reaction (PCR) to amplify specific rRNA genes, this technique may obviate the need for large numbers of parasites required for, e.g., an analysis of restriction fragment length polymorphism (RFLP). Vossbrinck et al. (1993) used a ~1,350 base pair fragment that extended from the small to the large rDNA region to amplify rDNA sequences that were then either sequenced or subjected to RFLP analysis. This approach allowed the authors to distinguish between two species of *Vairimorpha* (*V. necatrix* and *V. lymantriae*) and *Encephalitozoon* (*E. hellem* and *E. cuniculi*). There were minor differences (0.097 percent overall) in nucleotide sequence between *V. necatrix* and *V. lymantriae,* while generic differences were larger (~0.5 to 0.6 percent).

## *Evidence for Strains*

*Nosema pyrausta* is presumed to have been accidentally introduced from Europe into North America, possibly along with *O. nubilalis* or its parasitoids. A contemporary comparison between European and American isolates of *N. pyrausta* is needed to confirm this presumption. However, we were unable to isolate any microsporidia from corn borer larvae collected from ten locations in Europe (unpublished). The variable effects of different geographical isolates of *N. pyrausta* on fecundity as judged by the number of eggs per egg mass have been attributed to strain differences. Siegel et al. (1986) found no differences between sizes of egg masses from infected and uninfected females, whereas a reduction of up to 52 percent was reported by Kramer (1959b). It has also been suggested that the number of transovarial passages through *O. nubilalis* influenced the shedding of *N. pyrausta* spores in the frass (Solter et al. 1990). Unfortunately, in studies using colonies of transovarially infected insects, the levels of infection were rarely quantified. Furthermore, parameters such as the insect strain, methods of bioassay, and environmental conditions were not uniform. Thus, there is presently no unequivocal evidence for the existence of strains in *N. pyrausta.* In fact, Siegel et al. (1986) found two isolates of *N. pyrausta* from the same county in Illinois to have similar $IC_{50}$ values in laboratory bioassays. Vossbrinck et al. (1993) found that three different isolates of *Encephalitozoon hellem,* an opportunistic microsporidia infecting AIDS patients, had identical rDNA sequences in a hypervariable region. The best evidence available for the existence of variant strains of the same species is with *N. bombycis.* Microsporidia have been isolated from the lawn cutworm, *Spodoptera depravata,* and the beet armyworm, *Spodoptera exigua.* The microsporidia were identified as *N. bombycis* on the basis of infectivity for silkworms, the tissues they infected, reactivity with *N. bombycis* specific monoclonal and polyclonal antisera, patterns of growth and development in vitro and stimuli needed to induce spore germination (Iwano 1987; Ishihara and Iwano 1991; Iwano and Ishihara 1991a; Yasunaga et al. 1992). However, the spores of the lawn cutworm strain differed in shape and size, as well as virulence for silkworms, from the reference strain of *N. bombycis* (Iwano and Ishihara 1991a). On the other hand, Malone and McIvor (1993) found that a *V. oncoperae* isolate from grass grubs (*Costelytra zealandica*) had fourteen chromosomal DNA bands while another from porina larvae (*Wiseana* spp.) had sixteen. It is possible, though, that these results are due to chromosomal polymorphism in *V. oncoperae.*

Changes in virulence can result from spontaneous emergence of variants within a pathogen population. Extensive passage in vitro (Sohi and Wilson 1976) or through a different insect host may represent selective forces that favor existing, minor subpopulations of the pathogen, resulting in an overall change in virulence. When *N. furnacalis* is transferred extensively in cell cultures of *H. zea* (i.e., < 50 times), there is a decline in its infectivity and virulence for *O. nubilalis* (Kurtti et al. 1994). Clonal analysis of a pathogen population can yield information about the occurrence of variants and thus provides a powerful tool for the improvement of isolates. Cell culture–based cloning strategies have been successfully exploited in other obligate intracellular pathogens, most notably viruses, but are not yet available for microsporidia. We are in the process of developing cell culture cloning techniques for microsporidia. These may allow us to isolate variants (clones) that differ in virulence or transmission potential. Our aim is to determine if these traits can be manipulated to isolate microsporidia with desired characteristics for a given control project.

## Prospects for Genetic Manipulation of Microsporidia

Microsporidia are equipped to gain access to and maintain a foothold within a specific host insect. Genes involved with these processes are likely determinants of transmission potential and pathogenesis, and would therefore constitute desirable targets for manipulation to make microsporidia more effective biological control agents. As we have discussed here, rRNA genes are the only ones described and analyzed in microsporidia to any extent. Before such complex genetic traits as virulence, likely to be determined by a multigene family, can be tackled, an easily measurable trait that is directly linked with it must be found. Plaque size in susceptible monolayer cell cultures is one such measurable parameter known to be linked to infectivity and virulence in viruses. This awaits the development of a reliable plaque assay for microsporidia.

The principal issue for any microbial control agent is the apparent conflict between maximizing its virulence so that it can regulate population numbers below economic thresholds and at the same time optimizing transmission to retain its potential to recycle. This dilemma presents itself not only in the field but also in the laboratory when a pathogen is mass-produced for field release. To address this apparent dichotomy requires greater knowledge about the basic genetic organization of microsporidia and how to manipulate them in the laboratory. In this context, it remains to be determined by what means microsporidia can be genetically manipulated. A systematic approach to screening for or reproduction of entomopathogenic microsporidia has not been developed and is hindered by the lack of standard strain isolation and manipulation techniques. There are presently no general guidelines for the identification of more effective strains nor any simple procedures for selecting promising ones. The avenues for genetic improvement could include strain isolation, clonal selection and hybridization. Although evidence for the formation of sexual stages, or gametocytes, has been documented in *Amblyospora* species (Canning 1988), no similar observations have been made in *N. pyrausta, N. furnacalis,* or *V. necatrix.* This would curtail the potential for natural recombination between genetic variants. Present approaches to isolating genetic variants focus on geographical isolates of the same species, e.g., *N. pyrausta* from Europe and the USA, or isolating microsporidia from closely related insect species, such as *N. furnacalis* from the Asian corn borer. Direct genetic manipulation offers the possibility of introducing a desirable trait into a microsporidian by transferring a cloned gene from another pathogen, e.g., the delta endotoxin gene of *Bacillus thuringiensis.* In the absence of any known vectors or insertion sequences, and because the only viable extracelluar stage is the spore, this approach will have to overcome unique and considerable obstacles.

Genetic studies of obligate intracellular pathogens are more easily conducted using in vitro systems than the host. To this end, we have developed methods for the culture of *V. necatrix* (Kurtti et al. 1990), *N. furnacalis* (Kurtti et al. 1994) and *N. pyrausta* (Sagers et al. 1996). One approach to studying determinants of virulence is to compare virulent and avirulent strains. Although natural populations are believed to comprise variant forms, these may not be detected unless they represent a substantial subset. Most field isolates are considered to be genetically diverse, with the overall virulence or transmission potential of the population being lower than that of the most virulent or transmissible genotype present. This assumption has provided the rationale to seek pathogen clones with outstanding biological control potential from feral populations. However, most pathogen populations, especially bacteria, appear to be clonal in structure, as isolates from even large geographical areas are phenotypically similar (Selander and Musser 1990). Preliminary results indicate that this may also be true for microsporidia. A clone represents a population that is derived from a single organism by asexual reproduction and that is genetically homogeneous. Thus, a prerequisite to the isolation of microsporidia clones is their ability to multiply by cross infecting host cells. Because *N. furnacalis* was able to spread to uninfected cells in vitro, we were able to isolate several clones using medium solidified with agarose to restrict the movement of infected cells and parasites (Kurtti et al. unpublished). The method is similar to that used to plaque-purify baculoviruses and permits the selection of variants on the basis of colony morphology or size, which can serve as genetic markers linked to other phenotypic traits.

Once established, clones or strains can be subject to further molecular analysis. The wide size range of *N. furnacalis* chromosomes led us to examine whether chromosomal polymorphism or a heterogeneous mixture of organisms was responsible for these patterns. The DNA banding pattern of two clones, resolved by TAFE, was indistinguishable from that of the uncloned parent line, indicating that a single organism contained all thirteen chromosomes, and that at least those that could be cloned represented the same genotype. Hayasaka and Kawarabata (1990) isolated two clones of a strain of *N. bombycis* that had been maintained in the laboratory for over fifty years. Although their spores differed in size, immunological tests using monoclonal and polyclonal antibodies to spore surface antigens did not reveal any differences between the two clones. Considering that spore size is not a stably expressed microsporidia characteristic, these results may not be contradictory. The lack of marked differences among clones of a single microsporidian isolate emphasizes the need to compare and clone fresh isolates, particularly those from different geographical regions or insect hosts. Malone and McIvor (1993) found differences in the chromosomes of two isolates of *V. oncoperae.* An isolate from grass grubs (*Costelytra zealandica*) had fourteen chromosomes while one from porina larvae (*Wiseana* spp.) had sixteen, indicating the existence of chromosomal polymorphism in this species. The lack of variability in a single characteristic chosen for comparison must not necessarily be taken to imply identity in all other respects. Nevertheless, Vossbrinck et al. (1993) sequenced a 1,350 bp sized PCR fragment containing rDNA from both the small and the large subunit, including a hypervariable spacer region, and found it to be identical in three separate isolates of *E. hellem,* but to differ in sequence from four other microsporidia.. This fragment appears well suited for species identification, but may not be useful for the delineation of strains.

Isozyme analyses, using multilocus enzyme electrophoresis, can be used to measure allelic variation in strains of microorganisms and to determine how many gene loci are responsible for the observed electrophoretic pattern. Such data can also be used to detect possible recombinants that arose from strain crossings. Little has been done with the isozymes of microsporidia. Except for a brief description of the isozymes of several species by Joslyn et al. (1979), this tool has not been exploited to analyze the variations that might exist between strains or geographical isolates.

## Conclusions

We have identified several areas of research that need to be addressed before microsporidia can be genetically manipulated and reliably used in the biological control of European corn borer.

Microsporidian transmission in European corn borer is accomplished through both horizontal and vertical infection. Vertical transmission of *N. pyrausta* from overwintering infected corn borers to their progeny plays a major role in the infection of first-generation larvae, but is less important in transmitting *N. pyrausta* to the second generation. Horizontal infection occurs mainly during the second generation when larval population densities are high and feeding chambers are inhabited by both infected and uninfected larvae. Laboratory studies using both colonized and feral *O. nubilalis* infected with *N. pyrausta* indicate that mortality is greatest in transovarially infected larvae subjected to environmental stress, such as overwintering. Whether the epizootiological situation is similar in *N. furnacalis* needs to be determined, but it is certain to differ from a pathogen that is transmitted purely horizontally, as is *V. necatrix.* The interaction between introduced microsporidia and other biological control agents, such as parasitoids, has been examined only to a limited extent, in part because of the difficulties in identifying microsporidia species unequivocally. It is important that natural pathogen populations not be displaced by the artificial introduction of other biocontrol agents. This means that exotic species should be scrutinized not only with respect to their potential host range, but also for their impact on indigenous pathogens. The epizootiological models related to microsporidia need additional refinement and will serve to guide continued research in this area. As Onstad and Carruthers (1990) have pointed out, "empirical research provides underlying data for the development and validation of epizootiological models. Models are useful in guiding empirical research." This relationship must not be reversed.

A major problem is to obtain a definitive identification of the microsporidia being employed for biological control and to avoid working with a mixed infection. Several species may actually be present in what appears to be an epizootic or infection caused by one species. Most of the original species descriptions were made using light microscopy and ultrastructural or molecular details are often not available. The species descriptions have to be expanded to include molecular (Vossbrinck et al. 1993) or serological markers. Because the necessary equipment and reagents may not be available to all researchers active in microsporidiology, the establishment of national reference centers, able to provide identification or certified strains, would be a great advantage.

The relatively recent finding that certain species of *Nosema* may form two kinds of spores reveals that our understanding of sporogony and microsporidia development is incomplete. This is especially true for those that are transovarially transmitted. The limited information that is available indicates that microsporidia may undergo changes when they pass through the egg and developing embryo. Causing rather benign diseases after oral infection, *N. pyrausta* can cause high mortality in the progeny of infected European corn borer females. Continued research in this area may reveal important determinants of pathogenicity.

Before any attempts are made to identify infectivity or virulence factors, it is necessary to establish the degree of variation in these parameters that actually exists. Furthermore, host specificity and transmission patterns of the microsporidia that infect European corn borer should be investigated in several insect species. Conclusive evidence for the existence of strains in the three species discussed here is lacking. However, strains have been identified in other microsporidia, and there are no a priori reasons to expect that *N. pyrausta, N. furnacalis,* or *V. necatrix* should be an exception. Geographical isolates of the same species, most notably *N. pyrausta* from Europe, should be obtained for comparison. The variation that exists within isolates from a given geographical region should be examined in more detail to determine the degree of heterogeneity within a population. This information should provide a basis on which to decide whether native parasite populations should be screened for strains with better biological control potential or whether exotic isolates should be sought. Because of the apparent clonal nature of microsporidia that have been examined to date, it may be necessary to mutagenize microsporidia, as has been done with nuclear polyhedrosis viruses (Wood et al. 1981). This approach could yield variants with greater virulence, wider host range, and better transmission potential or persistence. As there is every reason to believe that sexual recombination does not occur in *Vairimorpha* and *Nosema,* it will be difficult, if not impossible, to obtain crosses by conventional means that display improved biological control characteristics. Alternative methods based on the introduction of foreign genes into microsporidia may be needed.

## References


Andreadis, T. G. 1980. *Nosema pyrausta* infection in *Macrocentrus grandii,* a braconid parasite of the European corn borer, *Ostrinia nubilalis. Journal of Invertebrate Pathology* 35:229–33.

______. 1984. Epizootiology of *Nosema pyrausta* in field populations of the European corn borer (Lepidoptera: Pyralidae). *Environmental Entomology* 13:882–87.

______. 1986. Dissemination of *Nosema pyrausta* in feral populations of the European corn borer, *Ostrinia nubilalis. Journal of Invertebrate Pathology* 48:335–43.

______. 1987. Horizontal transmission of *Nosema pyrausta* (Microsporida: Nosematidae) in the European corn borer, *Ostrinia nubilalis* (Lepidoptera: Pyralidae). *Environmental Entomology.* 16:1124–29.

Avery, S. W., and D. W. Anthony. 1983. Ultrastructural study of early development of *Nosema algerae* in *Anopheles albimanus. Journal of Invertebrate Pathology* 42:87–95.

Baig, M., R. K. Datta, B. Nataraju, M. V. Samson, and V. Sivaprasad. 1992. Protein A–linked latex antisera test for the detection of *Nosema bombycis* Naegeli spores. *Journal of Invertebrate Pathology* 60:312–13.

Canning, E. U. 1988. Nuclear division and chromosome cycle in microsporidia. *BioSystems* 21:333–40.

Curgy, J.-J., J. Vavra, and C. Vivares. 1980. Presence of ribosomal RNAs with prokaryotic properties in microsporidia, eukaryotic organisms. *Biologie Cellulaire* 38:49–51.

Greenstone, M. H. 1983. An enzyme-linked immunosorbent assay for the *Amblyospora* sp. of *Culex salinarius* (Microspora: Amblyosporidae). *Journal of Invertebrate Pathology* 41: 250–55.

______. 1986. The ELISA for *Amblyospora* sp.: reproducibility, sensitivity, and cross-reactivity with other microsporidia species. *Journal of the Kansas Entomological Society* 59: 658–65.

Hall, I. M. 1952. Observations on *Perezia pyraustae* Paillot, a microsporidia parasite of the European corn borer. *Journal of Parasitology* 38:48–52.

Hayasaka, S., and C. Ayuzawa. 1987. Diagnosis of microsporidians *Nosema bombycis* and closely related species by antibody-sensitized latex. *Journal of Sericultural Science, Japan* 56: 169–70.

Hayasaka, S., and T. Kawarabata. 1990. Cloning of a microsporidia, *Nosema bombycis* (Microsporida: Nosematidae), in insect cell cultures by a limiting dilution method. *Journal of Invertebrate Pathology* 55:35–40.

Hazard, E. I., E. A. Ellis, and D. J. Joslyn. 1981. Identification of microsporidia. In *Microbial control of pests and plant diseases 1970–1980,* ed. H. D. Burgess, 163–82. New York: Academic Press.

Herre, E. H. 1993. Population structure and the evolution of virulence in nematode parasites of fig wasps. *Science* 259:1442–45.

Hill, R. E., D. P. Carpino, and Z. B. Mayo. 1978. Insect parasites of the European corn borer, *Ostrinia nubilalis,* in Nebraska from 1948–1976. *Environmental Entomology* 7: 249–53.

Irby, W. S., Y. S. Huang, C. Y. Kawanishi, and W. M. Brooks. 1986. Immunoblot analysis of exospore polypeptides from some entomophilic microsporidia. *Journal of Protozoology* 33: 14–20.

Ishihara, R., and Y. Hayashi. 1968. Some properties of ribosomes from the sporoplasm of *Nosema bombycis. Journal of Invertebrate Pathology* 11:377–85.

Ishihara, R., and H. Iwano. 1991. The lawn grass cutworm, *Spodoptera depravata* Butler, as a natural reservoir of *Nosema bombycis* Naegeli. *Journal of Sericultural Science, Japan* 60: 326–27.

Iwano, H. 1987. Seasonal occurrence of microsporidia in the field populations of the lawn grass cutworm, *Spodoptera depravata* Butler. *Japanese Journal of Applied Entomology and Zoology* 31:321–27. (In Japanese with English summary)

Iwano, H., and R. Ishihara. 1989. Intracellular germination of spores of a *Nosema* sp. immediately after their formation in cultured cell. *Journal of Invertebrate Pathology* 54:125–27.

______. 1991a. Isolation of *Nosema bombycis* from moths of the lawn grass cutworm, *Spodoptera depravata* Butler. *Journal of Sericultural Science, Japan* 60:279–87. (In Japanese with English summary)

______. 1991b. Dimorphism of spores of *Nosema* spp. in cultured cell. *Journal of Invertebrate Pathology* 57:211–19.

______. 1991c. Dimorphic development of *Nosema bombycis* spores in gut epithelium of larvae of the silkworm, *Bombyx mori. Journal of Sericultural Science, Japan* 60:249–56.

Iwano, H., and T. J. Kurtti. 1995. Identification and isolation of dimorphic spores from *Nosema furnacalis* (Microspora:Nosematidae). *Journal of Invertebrate Pathology* 65:230–6.

Joslyn, D. J., J. F. Kelly, J. D. Knell, and C. R. Dillard. 1979. Isozymes in microsporidia. *Isozyme Bulletin* 12:60.

Kawakami, Y., T. Inoue, M. Kikuchi, M. Takayanagi, M. Sunairi, T. Ando, and R. Ishihara. 1992. Primary and secondary structures of 5S ribosomal RNA of *Nosema bombycis* (Nosematidae, Microsporidia). *Journal of Sericultural Science, Japan* 61:321–27.

Kawarabata, T., and S. Hayasaka. 1987. An enzyme-linked immunosorbent assay to detect alkali-soluble spore surface antigens of strains of *Nosema bombycis* (Microspora: Nosematidae). *Journal of Invertebrate Pathology* 50:118–23.

Ke, Z., W. Xie, X. Wang, Q. Long, and Z. Pu. 1990. A monoclonal antibody to *Nosema bombycis* and its use for identification of microsporidia spores. *Journal of Invertebrate Pathology* 56:395–400.

Kramer, J. P. 1959a. Studies on the morphology and life history of *Perezia pyraustae* Paillot (Microsporidia: Nosematidae). *Transactions of the American Microscopical Society* 78:336–42.

______. 1959b. Some relationships between *Perezia pyraustae* Paillot (Sporozoa, Nosematidae) and *Pyrausta nubilalis* (Hübner) (Lepidoptera, Pyralidae). *Journal of Insect Pathology* 1:25–33.

______. 1959c. Observations on the seasonal incidence of microsporidiosis in European corn borer populations in Illinois. *Entomophaga* 4:37–42.

Kurtti, T. J., U. G. Munderloh, and H. Noda. 1990. *Vairimorpha necatrix:* Infectivity for and development in a lepidopteran cell line. *Journal of Invertebrate Pathology* 55:61–68.

Kurtti, T. J., S. B. Ross, Y. Liu, and U. G. Munderloh. 1994. In vitro developmental biology and spore production in *Nosema furnacalis* (Microspora: Nosematidae). *Journal of Invertebrate Pathology* 63:188–96.

Laing, D. R., and R. P. Jaques. 1984. Microsporidia of the European corn borer (Lepidoptera: Pyralidae) in southwestern Ontario: Natural occurrence and effectiveness of microbial insecticides. *Proceedings of the Entomological Society of Ontario* 115:13–17.

Lewis, L. C. 1982. Persistence of *Nosema pyrausta* and *Vairimorpha necatrix* measured by microsporidiosis in the European corn borer. *Journal of Econonic Entomology* 75:670–74.

Lewis, L. C., J. E. Cossentine, and R. D. Gunnarson. 1983. Impact of two microsporidia, *Nosema pyrausta* and *Vairimorpha necatrix,* in *Nosema pyrausta* infected European corn borer larvae. *Canadian Journal of Zoology* 61:915–21.

Lewis, L. C., R. D. Gunnarson, and J. E. Cossentine. 1982. Pathogenicity of *Vairimorpha necatrix* (Microsporidia: Nosematidae) against *Ostrinia nubilalis* (Lepidoptera: Pyralidae). *Canada Entomologist* 114:599–603.

Lewis, L. C., and T. B. Johnson. 1982. Efficacy of two nuclear polyhedrosis viruses against *Ostrinia nubilalis* (Lep.: Pyralidae) in the laboratory and field. *Entomophaga* 27:33–38.

Lewis, L. C., and R. E. Lynch. 1970. Treatment of *Ostrinia nubilalis* larvae with Fumidil B to control infections caused by *Perezia pyraustae. Journal of Invertebrate Pathology* 15:43–48.

______. 1978. Foliar applications of *Nosema pyrausta* for suppression of populations of European corn borer. *Entomophaga* 23:83–88.

Lewis, L. C., R. E. Lynch, and J. J. Jackson. 1977. Pathology of a *Baculovirus* of the alfalfa looper, *Autographa californica,* in the European corn borer, *Ostrinia nubilalis. Environmental Entomology* 6:535–38.

Lublinkhof, J., and L. C. Lewis. 1980. Virulence of *Nosema pyrausta* to the European corn borer when used in combination with insecticides. *Environmental Entomology* 9:67–71.

Lublinkhof, J., L. C. Lewis, and E. C. Berry. 1979. Effectiveness of integrating insecticides with *Nosema pyrausta* for suppressing populations of the European corn borer. *Journal of Economic Entomology* 72:880–83.

MacVean, C. M., and J. L. Capinera. 1991. Pathogenicity and transmission potential of *Nosema locustae* and *Vairimorpha* sp. (Protozoa: Microsporida) in Mormon crickets (*Anabrus simplex;* Orthoptera: Tettigoniidae): A laboratory evaluation. *Journal of Invertebrate Pathology* 57:23–36.

______. 1992. Field evaluation of two microsporidia pathogens, an entomopathogenic nematode, and carbaryl for suppression of the Mormon cricket, *Anabrus simplex* Hald. (Orthoptera: Tettigoniidae). *Biological Control* 2:59–65.

Maddox, J. V. 1986. Possibilities for manipulating epizootics caused by protozoa: a representative case history of *Nosema pyrausta.* In *Fundamental and applied aspects of invertebrate pathology,* ed. R. A. Samson, J. M. Vlak, and D. Peters, 563–66. Wageningen: Foundation 4th International Colloquium of Invertebrate Pathology.

Maddox, J. V., and R. K. Sprenkel. 1978. Some enigmatic microsporidia of the genus *Nosema. Miscellaneous Publications of the Entomological Society of America* 11:65–84.

Maddox, J. V., M. L. McManus, M. R. Jeffords, and R. E. Webb. 1992. Exotic insect pathogens as classical biological control agents with an emphasis on regulatory considerations. In *Selection criteria and ecological consequences of importing natural enemies,* ed. W. C. Kauffman and J. R. Nechols, 27–52. Lanham, Md.: Entomological Society of America.

Malone, L. A., and C. A. McIvor. 1993. Pulsed field gel electrophoresis of DNA from four microsporidia isolates. *Journal of Invertebrate Pathology* 61:203–5.

Moore, C. B., and W. M. Brooks. 1992. An ultrastructural study of *Vairimorpha necatrix* (Microspora, Microsporida) with particular reference to episporontal inclusions during octosporogony. *Journal of Protozoology* 39:392–98.

Munderloh, U. G., T. J. Kurtti, and S. E. Ross. 1990. Electrophoretic characterization of chromosomal DNA from two microsporidia. *Journal of Invertebrate Pathology* 56:243–48.

Odindo, M. O., and W.G.Z.O. Jura. 1992. Ultrastructure of *Nosema marucae* sp. n. (Microspora, Nosematidae), a pathogen of *Maruca testulalis* (Lepidoptera, Pyralidae). *Current Microbiology* 25:319–25.

Oien, C. T., and D. W. Ragsdale. 1992. A species-specific enzyme-linked immunosorbent assay for *Nosema furnacalis* (Microspora: Nosematidae). *Journal of Invertebrate Pathology* 60: 84–88.

Onstad, D. W., and R. I. Carruthers. 1990. Epizootiological models of insect diseases. *Annual Review of Entomology* 35:399–419.

Onstad, D. W., and J. V. Maddox. 1989. Modeling the effects of the microsporidium, *Nosema pyrausta,* on the population dynamics of the insect, *Ostrinia nubilalis. Journal of Invertebrate Pathology* 53:410–21.

Pilley, B. M. 1976. A new genus, *Vairimorpha necatrix* (Protozoa: Microsporida), for *Nosema necatrix* Kramer 1965: Pathogenicity and life cycle in *Spodoptera exempta* (Lepidoptera: Noctuidae). *Journal of Invertebrate Pathology* 28:177–83.

Sagers, J. B., U. G. Munderloh, and T. J. Kurtti. 1996. Early events in the infection of a *Helicoverpa zea* cell line by *Nosema furnacalis* and *Nosema pyrausta* (Microspora: Nosematidae). *Journal of Invertebrate Pathology* 67:28–34.

Selander, R. K., and J. M. Musser. 1990. Population genetics of bacterial pathogenesis. In *The bacteria XI: Molecular basis of bacterial pathogenesis,* ed. B. H. Iglewski and V. L. Clark, 11–36. San Diego: Academic Press.

Siegel, J. P., J. V. Maddox, and W. G. Ruesink. 1986. Lethal and sublethal effects of *Nosema pyrausta* on the European corn borer (*Ostrinia nubilalis*) in central Illinois. *Journal of Invertebrate Pathology* 48:167–73.

______. 1987. Survivorship of the European corn borer, *Ostrinia nubilalis* (Hübner) (Lepidoptera: Pyralidae), in central Illinois. *Environmental Entomology* 16:1071–75.

Sohi, S. S., and G. G. Wilson. 1976. Persistent infection of *Malacosoma disstria* (Lepidoptera: Lasiocampidae) cell cultures with *Nosema* (*Glugea*) *disstriae* (Microsporida: Nosematidae). *Canadian Journal of Zoology* 54:336–42.

Solter, L. F., J. V. Maddox, and D. W. Onstad. 1991. Transmission of *Nosema pyrausta* in adult corn borers. *Journal of Invertebrate Pathology* 57:220–26.

Solter, L. F., D. W. Onstad, and J. V. Maddox. 1990. Timing of disease-influenced processes in the life cycle of *Ostrinia nubilalis* infected with *Nosema pyrausta. Journal of Invertebrate Pathology* 55:337–41.

Sprague, V., J. J. Becnel, and E. I. Hazard. 1992. Taxonomy of phylum Microspora. *Critical Reviews in Microbiology* 18:285–395.

Van Denburgh, R. S., and P. P. Burbutis. 1962. The host-parasite relationship of the European corn borer, *Ostrinia nubilalis,* and the protozoan, *Perezia pyraustae,* in Delaware. *Journal of Economic Entomology* 55:65–67.

Vavra, J., and A. H. Undeen. 1970. *Nosema algerae* n. sp. (Cnidospora, Microsporida), a pathogen in a laboratory colony of *Anopheles stephensi* Liston (Diptera, Culicidae). *Journal of Protozoology* 17:240–49.

Vossbrinck, C. R., and C. R. Woese. 1986. Eukaryotic ribosomes that lack a 5.8S RNA. *Nature* 320:287–88.

Vossbrinck, C. R., M. D. Baker, E. S. Didier, B. A. Debrunner-Vossbrinck, and J. A. Shadduck. 1993. Ribosomal DNA sequences of *Encephalitozoon hellem* and *Encephalitozoon cuniculi*: Species identification and phylogenetic construction. *Journal of Eukaryote Microbiology* 40:354–62.

Vossbrinck, C. R., J. V. Maddox, S. Friedman, B. A. Debrunner-Vossbrinck, and C. R. Woese. 1987. Ribosomal sequence suggests microsporidia are extremely ancient eukaryotes. *Nature* 326:411–14.

Wen, J.-z., and C.-x. Sun. 1988. Two new species of *Nosema* (Microspora: Nosematidae). *Acta Zootaxonomica Sinica* 13:105–11. (In Chinese with English abstract; translated by Dr. Shu)

Windels, M. B., H. C. Chiang, and B. Furgala. 1976. Effects of *Nosema pyrausta* on pupa and adult stages of the European corn borer *Ostrinia nubilalis. Journal of Invertebrate Pathology* 27:239–42.

Wood, H. A., P. R. Hughes, L. B. Johnston, and W.H.R. Langridges. 1981. Increased virulence of *Autographa californica* polyhedrosis virus by mutagenesis. *Journal of Invertebrate Pathology* 38:236–41.

Yasunaga, C., M. Funakoshi, T. Kawarabata, Y. Aratake, and H. Iwano. 1992. Isolation and characterization of *Nosema bombycis* (Microsporida: Nosematidae) from larvae of beet armyworm, *Spodoptera exigua. Japanese Journal of Applied Entomology and Zoology* 36:127–34. (In Japanese with English summary)

Zimmack, H. L., and T. A. Brindley. 1957. The effect of the protozoan parasite *Perezia pyraustae* Paillot on the European corn borer. *Journal of Economic Entomology.* 50:637–40.

# 13

# Monitoring and Impact of Weed Biological Control Agents

*Peter Harris*

Most weed biocontrol studies end with a report on the weed tissue destroyed and the depression of the weed population by the agent released. This information is not, however, directly related to the social and economic benefits such as increased forage yield, protection of a native plant community, or decreasing herbicide use, which are usually not identified as project goals. Indeed, weed reduction may not produce any benefits until it is below a threshold level. For example, cattle avoid grazing pasture with as little as 10 percent spurge cover (Lym and Kirby 1987) because the spurge latex blisters their mouths, but at 5 percent cover they graze around the stems. Thus, reduction of spurge cover to 5 percent is the goal of the Canadian spurge biocontrol project. It is rash to start a biocontrol project that may cost $4 million and take 20 years without an explicit statement of such a goal. Without a specific goal, there can be no assessment of success, or of whether benefits are likely to be achieved; there is no end point to the project; and because expectations are likely to differ, disappointment is inevitable. Establishing a meaningful goal in terms of a threshold weed population requires a preliminary study, however, so it is usually not done.

Conventional wisdom says that the vegetation should be sampled at the time and place of agent release. Unfortunately, this has a low benefit-cost return since about a third of the species fail to establish and most of the releases of successful species also fail or do not have an impact on the weed within five to ten years of release. A better approach is to monitor, in sequence, the following steps, each with explicit goals: agent establishment, intensity of agent attack, agent impact, and project benefits (Harris 1991). This process allows the selection of appropriate sampling sites and methods and avoids the collection of detailed data over long periods when there is no impact. If other major effects are noticed in the course of the program, the suggested approach is flexible enough to allow for investigation of them.

## Monitoring Agent Establishment

The first goal is to determine whether the agent is established, which I define as field survival for two years. Unfortunately, approximately a third of agents fail to become established (37 percent internationally [Julien 1989] and 34 percent in Canada), and many of the initial releases of those that do become established also fail. For example, of fourteen colonies of European stock of the cinnabar moth (*Tyria jacobaeae*) released in Canada, only two established. There were similar difficulties with spurge beetles in the genus *Aphthona.* Two colonies of *A. nigriscutis* and one of *A. cyparissiae* were established in Canada from fifteen releases of European stock. Even worse, only one of twelve releases of *A. flava* became established. All of these insects eventually became common and widespread, and the *Aphthona* species are now effective biocontrol agents in both Canada and the United States. Given the high failure rate of eventually successful species, monitoring agent establishment requires a rapid sampling method, which for *Aphthona* spp. is sweep netting the release site on a warm summer day.

The failures can be attributed in part to the poor climatic adaptation and the narrow ecological requirements of many biocontrol agents. There is little point in spending time on a species whose degree-day requirements greatly exceed that of the release area, but natural selection may accommodate a small difference. Another problem is not meeting an insect's habitat requirements. For example, the spurge beetle, *A. nigriscutis,* requires dry, open sites on coarse soils on the Canadian prairies, and some failures arose from a misplaced attempt to be kind to the beetle by releasing it on lush, robust stands of spurge. Undoubtedly, failures can be reduced by more preliminary investigation, but some are inevitable in determining the site requirements of an agent in a new region. For example, *A. nigriscutis* is thriving under the partial shade of open ponderosa pine stands in the summer-warm Okanagan of British Columbia. In the cooler Saskatchewan summer, however, it is intolerant of shade. Thus, there is some trial and error—and hence failed establishments—in determining these differences.

## Monitoring Intensity of Agent Attack

The goal for monitoring biocontrol agents that become established is to determine the intensity of attack, that is, the proportion of the resource exploited by the agent, such as the degree of defoliation. It is relatively easy to determine, and the result is publishable. McClay (1995) found 60 percent of biocontrol studies reported exploitation levels, but many of the researchers, at least by implication, seemed to equate this with benefit. Clearly, a low level of attack is unlikely to control the weed, but a high level also does not necessarily result in any benefit and may merely indicate that an agent that attacks another part of the plant is necessary to achieve weed control. Thus I normally do not start monitoring the impact on the weed population if less than 20 percent of the seed heads or other resource is exploited by the agent.

The Canadian record to 1990 is that twenty-five of the fifty-three agent species released in Canada exploit over 20 percent of their resource. Thus, less than half the species released in Canada justify impact studies. Julien (1989) calculated that, internationally, 24 percent were "effective" although it is difficult to compare the two figures since effectiveness was not defined. However, low-density agents need to be monitored, but with a minimum effort, since they can increase to become effective as a result of adaptation after ten or more years. Three examples are given, although in only one were the adaptive changes investigated: (1) Colonies of the beetle *Chrysolina quadrigemina* released in British Columbia remained at low densities for eight to thirteen years. They then increased rapidly to depress the weed to about 1 percent of its former density over the following two to three years (Harris 1962; Harris et al. 1969). The reason for the initial poor performance was that the original beetles, which were obtained from southern France via Australia and California, remained on the foliage in the fall until they were killed by early frost. The present population seeks shelter at 4°C (Peschken 1972) and reemerges to oviposit on the next warm day. (2) The cinnabar moth became established in eastern Canada from two pairs of moths surviving from the release of several thousand larvae. The population remained small for four years, increased in year five to achieve 11 percent defoliation and in year six stripped the weed of foliage and flowers (Harris et al. 1978). After nine years of field selection, larvae from this colony established at eight of nine releases (Harris et al. 1975). (3) The spurge beetle, *A. cyparissiae,* remained at a low density for six years, but stock from this colony now depresses the spurge at the release point within one or two years in two-thirds of the releases.

## Monitoring Agent Impact on the Weed

My purpose in monitoring agent impact on weed density, seed production, longevity, or other population measure is to determine whether other agents need to be screened and for what sites. Thus, I want to see developing trends as quickly as possible, but there is no point in looking for them until an agent significantly exploits the weed. This advice was not followed on the Canadian project for the biocontrol of Saint-John's-wort (*Hypericum perforatum*) with the defoliating beetle *C. quadrigemina.* No project goal was established, but the weed and all other plants were monitored annually in permanent plots from the time of release. By the time colonies started to increase, in eight to thirteen years, management, and consequently the flora, had changed on many sites and several were about to be flooded by the construction of a dam. Even on the undisturbed plots, the data were not used, as it was decided that the project goal was forage yield, which was more easily determined in other ways.

Often impact is measured by comparing the weed in and outside the release area. This is easy for *A. nigriscutis,* which tends to build up dense populations that control spurge in an annually increasing area around the release point, but with ingenuity similar data can be obtained for most insects. In 28 percent of the biocontrol studies reviewed by McClay (1995), impact was determined by comparing areas with and without the agent, although insecticide or exclusion cages usually were used to obtain agent-free samples. Most comparisons were based on the dry weight of the weed and other plants from clipped plots, but I question whether this is the best choice. Rangeland vegetation tends to be highly variable and sorting clipped samples by plant species is tedious and time consuming, so it is difficult to process enough samples to obtain a satisfactory standard error. Similar results can be obtained with less work using Daubenmire's approach for measuring plant cover, or in grazed pastures by point sampling (see Mueller-Dombois and Ellenberg 1974).

We have compared several methods for monitoring the impact of biocontrol agents on knapweed. A succession of annual knapweed seed head samples showed that the number of knapweed capitula at Chase, B.C., declined to a lower equilibrium following the establishment of *Urophora affinis* and *U. quadrifasciata* (Figure 13.1). Unfortunately, this equilibrium is above the threshold needed for control since knapweed has remained the dominant plant; however, as the site is subject to heavy use and churning by cattle, control might be achieved more cheaply by moving the salt licks or building a drift fence than by establishing another agent.

The addition in 1976 of a root insect, *Sphenoptera jugoslavica,* to the two *Urophora* species attacking diffuse knapweed at White Lake, B.C., decreased total seed production by about 95 percent. The various parameters sampled gave different early indications of the effect of the seed reduction on the weed population. Life table studies of the recruitment and survival of all knapweed plants in permanent plots in a cattle exclosure (Powell and Myers 1988) suggested that the three insects had reduced seed production to slightly below the knapweed replacement level. This implied that another agent was unnecessary. In contradiction, weed density (Figure 13.2) suggested that another agent was needed since there was little decline. The number of capitula per plant was encouraging until 1988 when the low-density knapweed produced many capitula, and the converse occurred in 1991 (Figure 13.2). Evidently, there can be many small plants or a few large ones. A more satisfactory indication of the trend was the number of capitula per square meter (Figure 13.3). Indeed, although the number of seeds per capitulum has declined, the drop in seed production from about 35,000/$m^2$ in 1976 to just under 1,300/$m^2$ between 1987 and 1991 has proven to be a better measurement than capitula density of impact on the weed.

If time is not limiting, which rarely seems to be the case, I would prefer the life table study as the most sensitive measure of the knapweed population trend. The next most sensitive parameter, and much easier to assess, was seed production per unit area, followed by capitulum density, dry weight and plant cover; the least sensitive was plant density. I suspect most plants behave similarly, so regardless of the plant organ attacked, I prefer to measure reproduction rather than plant density as a means of getting an early indication of the weed population trend. The decision not to screen another biocontrol agent at a cost of two scientist years ($700,000) was made in 1992 and appears to have been justified, because the weed continued to decline in 1993 and 1994.

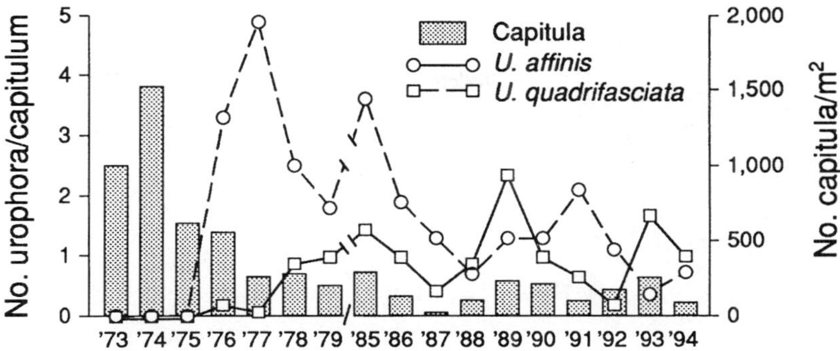


Figure 13.1. Number of spotted knapweed capitula/m², *Urophora affinis* capitulum, and *U. quadrifasciata* capitulum at Chase, British Columbia.

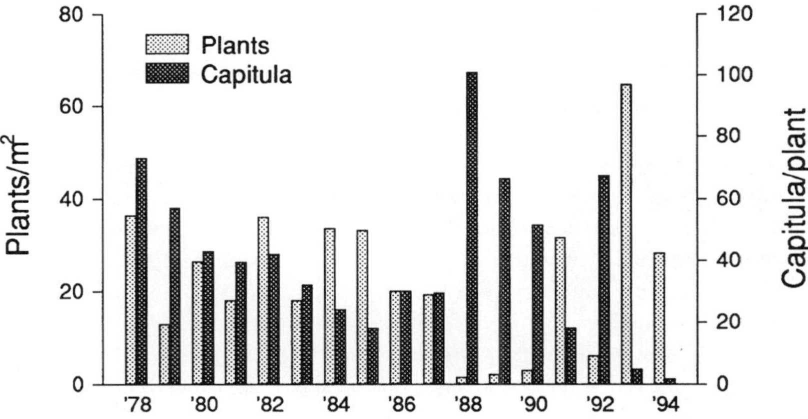


Figure 13.2. Number of diffuse knapweed plants/m² and capitula/plant in an exclosure at White Lake, British Columbia.

## Determining Project Benefits

Weed control is an impact threshold determined by human values. Goeden and Ricker (1981) suggested that convincing evidence of control involving a major vegetation change is provided by sequential photographs. However, obtaining suitable photographs for the knapweed project has been difficult since the greatest effects have tended to occur off the release sites. Another problem with photographs is that many scientific journals are reluctant to accept a pictorial story.

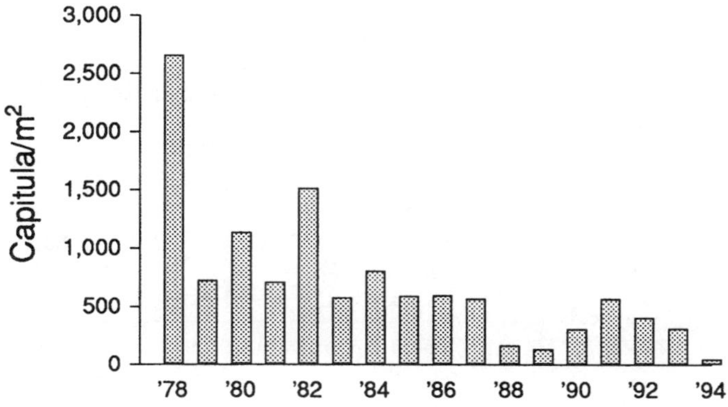


Figure 13.3. Number of diffuse knapweed capitula/$m^2$ at White Lake, British Columbia.

A slow response to knapweed control by seed feeders was expected as there is a reservoir of seed in the soil, which for spotted knapweed without any further input requires twelve years to decline by over 99 percent (Lacey 1985). The important thing is that the site now has some grass: in 1992 the clipped samples had 22 g/$m^2$ dry weight of forage and 28 g/$m^2$ knapweed. Spotted knapweed displaces about 1.16 times its weight in grass (Lacey et al. 1989), so if the figure for diffuse knapweed is similar, the site now realizes about 40 percent of its capacity to produce grass. The results obtained with the Daubenmire cover scale in 1992 were 24 percent grass and 46 percent knapweed (A. Sturko, personal communication, 1992). This is a less favorable ratio than that obtained by clipping the vegetation, but it shows the trend toward increased grass cover and it was obtained with less effort.

A surprise to me is that the best knapweed control is being achieved in the area with moderate grazing in the fall rather than in the cattle exclosure. Fortunately, it has been possible to monitor this with Daubenmire transects, which, as they involve little time, can be established in many places on a speculative basis. Outside the exclosure, knapweed cover declined from 52.2 percent in 1986 to 13.7 percent in 1992 (Sturko, personal communication, 1992), so approximately 82 percent of the site capacity to produce grass has been realized compared with 40 percent inside the exclosure.

The Saint-John's-wort beetle, *C. quadrigemina,* increased forage yields in an infested Ontario pasture by almost 2.5 times and the weed declined by 99.8 percent between 1970 and 1974 (Dougan 1975). This resulted in the removal of the weed from the Ontario noxious weed list, which means that the weed can no longer be used by municipalities as a reason for mowing private property at the owner's expense, and so one benefit is decreased social confrontation.

I have tended to discuss weed biocontrol in terms of forage yields because it is mainly the interests of the British Columbia Cattlemen's Association that have been responsible for several of the programs, but there are other parameters that are equally valid measures of the success of biocontrol programs. For example, knapweed increases surface runoff and stream sedimentation (Lacey et al. 1989), and there is a suggestion that it reduces ponderosa pine regeneration. Both of these could have been used as project objectives. A concern with leafy spurge is that it is displacing the western prairie fringed orchid (*Plantanthera praeclara*), of which there are about 2,000 plants in North Dakota (Federal Register 1989). Hence the objective of the U.S. Forest Service group responsible for managing the main orchid populations should be to maintain or increase it. The Canadian Department of National Defense is concerned with spurge at Shilo, Manitoba, because it does not bind the sandy soil, so the area is subject to wind erosion after tank exercises. Biocontrol success would be an increase in the root mat near the surface as grass replaces spurge. Military authorities are also concerned with the maintenance of native plant communities and being a good neighbor to surrounding farmers, both of which have to be measured in other terms. One of the government of Saskatchewan's goals for the leafy spurge biocontrol program is the elimination of an annual $150,000 subsidy for its chemical control; the results achieved are sufficiently encouraging that this subsidy has already been withdrawn, so this goal has been met. The point is that each of these objectives needs to be measured in terms other than weed reduction, and many weed biocontrol projects are not taking this final step.

## Summary

The impact of biocontrol agents—such as degree of defoliation or reduction in weed population vigor—on their target weed is not linearly related to the benefits of weed biocontrol, such as forage yield. Indeed, the benefits are usually not studied because they are not established because project goals.

Monitoring weed biocontrol projects is difficult because many parameters must be followed and the effects do not necessarily occur where and when they are expected. The system suggested is to divide the project into the following steps, each with its own goal, to be investigated in sequence: agent establishment, intensity of agent attack, impact of the agent on the weed, and project benefits. Many projects are terminated with an assessment of the agent impact on the weed, which is done by clipping or sorting vegetation inside and outside the agent area. However, satisfactory results can usually be obtained with much less effort with Daubenmire's approach for assessing plant species cover. This is not the end of a project, since the benefits of reducing weed cover have to be measured in other terms, such as forage yield, reduced erosion, reduced herbicide use, and the protection of rare species. These are the benefits that justify starting a biocontrol project, and they vary with each weed species.

## Acknowledgments

I gratefully acknowledge the assistance of Alan Sturko of the British Columbia Forest Service for collecting samples and providing data from Daubenmire transects, and the following from Agriculture and Agri-Food Canada: Alison Paton and Randall Brandt for processing many vegetation samples, Helen McMenamin for suggestions on the manuscript, and Sheila Torgunrud for preparing the figures. L.R.C. Contribution No. 3879530.

## References

Cranston, R. S., A. H. Bawtree, and G. R. Keay. 1983. *Knapweed in British Columbia. A description of the problem, efforts to contain its spread and the economics of control.* Kamloops: British Columbia Ministry of Agriculture and Food.

Dougan, J. M. 1975. *The ecology and biological control of* Hypericum perforatum *L. (Hypericaceae) in Ontario.* M.Sc. thesis, University of Guelph, Ontario.

Federal Register (U.S.). 1989. Endangered and threatened wildlife and plants: Determination of threatened status for eastern and western prairie fringed orchids. *Rules and regulations,* Department of the Interior, Washington, D.C. No 39857, 50 CFR part 17.

Goeden, R. D., and D. W. Ricker. 1981. Santa Cruz Island—revisited. Sequential photography records the causation, rates of progress, and lasting benefits of successful biological weed control. In *Proceedings of the V International Symposium on Biological Control of Weeds,* ed. E. S. Delfosse, 355–65. Melbourne: CSIRO.

Harris, P. 1962. Effect of temperature on fecundity and survival of *Chysolina quadrigemina* (Suffr.) and *C. hyperici* (Forst.) (Coleoptera: Chrysomelidae). *Canadian Entomologist* 94:774–80.

______. 1991. Classical biocontrol of weeds: Its definition, selection of effective agents, and administrative-political problems. *Canadian Entomologist* 123:827–49.

Harris, P., D. Peschken, and J. Milroy. 1969. The status of biological control of the weed *Hypericum perforatum* in British Columbia. *Canadian Entomologist* 101:1–15.

Harris, P., A.T.S. Wilkinson, M. E. Neary, L. S. Thompson, and D. Finnamore. 1975. Establishment in Canada of the cinnabar moth, *Tyria jacobaeae* (Lepidoptera: Arctiidae), for controlling the weed *Senecio jacobaea. Canadian Entomologist* 107:913–17.

Harris, P., A.T.S. Wilkinson, L. S. Thompson, and M. Neary. 1978. Interaction between the cinnabar moth, *Tyria jacobaeae* L. (Lep:Arctiidae), and ragwort, *Senecio jacobaea* L. on tansy ragwort (Compositae) in Canada. In *Proceedings IV International Symposium on Biological Control of Weeds,* ed. T. E. Freeman, 163–73. Gainesville: University of Florida.

Julien, M. H. 1989. Biological control of weeds worldwide: Trends, rates of success and the future. *Biocontrol News and Information* 10:299–306.

Lacey, C. A. 1985. *A weed education program, and the biology and control of spotted knapweed (*Centaurea maculosa *Lam.) in Montana.* M.Sc. thesis, Montana State University, Bozeman.

Lacey, J. R., C. B. Marlow, and J. R. Lane. 1989. Influence of spotted knapweed (*Centaurea maculosa*) on surface runoff and sediment yield. *Weed Technology* 3:627–31.

Lym, R. G., and D. R. Kirby. 1987. Cattle foraging behaviour in leafy spurge (*Euphorbia esula*) infested range. *Weed Technology* 1:314–18.

McClay, A. S. 1995. Beyond "before-and-after": Experimental design and evaluation in classical weed biocontrol. In *Proceedings of the VIII International Symposium on Biological Control of Weeds,* ed. E. S. Delfosse and R. R. Scott, 213–19. Canterbury, New Zealand: Lincoln University.

Mueller-Dombois, D., and H. Ellenberg. 1974. *Aims and methods of vegetation ecology.* Toronto: Wiley.

Peschken, D. P. 1972. *Chrysolina quadrigemina* (Coleoptera: Chrysomelidae) introduced from California to British Columbia against the weed *Hypericum perforatum*: Comparison of behaviour, physiology, and colour in association with post colonization adaptation. *Canadian Entomologist* 104:1689–98.

Powell, R. D., and J. H. Myers. 1988. The effect of *Sphenoptera jugoslavica* Obenb. (Col., Buprestidae) on its host plant *Centaurea diffusa* Lam. (Compositae). *Journal of Applied Entomology* 106:25–45.

# 14

# Biological Control of Plant Disease Using Antagonistic *Streptomyces*

*Daqun Liu, Linda L. Kinkel, Eric C. Eckwall, Neil A. Anderson, and Janet L. Schottel*

## Introduction

*Streptomyces* spp. are common soil-borne microorganisms that are noted for their abilities to produce antibiotics and other secondary metabolites. About two-thirds of the naturally occurring antibiotics are produced by *Streptomyces* spp. and other actinomycetes (Chater and Hopwood 1989). Streptomycetes are mainly saprophytic and utilize insoluble organic debris by producing extracellular hydrolytic enzymes such as cellulases, hemicellulases, proteases, amylases, and nucleases. They also adapt to their environments by forming hyphae that penetrate substrates and allow enzymes to be secreted with a relatively high local concentration (Chater and Hopwood 1989).

Because of their competence in colonizing the soil and their proficiency in producing antimicrobial compounds, *Streptomyces* strains are attractive as potential biocontrol agents. Recently, members of this genus have been investigated for their potential to control a wide range of plant pests in a number of different systems. In our work at the University of Minnesota, *Streptomyces* strains have been extensively investigated in field, greenhouse, and laboratory studies for their ability to control *Streptomyces scabies* (causal agent of potato scab) and other plant pathogens. In addition to practical studies on biocontrol, we have focused research on the ecology, biochemistry, taxonomy, and physiology of the naturally occurring pathogen-suppressive *Streptomyces* strains in efforts to enhance biological control. In this chapter we will briefly survey systems in which streptomycetes have been investigated for their ability to act as biocontrol agents. Our primary focus will be to summarize the information obtained in our programs on the use of *Streptomyces* spp. for the biological control of potato scab.

## Streptomycetes as Natural Antagonists and Biocontrol Agents

*Streptomyces* spp. and other actinomycetes have been implicated in antagonism of a wide variety of plant pathogens. These pathogens include plant pathogenic fungi causing root diseases (Bolton 1980; Broadbent et al. 1971; Crawford et al. 1993; Filnow and Lockwood 1985; Rothrock and Gottlieb 1984; Sutherland and Papavizas 1991; Tahvonen 1982; Tu 1986; Turhan and Grossmann 1986; Wadi and Easton 1985), fungi causing leaf diseases (Hodges et al. 1993; Zaher et al. 1985), bacterial plant pathogens (El-Abyad et al. 1993; Liu 1992; Tanii et al. 1990), and plant pathogenic nematodes (Dicklow et al. 1993; Walker et al. 1966). In some systems, the development of a naturally occurring pathogen or disease-suppressive soil is associated with the buildup of *Streptomyces* spp. in that soil (Dicklow et al. 1993; Lorang et al. 1989; L. Kinkel, unpublished data). Use of green manure crops that lead to disease suppression has also been correlated with an enhancement of soil-borne streptomycete populations (Papavizas and Davey 1960; Weinhold et al. 1964).

In addition to alterations of the soil-borne microbial community through management practices, direct inoculation of *Streptomyces* spp. into the soil has been attempted for the control of a number of soil-borne pathogens (see above). Also, foliar application of streptomycetes has been used for the management of above-ground pathogens of wheat and aspen (Hodges et al. 1993; Ostry, Shimizu and Anderson, unpublished data). Finally, added to their ability to inhibit a diversity of fungal, bacterial, and nematode pathogens, streptomycetes have been shown by a number of researchers to have the potential to inhibit potato scab, caused by closely related *Streptomyces* spp. (Alivizatos and Pantazis 1992; Hayashida et al. 1989; Liu et al. 1995; Lorang et al. 1995; Tanii et al. 1990).

## Potato Scab Disease

Common scab of potato (*Solanum tuberosum*) is caused primarily by *Streptomyces scabies* (Elesawy and Szabo 1979; Faucher et al. 1992), although other species of *Streptomyces* have been reported to cause potato scab (Doering-Saad et al. 1992; Faucher et al. 1993; Healy and Lambert 1991; Lambert and Loria 1989a, 1989b; Lorang et al. 1995). The pathogen infects actively growing tubers through young lenticels, wounds, and stomata of the epidermis (Adams and Lapwood 1978). Insect larval feeding can also help penetration of this pathogen into growing tubers (Hooker 1981). Soil arthropods such as springtails and mites that carry the pathogen both on and in their bodies (Storch et al. 1978) may also play a role in acid scab development (Manzer et al. 1984).

Scab is present to some degree in most areas where potatoes are grown, and it is a major production problem affecting both grade quality and market value (Slack 1992). Potato scab has been reported to reduce potato yield in some studies (Adams and Hide 1981; Bång 1979) although other researchers report no effect (Hooker

1981). *Streptomyces* spp. are pathogens on other crops such as radish, sweet potato, peanut, table beet, sugar beet, carrot, rutabaga, parsnip, and turnip (Hooker 1981). In addition, *Streptomyces* spp. have also been reported to cause root tumors on melons (Kobayashi et al. 1987).

## Survey of Control Methods

Cultural practices used for control of potato scab and mechanisms for scab control were reviewed by Trehan and Grewal (1980). A general description of the effects of organic amendments and crop rotation influences on soil-borne plant pathogens was given by Huber and Watson (1970) and Windels (Chapter 18 in this volume). Some scab disease control has been obtained by application of green manures (Millard and Taylor 1927; Oswald and Lorenz 1956; Rouatt and Atkinson 1950; Sandar and Nelson 1968; Weinhold et al. 1964); by use of organic amendments (Hooker 1981; Huber and Watson 1970; Rogers 1969; Weinhold et al. 1964); by use of potato rotation crops (Hooker 1981); and by keeping soil moisture levels high during tuber formation (Davis et al. 1974b, 1976b; Lapwood and Hering 1970; Lapwood et al. 1970; Lewis 1970). Studies have suggested that increases in scab are related to increases in soil pH (Doyle and MacLean 1960) and that maintaining low soil pH may decrease the disease (Davis et al. 1974a; Hooker 1981). Lambert and Manzer (1991) reported that scab incidence was correlated with soil pH but was not correlated with calcium concentration in soil, healthy tuber periderm, or healthy medulla tissue, which is in contrast to the results of other studies (Davis et al. 1976a; Horsfall et al. 1954).

Effects of NPK fertilizers on potato scab incidence are not conclusive since some studies have suggested that the fertilizers decreased the disease in potato (Sandar and Nelson 1968) while others have indicated that the fertilizers did not affect disease incidence (Davis et al. 1976a; Lapwood and Dyson 1966). Sulphur application (Davis et al. 1974a; Muncie et al. 1944), the addition of copper (Mader and Blodgett 1935; Mortvedt et al. 1963), and the application of manganese (McGregor and Wilson 1964, 1966; Mortvedt et al. 1961, 1963) have been reported to reduce potato scab. Scab lesion diameter was inversely correlated with tuber magnesium and manganese concentrations (Lambert and Manzer 1991). Davis et al. (1974a) and Nuget (1956) have suggested that potato scab could be controlled by increasing soil acidity while applying fungicides to the soil. Chemical control (Davis et al. 1974a; Hooker 1981), soil fumigation (Davis et al. 1974a; Hooker 1981; Nuget 1956), foliar sprays (Burrell 1981; Cetas and Sawyer 1962; McIntosh et al. 1981, 1985, 1988), and seed piece treatments with fungicides (Averre and Jones 1985; Johnson 1991) also were reported to reduce scab. Soil-applied systemic insecticides (aldicarb, carbofuran, disulfoton, and thiofanox) showed effective control of acid scab (Manzer et al. 1979, 1980, 1981), presumably by reducing the field population of springtails and mites (Storch 1981). Co-inoculation of virulent and saprophytic isolates of *Streptomyces* on detached potato tubers in sterile soil reduced scab

(Alivizatos and Pantazis 1992). Hayashida et al. (1988, 1989) reported that an antibiotic biofertilizer (swine feces) containing *Streptomyces albidoflavus* strain CH-33 controlled potato common scab in both greenhouse and field tests. Tanii et al. (1990) have used various species of bacteria in a process called potato seed tuber bacterization and claimed a 30 percent disease decrease. Daines (1937) reported an antagonistic effect of *Trichoderma* on the scab pathogen. However, the use of scab-resistant potato cultivars still remains the most reliable method for scab control (Keinath and Loria 1991). Unfortunately, many potato lines lack resistance to scab, and approximately 70 percent of the potato clones evaluated over the past fifteen years in the University of Minnesota potato breeding program have been susceptible to potato scab and consequently dropped from further development. Thus, the development of alternative control strategies for potato scab will present a valuable addition to potato crop management.

## History of Potato Scab–Suppressive Soils

Naturally occurring disease-suppressive soils have been investigated for a number of pathogen-host systems (Schneider 1982). In many cases, these soils have developed following continuous long-term cultivation of a single crop plant species. Disease suppression has often been shown to be dependent upon microbial interactions because heat-sterilized soils do not show suppression. From a practical disease management perspective, disease-suppressive soils offer an ideal biocontrol strategy: stable, long-lasting, nonchemical, and minimal-input–based disease control.

Scab disease decline has been observed in some soils that have been maintained in continuous potato production for extended periods. Menzies (1959) first demonstrated that a biological factor was responsible for scab decline in suppressive soils. In field trials, Menzies (1959) showed that application of 1 percent suppressive soil plus 1 percent alfalfa meal into scab-conducive soil reduced potato scab. Earlier reports showed that naturally occurring saprophytic actinomycetes could be enhanced following amendment of organic residues into the soil and that such treatments could reduce potato scab (Basu-Chaudhary 1967; Millard and Taylor 1927). Menzies (1959) noted that these results might provide a possible model for understanding the mechanism by which scab was reduced in the suppressive soil, but specific organisms responsible for scab suppression were not isolated.

A scab decline phenomenon was observed in a plot established in 1943 for evaluating resistance of potato breeding lines to scab at Grand Rapids, Minnesota (Lorang et al. 1989). After twenty-eight years of potato monoculture, scab levels on susceptible control lines planted in the plot began to decline. In 1972, after thirty years of continuous potato production, the plot was abandoned for the evaluation of breeding lines because attempts to restore high levels of scab had failed. The soil had become suppressive to the development of potato scab. More recently, a second breeding plot was established in scab-conducive soil at Grand Rapids in 1985 for screening potato scab–resistant lines. After eight years of potato monoculture, the plot had become nearly scab-suppressive. Type 3 lesions (broken periderm of potato tubers) were scarce on the tubers, and overall disease levels had declined drastically.

To study the causes of scab decline in Minnesota, scab-susceptible potato cultivars were grown in these suppressive plots. No typical scab lesions were noted, but raised tissue around the lenticels of tubers (potato scab–susceptible cultivar Pontiac) was observed, and suppressive strains were isolated from these lenticels. So far, hundreds of *Streptomyces* strains have been isolated from the suppressive Minnesota plots and from the suppressive soil studied by Menzies (1959). Laboratory studies have indicated that many of these strains produce antibiotics that inhibit the growth of pathogenic isolates of *S. scabies.* Suppressive strains are not pathogenic, and at least twenty of these isolates effectively reduce potato scab when they are co-inoculated with virulent strains on leaf-bud microtubers (Lauer 1977), on radish or potato in greenhouse tests, or with potato tubers in field tests (Fig. 14.1) (Liu 1992; Liu and Anderson 1992a, 1992b; Liu et al. 1994, 1995; Lorang 1988; Lorang et al. 1995). In addition, many of these naturally occurring antagonists have shown potential in laboratory or controlled-environment assays to inhibit other plant pathogens, including *Verticillium dahliae, V. albo-atrum, Rhizoctonia solani* AG3 and AG4, *Cylindrocladium floridanum, Clavibacter michiganense* subsp. *sepedonicum, Phytophthora megasperma* pv. *glycinea* race 1, *Fusarium solani, Fusarium* spp. associated with white pine root rot, *Sclerotinia* spp., *Helminthosporium solani,* and *Septoria* spp. (Liu and Anderson, unpublished data).

## Taxonomy

Insight into the taxonomic, phylogenetic, and ecological relationships among *Streptomyces* strains is critical to understanding the potential for interactions among pathogenic and antagonistic *Streptomyces* isolates in the rhizosphere. Unfortunately, the taxonomy and phylogeny of this group are poorly understood. Several different species of *Streptomyces* have been reported to cause scab on a variety of underground vegetables. Thaxter was the first to isolate a microorganism associated with scab lesions, in 1890, and he named this bacterium *Oospora scabies* (Thaxter 1891, 1892). Gussow (1914) renamed the organism *Actinomyces scabies,* and Waksman and Henrici (1948) renamed the organism *Streptomyces scabies.* In 1961 Waksman redescribed the species, but unfortunately designated a nonscabieslike isolate (IMR 3018 = ISP 5078) as the neotype strain (Waksman 1961). Although pathogenic, this strain did not produce spiral spore chains and melanin pigment as described by Thaxter. Partly in consequence of this, a disproportionate number of putative *S. scabies* reference strains isolated from potato scab lesions have been placed in different species since the 1960s (Lambert and Loria 1989b). As a result, *S. scabies* was listed in *Bergey's Manual of Determinative Bacteriology* (Holt et al. 1994) as an uncertain species.

Does the species *S. scabies* represent a valid taxonomic entity, and does this designation reflect any useful information about the ecology of a strain? Elesawy and Szabo (1979) presented a neotype strain of *S. scabies* (ATCC 33282) that had the

Figure 14.1. Biocontrol results. *Notes:* On the left are tubers from plots treated with 10 percent oatmeal-vermiculite PonSSII inoculum. On the right are tubers from the untreated check plot. Note the reduction in numbers of scab lesions on tubers from the PonSSII-treated soil.

characteristics described by Thaxter in addition to smooth spores and the use of all ISP sugars. Lambert and Loria (1989b) studied pathogenic *Streptomyces* isolates from common scab lesions on tubers collected from Hungary, Canada, and the eastern United States. They concluded that the majority of the isolates were similar to the original *S. scabies* description and suggested that the name *S. scabies* be redesignated to describe this group of organisms. However, a more recent study has shown that fewer than 6 percent of isolates that are taxonomically designated as *S. scabies* are pathogenic (Keinath and Loria 1989). Thus, *S. scabies* may represent a valid taxonomic entity, but classification of a strain as *S. scabies* does not necessarily indicate pathogenicity.

If not all *S. scabies* are pathogens, can non–*S. scabies* strains cause scab? Millard and Burr (1926) described eleven different *Streptomyces* species that cause scab. Harrison (1962) isolated non–*S. scabies* strains of *Streptomyces* that cause russet scab of tubers in the alkaline (pH 8) Red River Valley soils of Minnesota and North Dakota, and russet-scab–causing strains also have been distinguished from *S. scabies* by Faucher et al. (1993). The streptomycetes that infect tubers in highly acidic soils (pH 5.2 or below) of northern Maine have been described as *S. acidiscabies* (Lambert and Loria 1989a; Bonde and McIntyre 1968). In addition, six species of *Strep-*

*tomyces* have been reported to cause pit scab of potato in the Pacific Northwest (Archuleta and Easton 1981). A recent study in Israel indicated that 57 percent of *Streptomyces* isolates causing scab lesions were *S. violaceous* and 22 percent were *S. griseus* (Doering-Saad et al. 1992). Healy and Lambert (1991) reported that *S. scabies, S. acidiscabies,* and the *S. albidoflavus* cluster are the major scab-causing organisms. Overall, numerical taxonomy, RFLP analysis, and Southern hybridizations have all indicated that there is a great diversity among scab-causing streptomycetes.

If not all *S. scabies* cause scab, and many non–*S. scabies Streptomyces* can cause scab, where do the pathogen-suppressive *Streptomyces* fit in? We have found that most of the suppressive *Streptomyces* strains isolated from the Grand Rapids suppressive soil are taxonomically *S. scabies.* These strains are predominantly nonpathogenic. In fact, the ability of the suppressive strains to control potato scab in the field may be partly a function of the fairly close relationship between the pathogenic and suppressive *Streptomyces* strains. We have found that inoculation of suppressive *Streptomyces* into disease-conducive soil results in substantial reductions in the pathogenic *Streptomyces* population in that soil, though the suppressive strains have no detectable impact on the saprophytic *Streptomyces* community (Bowers et al. 1996). Thus, the taxonomic identity of these strains may provide some index to their ecological niche such that *S. scabies* isolates are more likely to interact with other *S. scabies* isolates than with non–*S. scabies* isolates as a function of their preferred colonization sites or resource needs in the rhizosphere. However, despite their taxonomic similarity, fatty acid analysis of *Streptomyces* strains isolated from Minnesota soils can distinguish the pathogens from the nonpathogens (Ndowora et al. 1996; Bowers et al. 1996). This suggests that even though these organisms are taxonomically similar, they are distinct groups. Investigation of the ability of the suppressive *Streptomyces* strains to control scab caused by non–*S. scabies* strains may provide further insight into the utility of traditional taxonomic criteria in shedding light on the ecology of *Streptomyces* populations in soil.

## Use of the Naturally Occurring Antagonistic *Streptomyces* Strains in Disease Biocontrol

A four-year field plot experiment at Becker, Minnesota, was carried out to investigate the biocontrol potential of two suppressive strains of *Streptomyces* against potato scab (Liu and Anderson 1993; Liu et al. 1995). PonSSII (*S. diastatochromogenes*) and PonR (*S. scabies*) were isolated from the Grand Rapids suppressive soil and from a nonsuppressive soil, respectively, and were both shown to produce antibiotics against the potato scab pathogen in vitro (Lorang et al. 1995). These strains were grown individually on a vermiculite-oatmeal broth base and inoculated into scab-conducive soil at 1 percent, 5 percent, and 10 percent volume to volume ratios during the first year of the experiment.

Over all four years of the field pot trial, both PonR and PonSSII at all three inoculum levels significantly reduced potato scab (potato cv. Norchip) (Figure 14.1;

Liu et al. 1995). Expressed as percent disease control relative to the nonamended control, biocontrol by PonR and PonSSII was stable or increased in the various inoculum doses over the four years of the experiment. There were significant differences in the level of disease control among the three different inoculum doses in year one of the experiment, with the greatest level of disease control obtained using the highest inoculum dose. After the first year of the trial, there were no significant differences in the level of disease control among the inoculum doses. The level of scab control achieved with PonSSII at the 1 percent inoculum dose increased every year during the four-year trial, while the percent disease control obtained following inoculation with 10 percent PonSSII was stable throughout the field trial. The suppressive strains were reisolated from the soil or tubers in every year of the experiment (Liu et al. 1995). These data provide evidence that the suppressive strains effectively colonized the soil and that a single inoculation can provide relatively long-term disease control. Additionally, a smaller inoculum dose can ultimately provide the same level of disease control as the larger dose, given time for the suppressive strain population to increase in the soil.

Further field trials on the efficacy of the suppressive *Streptomyces* spp. in potato scab biocontrol were conducted on a large (2.5 acre) field plot maintained under standard potato production conditions (Liu et al. 1994, 1997). Over four years of evaluation, the suppressive strains PonSSII and PonR, as well as additional suppressive strains, were evaluated for their scab control potential when inoculated at different doses and using a variety of inoculation strategies. As before, PonSSII and PonR were grown on vermiculite plus oatmeal broth to provide one form of inoculum. Vermiculite inoculum was incorporated in-furrow at planting at a dose of 15 or 45 cc per seed tuber. Strains were inoculated both singly or in combination. As a second inoculum formulation, strains were grown individually on oatmeal agar plates, and the strains plus agar were macerated in a Waring blender with sterile water. The resulting macerate was used to coat seed tubers immediately prior to planting. In the second year of this study, both strains were reinoculated into half of the microplots that also had been inoculated the previous year.

The field results show that both inoculum dose and inoculation strategy had a significant influence on the percent potato scab reduction achieved in inoculated pots (Liu et al. 1997). Larger inoculum doses provided a greater level of disease control, though tripling the amount of inoculum did not triple the level of control. Addition of inoculum in two successive seasons provided greater disease reduction than inoculation in a single season, though the increase in disease reduction with a second inoculum application was sometimes small.

Results from the field study suggest that the differences in biocontrol effectiveness between strains PonR and PonSSII are partly a function of inoculum density. In particular, PonR provided superior scab control compared to PonSSII at low inoculum densities, and PonSSII provided better control at high densities. This may reflect the differences in colonization ability between the two strains. PonR tends to achieve greater population densities in the soil than PonSSII (Bowers et al. 1993,

1996). Thus, at low inoculum densities, PonR may be better able than PonSSII to achieve population levels necessary for optimal biocontrol. However, PonSSII shows greater inhibition of pathogenic *S. scabies* in the antibiotic inhibition assay, suggesting either that PonSSII produces more antibiotic, a more toxic antibiotic, or a more diffusible antibiotic than strain PonR (data not shown). Additionally, analysis of whole cell fatty acids shows that strain PonSSII is more closely related to the pathogenic *S. scabies* group than is PonR (Bowers et al. 1996; Ndowora et al. 1996). These factors may explain the superior biocontrol offered by strain PonSSII in cases where it is not input-limited (that is, at high inoculum densities).

Finally, the field results demonstrated that the two strains in combination did not control scab as well as either strain individually (Liu et al. 1995). This suggests that competitive or antagonistic interactions between the suppressive strains in the rhizosphere influenced their ability to inhibit the pathogen population. Interestingly, neither PonR nor PonSSII are completely resistant to the antibiotic produced by the other strain. In related work, the potential for the scab-suppressive *Streptomyces* strains to reduce disease in alfalfa has been investigated. Specifically, field, greenhouse, and controlled-environment studies have been used to investigate the ability of the disease-suppressive strains to enhance alfalfa plant vigor (Jones and Samac 1994a, 1994b). In field trials in *Phytophthora*-infested soil, treatment with *Streptomyces* spp. enhanced alfalfa stand establishment and dry matter (g/m$^2$). Additionally, in controlled-environment assays with *Pythium ultimum,* disease severity index was significantly reduced on *Streptomyces*-treated seedlings compared with nontreated seedlings.

## Mechanisms of Biocontrol

The field results raise the question of the mechanism(s) of interaction between the pathogen and suppressive strain that results in disease control. Mechanisms of interaction that have been implicated in the inhibition of pathogen populations or the infection process by microbial antagonists include antibiosis (Boudreau and Andrews 1987; Kerr 1980; Lumsden et al. 1992; Thomashow and Weller 1988), resource competition (Lindow 1988), and parasitism (Tapio and Pohto-Lahdenpera 1991). Effective biocontrol may depend on more than one mechanism of interaction, and a clear understanding of the relative importance of and interactions among different mechanisms of biological control is a fundamental prerequisite for the development and enhancement of biocontrol systems. Additionally, information on the mechanisms of interaction between the pathogen and suppressive populations that result in disease inhibition may provide insight into the means by which the naturally occurring disease-suppressive soil developed.

Our work on potato scab biocontrol has included studies of the role of antibiotic production by the suppressive *Streptomyces* spp. in biocontrol efficacy. Specifically, we have considered the correlation between in vitro levels of antibiotic inhibition of the pathogen and biocontrol efficacy. We have generated both non–antibiotic-producing mutants of the suppressive strains and antibiotic-resistant mutants of the pathogen strains. These mutants are being used to quantify the relative roles of antibiosis and resource competition in biocontrol and to determine the potential fitness of antibiotic-resistant strains of the pathogen. Additionally, the antibiotic produced by one of the suppressive strains has been partially purified and characterized (Eckwall 1994), and strategies for purifying antibiotics produced by other strains are under consideration. Finally, information on the diversity in antibiotics produced among isolates from the suppressive soil is being used in the development of suppressive strain combinations that model the naturally disease-suppressive soil.

## Summary

The use of naturally occurring pathogen-suppressive strains of *Streptomyces* offers tremendous promise for the control of potato scab and other plant diseases. In our trials, *Streptomyces* strains effectively colonized the soil, reduced pathogen populations in the soil, and significantly reduced levels of potato scab multiple years after inoculation. Future efforts to enhance the levels of biocontrol obtained using *Streptomyces* to control plant diseases will focus on the development of strain combinations for biocontrol, the integration of biocontrol with crop rotation and other crop management strategies, and the selection of a variety of strains that are optimized to different physical conditions. Additional work here and in other programs is investigating the abilities of new, recently isolated *Streptomyces* strains to control a large number of different pathogens on many different crops. As biological strategies for controlling plant disease become a more integral part of pest management, streptomycetes promise to play an important role.

## Acknowledgments

Support for this research was provided by the Minnesota Agricultural Experiment Station, the Legislative Commission on Minnesota Resources, the Red River Valley Potato Growers Association, and the USDA North Central Region Integrated Pest Management Program.

## References

Adams, M. J., and G. A. Hide. 1981. Effects of common scab (*Streptomyces scabies*) on potatoes. *Annals of Applied Biology* 98:211–16.

Adams, M. J., and D. H. Lapwood. 1978. Studies on the lenticel development, surface microflora and infection by common scab (*Streptomyces scabies*) of potato tubers growing in wet and dry soils. *Annals of Applied Biology* 90:335–43.

Alivizatos, A. S., and S. Pantazis. 1992. Preliminary studies on biological control of potato common scab caused by *Streptomyces*. In *Biological control of plant diseases,* ed. E. C. Tjamos, G. C. Papavizas, and R. J. Cook, 85–91. New York: Plenum.

Archuleta, J. G., and G. D. Easton. 1981. The cause of deep-pitted scab of potatoes. *American Potato Journal* 58:385–92.

Averre, C. W., and T. J. Jones. 1985. Potato seed/soil treatment for the control of seed-piece decay, tuber rots, and potato scab. *Fungicide and Nematicide Tests* 41:135–36.

Bång, H. O. 1979. Studies on potato russet scab. 2. Influence of infection on the production capacity of seed. *Potato Research* 22:203–28.

Basu-Chaudhary, K. C. 1967. The influence of organic soil amendment on common scab of potato. *Indian Phytopathology* 20:203–5.

Bolton, T. A. 1980. Control of *Pythium aphanidermatum* in poinsettia in a soilless culture by *Trichoderma viride* and a *Streptomyces* sp. *Canadian Journal of Plant Pathology* 2:93–95.

Bonde, M. R., and G. A. McIntyre. 1968. Isolation and biology of a *Streptomyces* sp. causing potato scab in soils below pH 5.0. *American Potato Journal* 45:273–78.

Boudreau, M. A., and J. H. Andrews. 1987. Factors influencing antagonism of *Chaetomium globosum* to *Venturia inaequalis:* A case study in failed biocontrol. *Phytopathology* 77:1470–75.

Bowers, J. H., L. L. Kinkel, and R. K. Jones. 1996. Influence of disease suppressive strains of *Streptomyces* on the native and *Streptomyces* community in soil as determined by the analysis of cellular fatty acids. *Canadian Journal of Microbiology* 42:27–37.

Bowers, J. H., L. L. Kinkel, R. K. Jones, and N. A. Anderson. 1993. Development of fatty acid libraries to track *Streptomyces* spp. in the soil. *Phytopathology* 83:1420.

Broadbent, P., K. F. Baker, and Y. Waterworth. 1971. Bacteria and actinomycetes antagonistic to fungal root pathogens in Australian soils. *Australian Journal of Biological Science* 24:925–44.

Burrell, M. M. 1981. The mode of action of ethionine foliar sprays against potato common scab (*Streptomyces scabies*). *Physiological Plant Pathology* 18:369–78.

Cetas, R. C., and R. L. Sawyer. 1962. Evaluation of uracide for the control of common scab on potatoes on Long Island. *American Potato Journal* 39:456–59.

Chater, K. F., and D. A. Hopwood. 1989. Antibiotic biosynthesis in *Streptomyces.* In *Genetics of bacterial diversity,* ed. D. A. Hopwood and K. F. Chater, 129–50. New York: Academic.

Crawford, D. L., J. M. Lynch, J. M. Whipps, and M. A. Ousley. 1993. Isolation and characterization of actinomycete antagonists of a fungal root pathogen. *Applied and Environmental Microbiology* 59:3899–905.

Daines, R. H. 1937. Antagonistic action of *Trichoderma* on *Actinomyces scabies* and *Rhizoctonia solani. American Potato Journal* 14:85–93.

Davis, J. R., J. G. Garner, and R. H. Callihan. 1974a. Effects of gypsum, sulfur, terraclor and terraclor super-x for potato scab control. *American Potato Journal* 51:35–43.

Davis, J. R., R. E. McDole, and R. H. Callihan. 1976a. Fertilizer effects on common scab of potato and the relation of calcium and phosphate-phosphorus. *Phytopathology* 66: 1236–41.

Davis, J. R., G. M. McMaster, R. H. Callihan, J. G. Garner, and R. E. McDole. 1974b. The relationship of irrigation timing and soil treatments to control potato scab. *Phytopathology* 64:1404–10.

Davis, J. R., G. M. McMaster, R. H. Callihan, F. H. Nissley, and J. J. Pavek. 1976b. Influence of soil moisture and fungicide treatments on common scab and mineral content of potatoes. *Phytopathology* 66:228–33.

Dicklow, M. B., N. Acosta, and B. M. Zuckerman. 1993. A novel *Streptomyces* species for controlling plant-parasitic nematodes. *Journal of Chemical Ecology* 19:159–73.

Doering-Saad, C., P. Kampfer, S. Manulis, G. Kritzman, J. Schneider, J. Zakrzewska-Czerwinska, H. Schrempf, and I. Barash. 1992. Diversity among *Streptomyces* strains causing potato scab. *Applied and Environmental Microbiology* 58:3932–40.

Doyle, J. J., and A. A. MacLean. 1960. Relationships between Ca: K ratio, pH, and prevalence of potato scab. *Canadian Journal of Plant Science* 40:616–19.

Eckwall, E. C. 1994. Production of secondary metabolites associated with potato scab disease. M.S. thesis, University of Minnesota, St. Paul.

El-Abyad, M. S., M. A. El-Sayed, A.-R. El-Shanshoury, and N. H. El-Batanouny. 1993. Inhibitory effects of UV mutants of *Streptomyces corchorusii* and *Streptomyces spiroverticillatus* on bean and banana wilt pathogens. *Canadian Journal of Botany* 71:1080–86.

Elesawy, A. A., and I. M. Szabo. 1979. Isolation and characterization of *Streptomyces scabies* strains from scab lesions of potato tubers. Designation of the neotype strain of *Streptomyces scabies. Acta Microbiologica Hungarica* 26:311–20.

Faucher, E., B. Otrysko, E. Paradis, N. C. Hodge, R. E. Stall, and C. Beaulieu. 1993. Characterization of streptomycetes causing russet scab in Quebec. *Plant Discovery* 77:1217–20.

Faucher, E., T. Savard, and C. Beaulieu. 1992. Characterization of actinomycetes isolated from common scab lesions on potato tubers. *Canadian Journal of Plant Pathology* 14: 197–202.

Filnow, A. B., and J. L. Lockwood. 1985. Evaluation of several actinomycetes and the fungus *Hypochytrium catenoides* as biocontrol agents of *Phytophthora* root rot of soybean. *Plant Discovery* 69:1033–36.

Gussow, H. T. 1914. The systematic position of the organism of the common potato scab. *Science* 39:431–32.

Harrison, M. D. 1962. Potato russet scab, its cause and factors affecting its development. *American Potato Journal* 39:368–87.

Hayashida, S., M.-Y Choi., N. Nanri, and M. Miyaguchi. 1988. Production of potato common scab-antagonistic biofertilizer from swine feces with *Streptomyces albidoflavus. Agricultural and Biological Chemistry* 52:2397–402.

Hayashida, S., M.-Y. Choi, N. Nanri, M. Yokoyama, and T. Uematsu. 1989. Control of potato common scab with an antibiotic biofertilizer produced from swine feces containing *Streptomyces albidoflavus* CH-33. *Agricultural and Biological Chemistry* 53:349–54.

Healy, F. G., and D. H. Lambert. 1991. Relationships among *Streptomyces* spp. causing potato scab. *International Journal of Systematic Bacteriology* 41:479–82.

Hodges, C. F., D. A. Campbell, and N. Christians. 1993. Evaluation of *Streptomyces* for biocontrol of *Bipolaris sorokiniana* and *Sclerotinia homoeocarpa* on the phylloplane of *Poa pratensis. Journal of Phytopathology* 139:103–9.

Holt, J. G., N. R. Kneg, P.H.A. Sneath, J. T. Staley, and S. T. Williams. 1994. *Bergey's manual of determinative bacteriology.* 9th ed. Baltimore: Williams and Wilkins.

Hooker, W. J. 1981. Common scab. In *Compendium of potato disease,* ed. W. J. Hooker, 33–34. St. Paul, Minn.: American Phytopathological Society.

Horsfall, J. G., J. P. Hollis, and H.G.M. Jacobson. 1954. Calcium and potato scab. *Phytopathology* 44:19–24.

Huber, D. M., and R. D. Watson. 1970. Effect of organic amendment on soil-borne plant pathogens. *Phytopathology* 60:22–26.

Johnson, S. B. 1991. Evaluation of potato seed piece fungicides. *Fungicide and Nematicide Tests* 47:110.

Jones, C. R., and D. A. Samac. 1994a. Biocontrol of alfalfa seedling damping off with a suppressive strain of *Streptomyces. Phytopathology* 84:1158.

______. 1994b. Biocontrol of Phytophtora root rot of alfalfa with a suppressive strain of *Streptomyces. Phytopathology* 84:1159.

Keinath, A. P., and R. Loria. 1989. Population dynamics of *Streptomyces scabies* and other actinomycetes as related to common scab of potato. *Phytopathology* 79:681–87.

______. 1991. Effects of inoculum density and cultivar resistance on common scab of potato and population dynamics of *Streptomyces scabies. American Potato Journal* 68:515–24.

Kerr, A. 1980. Biological control of crown gall through production of Agrocin 84. *Plant Discovery* 64:25–30.

Kobayashi, K., M. Yoshida, T. Nakayama, and S. Koga. 1987. Root tumor of melon caused by *Actinomycetales. Annals of the Phytopathological Society of Japan* 53:562–65.

Lambert, D. H., and R. Loria. 1989a. *Streptomyces acidiscabies* sp. nov. *International Journal Systematic Bacteriology* 39:393–96.

______. 1989b. *Streptomyces scabies* sp. nov., nom. rev. *International Journal of Systematic Bacteriology* 39:387–92.

Lambert, D. H., and F. E. Manzer. 1991. Relationship of calcium to potato scab. *Phytopathology* 81:632–36.

Lapwood, D. H., and P. W. Dyson. 1966. An effect of nitrogen on the formation of potato tubers and the incidence of common scab (*Streptomyces scabies*). *Plant Pathology* 15:9–14.

Lapwood, D. H., and T. F. Hering. 1970. Soil moisture and the infection of young potato tubers by *Streptomyces scabies* (common scab). *Potato Research* 13:296–304.

Lapwood, D. H., L. W. Wellings, and W. R. Rosser. 1970. The control of common scab of potatoes by irrigation. *Annals of Applied Biology* 66:397–405.

Lauer, F. I. 1977. Tubers from leaf-bud cuttings: A tool for potato seed certification and breeding programs. *American Potato Journal* 54:457–64.

Lewis, B. G. 1970. Effects of water potential on the infection of potato tubers by *Streptomyces scabies* in soil. *Annals of Applied Biology* 66:83–88.

Lindow, S. E. 1988. Lack of correlation of in vitro antibiosis with antagonism of ice nucleation active bacteria on leaf surfaces by non-ice nucleation active bacteria. *Phytopathology* 78:444–50.

Liu, D. 1992. Biological control of *Streptomyces scabies* and other plant pathogens. Ph.D. thesis, University of Minnesota, St. Paul.

Liu, D., and N. A. Anderson. 1992a. Biological control of potato scab with suppressive isolates of *Streptomyces. Phytopathology* 82:1108.

______. 1992b. Selecting isolates of *Streptomyces* suppressive to *Streptomyces scabies. Phytopathology* 82:1108.

______. 1993. Biological control of potato scab by suppressive strains of *Streptomyces. American Potato Journal* 70:823–24.

Liu, D., N. A. Anderson, and L. L. Kinkel. 1994. Biocontrol of potato scab in field tests. *Phytopathology* 84:1114.

______. 1995. Biological control of potato scab in the field with antagonistic *Streptomyces scabies. Phytopathology* 85:827–31.

______. 1997. Effects of inoculum dose, inoculation strategy, and suppressive *Streptomyces* strains on potato scab biocontrol. *Phytopathology,* submitted.

Lorang, J. M. 1988. Heterokaryosis and inhibitory reactions among isolates of *Streptomyces scabies* causing scab on potato. Master's thesis. University of Minnesota, St. Paul.

Lorang, J. M., N. A. Anderson, F. I. Lauer, and D. K. Wildung. 1989. Disease decline in a Minnesota potato scab plot. *American Potato Journal* 66:531.

Lorang, J. M., D. Liu, N. A. Anderson, and J. L. Schottel. 1995. Identification of potato scab inducing and suppressive species of *Streptomyces. Phytopathology* 85:261–68.

Lumsden, R. D., J. C. Locke, S. T. Adkins, J. F. Walter, and C. J. Ridout. 1992. Isolation and localization of the antibiotic gliotoxin produced by *Gliocladium virens* from alginate prill in soil and soilless media. *Phytopathology* 82:230–35.

Mader, E. O., and F. M. Blodgett. 1935. Potato spraying and potato scab. *American Potato Journal* 12:137–42.

Manzer, F. E., A. E. Giggie, R. H. Storch, and G. H. Sewell. 1981. Control of acid scab with seedpiece and systemic insecticide treatments, 1980. *Fungicide and Nematicide Tests* 36:159.

Manzer, F. E., O. P. Smith, and A. E. Giggie. 1979. Control of acid scab with seedpiece and systemic insecticide treatments, 1978. *Fungicide and Nematicide Tests* 34:72.

Manzer, F.E.O.P. Smith, A. E. Giggie, R. H. Storch, and G. H. Sewell. 1980. Control of acid scab with seedpiece fungicide and systemic insecticide treatments, 1979. *Fungicide and Nematicide Tests* 35:186.

Manzer, F. E., R. H. Storch, and G. H. Sewell. 1984. Evidence for a relationship between certain soil arthropods and acid scab development. *American Potato Journal* 61:741–47.

McGregor, A. J., and G.C.S. Wilson. 1964. The effect of applications of manganese sulphate to a neutral soil upon the yield of tubers and the incidence of common scab in potatoes. *Plant and Soil* 20:59–64.

______. 1966. The influence of manganese on the development of potato scab. *Plant and Soil* 25:3–16.

McIntosh, A. H., G. L. Bateman, and K. Chamberlain. 1988. Substituted benzoic and picolinic acids as foliar sprays against potato common scab. *Annals of Applied Biology* 112: 397–401.

McIntosh, A. H., G. L. Bateman, K. Chamberlain, G. W. Dawson, and M. M. Burrell. 1981. Decreased severity of potato common scab after foliar sprays of 3, 5-dichlorophenoxyacetic acid, a possible antipathogenic agent. *Annals of Applied Biology* 99:275–81.

McIntosh, A. H., K. Chamberlain, and G. W. Dawson. 1985. Foliar sprays against potato common scab: Compounds related to 3,5-dichlorophenoxyacetic acid. *Crop Protection* 4:473–80.

Menzies, J. D. 1959. Occurrence and transfer of a biological factor in soil that suppresses potato scab. *Phytopathology* 49:648–52.

Millard, W. A., and S. Burr. 1926. A study of twenty-four strains of *Actinomyces* and their relation to types of common scab of potato. *Annals of Applied Biology* 13:580–644.

Millard, W. A., and C. B. Taylor. 1927. Antagonism of microorganisms as the controlling factor in the inhibition of scab by green manuring. *Annals of Applied Biology* 14:202–16.

Mortvedt, J. J., K. C. Berger, and H. M. Darling. 1963. Effect of manganese and copper on the growth of *Streptomyces scabies* and the incidence of potato scab. *American Potato Journal* 40:96–102.

Mortvedt, J. J., M. H. Fleischfresser, K. C. Berger, and H. M. Darling. 1961. The relation of soluble manganese to the incidence of common scab in potatoes. *American Potato Journal* 38:95–100.

Muncie, J. H., H. C. Moore, J. Tyson, and E. J. Wheeler. 1944. The effect of sulphur and acid fertilizer on incidence of potato scab. *American Potato Journal* 21:293–304.

Ndowora, T.C.R., L. L. Kinkel, R. K. Jones, and N. A. Anderson. 1996. Fatty acid analysis of pathogenic and suppressive strains of *Streptomyces* species isolated in Minnesota. *Phytopathology* 86:138–43.

Nuget, T. J. 1956. Soil treatments with PCNB (Terraclor) for control of potato scab. *Plant Disease Reporter* 40:428.

Oswald, J. W., and O. A. Lorenz. 1956. Soybeans as a green manure crop for the prevention of potato scab. *Phytopathology* 46:22.

Papavizas, G. C., and C. B. Davey. 1960. Rhizoctonia disease of bean as affected by decomposing green plant materials and associated microfloras. *Phytopathology* 50:516–22.

Rogers, P. F. 1969. Organic manuring for potato scab control and its relation to soil manganese. *Annals of Applied Biology* 63:371–78.

Rothrock, C. S., and D. Gottlieb. 1984. Role of antibiosis in antagonism of *Streptomyces hygroscopicus* var. *geldanus* to *Rhizoctonia solani* in soil. *Canadian Journal of Microbiology* 30: 1440–47.

Rouatt, J. W., and R. G. Atkinson. 1950. The effect of the incorporation of certain cover crops on the microbiological balance of potato scab infested soil. *Canadian Journal of Research C* 28:140–52.

Sandar, N., and D. C. Nelson. 1968. Effect of plant residues and nitrogen applications on yield, specific gravity, russet scab and silver scurf. *American Potato Journal* 45:327–34.

Schneider, R. W. 1982. *Suppressive soils and plant disease.* St. Paul, Minn.: American Phytopathological Society.

Slack, S. A. 1992. A look at potato leafroll virus and PVY: Past, present and future. *Valley Potato Grower* 57:35–39.

Storch, R. H. 1981. Effects of insecticides on collembola and mites in potato fields. *Proceedings of the Acadian Entomological Society* 41:12.

Storch, R. H., J. A. Frank, and F. E. Manzer. 1978. Fungi associated with collembola and mites isolated from scabby potatoes. *American Potato Journal* 55:197–201.

Sutherland, E. D., and G. C. Papavizas. 1991. Evaluation of oospore hyperparasites for the control of *Phytophthora* crown rot of pepper. *Journal of Phytopathology* 131:33–39.

Tahvonen, R. 1982. Preliminary experiments into the use of *Streptomyces* spp. isolated from peat in the biological control of soil and seed-borne diseases in peat culture. *Journal of the Scientific Agricultural Society of Finland* 54:357–69.

Tanii, A., T. Takeuchi, and H. Horita. 1990. Biological control of scab, black scurf and soft rot of potato by seed tuber bacterization. In *Biological control of soil-borne plant pathogens,* ed. D. Hornby, 143–64. Oxon, U.K.: CAB International.

Tapio, E., and A. Pohto-Lahdenpera. 1991. Scanning electron microscopy of hyphal interaction between *Streptomyces griseoviridis* and some plant pathogenic fungi. *Journal of Agricultural Science of Finland* 63:435–41.

Thaxter, R. 1891. Report of the mycologist. The potato "scab." *Connecticut Agricultural Experiment Station Report* 1890:80–95.

______. 1892. The potato scab. *Conneticut Agricultural Experiment Station Report* 1891: 153–60.

Thomashow, L. S., and D. W. Weller. 1988. Role of a phenazine antibiotic from *Pseudomonas fluorescens* in biological control of *Gaeumannomyces graminis* var. *tritici. Journal of Bacteriology* 170:3499–508.

Trehan, S. P., and J. S. Grewal. 1980. Cultural practices for the control of potato common scab—a review. *Agricultural Review* 1:103–13.

Tu, J. C. 1986. Hyperparasitism of *Streptomyces albus* on a destructive mycoparasite *Nectria inventa. Journal of Phytopathology* 117:71–76.

Turhan, G., and F. Grossmann. 1986. Investigation of a great number of actinomycete isolates on their antagonistic effects against soil-borne fungal plant pathogens by an improved method. *Journal of Phytopathology* 116:238–43.

Wadi, J. A., and G. D. Easton. 1985. Control of *Verticillium dahliae* by coating potato seed pieces with antagonistic bacteria. In *Ecology and management of soil-borne plant pathogens,* ed. C. A. Parker, A. D. Rovira, K. J. Moore, P.T.W. Wong, and J. F. Kollmorgen, 134–36. St. Paul, Minn.: American Phytopathological Society.

Waksman, S. A. 1961. *The Actinomycetes* Vol. 2. *Classification, identification and description of genera and species.* Baltimore: Williams and Wilkins.

Waksman, S. A., and A. T. Henrici. 1948. Family II. *Actinomycetaceae* Buchanan and family *Streptomycetaceae* Waksman and Henrici. In *Bergey's manual of determinative microbiology,* ed. R. S. Breed, E.G.D. Murray, and A. P. Hitchens, 892–980. Baltimore: Williams and Wilkins.

Walker, J. T., C. H. Specht, and J. F. Bekker. 1966. Nematicidal activity to *Pratylenchus penetrans* by culture fluids from actinomycetes and bacteria. *Canadian Journal of Microbiology* 12:347–51.

Weinhold, A. R., J. W. Oswald, T. Bowman, J. Bishop, and D. Wright. 1964. Influence of green manures and crop rotation on common scab of potato. *American Potato Journal* 41:265–73.

Zaher, E., F. M. Barakat, A. R. Osman, and M. I. El-Khaleely. 1985. Antagonism between phyllosphere bacteria and actinomycetes and *Ulocladium botrytis* Preuss., causing tomato leaf spot. *Egyptian Journal of Phytopathology* 17:15–22.

# 15

# Host Searching by *Trichogramma* and Its Implications for Quality Control and Release Techniques

*Franz Bigler, Bas P. Suverkropp, and Fabio Cerutti*

A variety of *Trichogramma* species are mass reared and used for inundative releases in different agricultural crops and forest systems. Each crop is a distinct ecosystem that changes during growth. Thus, the wasps are faced with ecologically diverse situations. Crop architecture (size, shape, density) and plant morphology (veins, hairs, shape of organs) affect a wasp's searching behavior and success. These ecological differences become obvious if we compare, for example, a pine forest, a corn or sugar cane field, and a cabbage crop. Moreover, host eggs may be laid on specific plant parts depending on the pest species, plant variety, growth stage, and climatic conditions. Thus, the distribution of host eggs within the crop, on the plant, and between plant parts affects the probability of host encounters.

In addition to plant and host attributes, host searching is determined by the female's intrinsic capacity for dispersal within the crop and foraging on the plant. While dispersal of *Trichogramma* is mainly a function of flight abilities, foraging on the plant is affected primarily by walking and perceiving chemical and physical cues from the plant and the host guiding the female to the host egg. A number of behavioral attributes are needed to perform host searching successfully.

Mass rearing of *Trichogramma* for inundative releases tends to select for populations that perform well in the laboratory under optimal conditions but not necessarily in the field. It is therefore important to develop and implement quality control procedures to detect changes in attributes that affect host searching. Experiments based on female behavior must be designed to determine the value of quality traits that contribute to the overall performance. The aim of inundative release of *Trichogramma* is to bring the females as close as possible to the host egg in order to increase host encounters with the least possible waste of time and energy. This is not always easy to achieve. The development of an optimal release system must take into consideration crop and pest status, climate, labor costs, and the availability of technical facilities. The requirements of searching behavior are not the same for adults emerging, for example, on the plant or on the ground, in an open or closed canopy, pointwise or evenly distributed, in a low-growing annual crop or in a pine forest.

It is the aim of this chapter to discuss the factors that influence host searching, to relate them to quality control measures in mass rearing, and to elucidate the impact of release techniques on host searching requirements.

## Host Searching: Dispersal

Flight by *Trichogramma* can be divided into short, local movements and long-distance flights. Movement between plants is assumed to be random (Franz and Voegelé 1974; Bigler et al. 1988). Dispersal by female *Trichogramma* and landing on other plants or plant parts increases the chance of finding hosts or host cues compared to intensive and lengthy searching in places where few or no hosts occur. Salmanova (1991) demonstrated that loss of females' ability for spontaneous flights was the main cause of decreased parasitism in the field. Low temperatures inhibit flight initiation by *T. brassicae* in corn and decrease parasitism (Bigler et al. 1982). Forsse et al. (1992), and Dutton and Bigler (1995) conclude from their flight experiments with *T. minutum* and *T. brassicae* respectively that temperature threshold for flight will have a major impact on parasitism. Therefore, flight initiation is considered an essential parameter for quality control. When temperatures are below 20°C, walking and short jumps are the major means of dispersal in the field (Pak et al. 1985; van der Schaaf et al. 1984). The denser the canopy, the more easily a parasitoid moves from plant to plant without flight. Van Heiningen et al. (1985) found that parasitism by *Trichogramma* spp. increased after the canopy of a cabbage crop closed. However, greenhouse experiments showed no effect of between-plant distance on dispersal even at low temperatures, i.e., when flight was inhibited (van Alebeek et al. 1986). Long-distance flights can be strongly influenced by wind. The maximum flight speed of *Trichogramma* is low, and it is unable to fly against winds stronger than 7 km/hour (Steenburgh 1934), so wasps are easily transported by wind and blown out of the crop where they were released. This can occur especially if the plots are small and the canopy not dense enough to break the wind. Bigler et al. (1990) showed that transportation by wind negatively affects parasitism in release fields in Switzerland, in areas where *T. brassicae* is commercially used against the European corn borer, *Ostrinia nubilalis* Hbn. Based on parasitism of sentinel egg masses inside and outside the release fields, it was concluded that more than 60 percent of the released *T. brassicae* population left the cornfields. Both dispersal speed and the strong correlation between dispersal and wind direction indicate that windborne flight was the main mechanism involved. A survey of cornfields, treated by farmers with *T. brassicae* or left untreated, showed that *T. brassicae* parasitized more that 25 percent (0–50 percent, $n = 35$) of egg masses in fields where no wasps were released (Bigler et al. 1990). It was concluded that a relatively high proportion of the released *T. brassicae* emigrated actively or were transported out of the release fields. These results are in accordance with those reported by Andow and Prokrym (1991), who concluded from parasitization data that *T. nubilale* females, released in corn, disappeared from small plots (140 $m^2$) at daily rates of 40 percent from the remaining population. Using sticky cards, Hendricks (1967) found that dispersal of *T. semifumatum* in cotton was influenced by wind. When they were released above the canopy, the wasps dispersed over longer distances, but only in the direction of the wind.

Apart from different flight propensities of species and strains of *Trichogramma,* the shape and density of the canopy will determine how strongly dispersal is influenced by wind. In tree habitats, dense foliage alternates with open spaces. Within apple tree foliage, dispersal of *T. minutum* and *T. pretiosum* was not influenced by wind direction, but the dispersal to other trees was partly wind-dependent (Yu et al. 1984). Schread (1932) found no effect of wind on within- or between-tree dispersal of *T. minutum* in peach orchards. *Trichogramma pallida* and *T. embriophagum* in apple trees were recaptured on sticky cards mostly within 1 meter from the release point (Kolmakova and Molchanova 1981). Kot (1964) found a maximum dispersal of 50 meters of *Trichogramma* in orchards. In pine forests, an effect of wind on dispersal was found for *T. minutum* (Smith 1988) and *T. dendrolimi* (Ma et al. 1988). An average dispersal of 4.3 meters from the release point was found by Smith (1988) for *T. minutum* in forest trees. Experiments in corn showed that parasitism of *O. nubilalis* eggs by *T. maidis* (= *brassicae*) declines within 8 meters of the release point (Bigler et al. 1988; Maini et al. 1991). Dispersal of *T. minutum* and *T. pretiosum* in cotton crops was very limited, and most of the wasps stayed in the release field (Fye and Larsen 1969; Keller and Lewis 1985).

With regard to dispersal, *Trichogramma* used in inundative release should have a low threshold for flight initiation and a high propensity for short-range flights or jumps, resulting in fast dispersion throughout the crop. However, the wasps should have a low propensity for long-distance flights to prevent emigration from the habitat where they are supposed to control pest insects.

### *Host Searching on Plants*

Landing patterns of *Trichogramma* on plants have been studied by direct observation only in corn. The largest proportion of released *T. brassicae* land on leaves in the middle part of the plant, and the next largest proportion on leaves in the top third of the plant. This proportion is greater than would be expected from surface area alone (Suverkropp 1994). Some authors found that parasitism of *O. nubilalis* egg masses by *T. maidis* (= *brassicae*) in corn was highest in the top part (Neuffer 1987; Milani et al. 1988), while *T. nubilale* (Burbutis et al. 1977) and other *T. maidis* strains (Milani et al. 1988) preferred the lower part. Hawlitzky et al. (1994) found no difference between parasitism rates at different heights. In apple trees, *T. pallida* and *T. embriophagum* were recaptured mainly in the lower part of the tree (Kolmakova and Molchanova 1981).

The effectiveness of *Trichogramma* used in inundative release depends largely on the distribution pattern within the crop and on the plant. If preferences for specific niches are observed, they should coincide with those of the host. Release techniques must be designed to bring *Trichogramma* as close as possible to the host eggs and to reach all parts of the crop where the pest occurs.

On the leaves, *Trichogramma* walks in a relatively straight course, interrupted by sharp turns (Gardner and van Lenteren 1986). Movement on the leaves is not random, since linear structures (veins and leaf edges) are often followed. *Trichogramma brassicae* females spend 18 to 24 percent of their time on the leaf following veins and edges (Suverkropp, in prep.). *Trichogramma evanescens* on brussels sprouts leaf spent a significant amount of time on the leaf edge (Noldus et al. 1991). *Trichogramma maidis* (= *brassicae*) spends 1 to 10 percent of total walking time on the upper surface of corn leaves following the central vein and 10 to 20 percent on the lower surface (Gass 1988). Simulation studies showed that this behavior had no effect on searching efficiency of *T. brassicae* on corn. Not enough is known about the relationship between walking patterns, plant spatial structure, and searching efficiency to use movement patterns (tortuosity, edge and vein following) as a quality trait.

The number of eggs encountered depends on the area searched per time unit (Skellam 1958). High walking velocity results in a higher probability of encountering hosts and cues that may lead the female to the host (Bieri et al. 1990). Bigler et al. (1988) detected a relationship between walking speed and parasitism of egg masses of *O. nubilalis* Hbn. in the field by different laboratory-reared strains of *T. maidis* (= *brassicae*), which indicates the importance of walking speed as a quality trait.

The area searched per time unit is also dependent on walking activity, which is the fraction of total time spent walking. At 20°C, walking activity during daylight is 80 percent for *T. minutum, T. evanescens, T. semifumatum* (Biever 1972), and *T. brassicae* (Pompanon et al. 1994). At lower temperatures, walking activity of *T. brassicae* declines rapidly to only 10 percent at 12°C (Suverkropp, in prep). Because walking activity depends heavily on temperature, quality assessments should take into account measurements performed under suboptimal conditions, i.e., at temperatures between 12°C and 20°C for species and strains used in temperate areas.

In several experiments, parasitism declined as leaf area of the crop increased during the season, because the parasitoids had to search larger areas to find their hosts. *Trichogramma pretiosum* parasitism of *Heliothis virescens* was much higher on small than on large cotton plants (Ables et al. 1980). The same was found for *Trichogramma* spp. parasitism of *O. nubilalis* on green pepper (Burbutis and Koepke 1981) and on corn (Need and Burbutis 1979; Kanour and Burbutis 1984; Maini et al. 1991). For *T. maidis* (= *brassicae*) on corn, a 9 percent increase of the leaf area index led to 12 percent decrease of parasitism of *O. nubilalis* egg masses (van den Heuvel 1986). If *Trichogramma* spp. are released into crops with changing leaf areas (e.g., annual crops) during the parasitization period, the increase should be taken into account by increasing the numbers of females released.

## Semiochemicals in Host Searching

Although *Trichogramma* reacts to volatile plant and host compounds (Noldus 1988; Kaiser et al. 1989), their significance in biological control remains unclear (see Lewis and Sheehan, Chapter 17 in this volume).

Long-distance volatiles are basically not needed in inundative applications where *Trichogramma* is released in the target crop. However, they can be used as kairomones by *Trichogramma* females if these compounds guide the wasps to the sites where eggs are laid. In olfactometer experiments, sex pheromones of several lepidopterous species seemed to attract *Trichogramma. Trichogramma pretiosum* reacted to sex pheromones of *Heliothis zea* (Noldus 1988) and *T. evanescens* to sex pheromones of *Mamestra brassicae* (Noldus and van Lenteren 1985a), *Pectinophora gossypiella, Erias insulana,* and *Spodoptera littoralis* (Zaki 1985). In olfactometers, only short-range movements are recorded. Whether attraction over longer distances occurs in field situations is less clear. Sex pheromones, emitted during the night, can arrest females the following day and increase their searching activity (Noldus et al. 1991b). However, wind tunnel experiments did not demonstrate attraction (Noldus et al. 1991a). Arrestment may certainly increase parasitism when host mating and oviposition occur in the same place. It may hinder parasitism if mating localities are different from oviposition sites. The latter is true for *O. nubilalis* in the one-generation distribution area of Europe where corn is grown generally in a crop rotation. The moths migrate between fields and between the center and edges of cornfields, and mating occurs most often outside the cornfields (Buechi 1981; Cordillot 1989).

Not only odors produced by adult hosts influence *Trichogramma.* Kaiser et al. (1989) found that *O. nubilalis* egg odor in combination with artificial sex pheromones and corn odor was attractive to *T. maidis* (= *brassicae*). Single components did not attract the wasps. *Trichogramma evanescens* is attracted by the odors of *Ephestia kuehniella* eggs (Ferreira et al. 1979). Again, these results were obtained in small olfactometers. In greenhouse experiments, it was shown that the presence of *O. nubilalis* eggs and host odors on corn plants does not increase landing by *T. brassicae* (Suverkropp 1994).

Short-range or contact chemical cues, e.g., those associated with moth scales, increase searching activity and time when they are encountered (Lewis et al. 1972). When *Trichogramma* encounters a host cue, klinokinesis increases, so the female wasp remains in the immediate vicinity and searches this area more intensively (Shu and Jones 1989). Such arrestment responses have been observed in *T. pretiosum* with *Heliothis zea* scales and scale extract (Beevers et al. 1981); in *T. evanescens* with *Cadra cautella* scales and scale extract (Lewis et al. 1972); in *T. achaeae* (Lewis et al. 1975) and *T. evanescens* (Jones et al. 1973) with *Heliothis zea* scale extract; in *T. nubilale* (Shu and Jones 1989) and *T. maidis* (= *brassicae*) (Kaiser et al. 1989) with *O. nubilalis* scales; and in *T. evanescens* with *Pieris brassicae* (Gardner and van Lenteren 1986) and *Pieris rapae* (Noldus and van Lenteren 1985b). When no eggs are found, the walking path will gradually become less tortuous until the normal walking pattern is resumed (Zaborski et al. 1987). In addition to initializing area-restricted search, kairomones also seem to stimulate the wasps to stay on the plant (Beevers et al. 1981; Shu and Jones 1989; Jones et al. 1973; Suverkropp et al., in prep.). In contrast to volatiles, contact cues can be used by *Trichogramma* only if their presence is associated with hosts. Although responses of *Trichogramma* to different kairomones can be quantified, it is not known what responses would optimize host finding in field situations. Because the role of host semiochemicals for searching success of *Trichogramma* under field situations is as yet unclear and certainly variable, it is not possible so far to assess quality attributes related to responses to semiochemicals.

## Vision in Host Searching

Vision plays an important role in host discovery at short distances. Pak et al. (1990a) found that *Trichogramma* can distinguish host eggs by shape and volume at a distance. Mean reactive distance by vision is approximately 1 to 3 mm for *T. evanescens* and *T. dendrolimi* depending on the size of the host eggs (Glas et al. 1981; Pak et al. 1990b). These authors measured differences of the reactive distance between species. *Trichogramma* inspects all egglike objects encountered. Wajnberg (1994) found that reactive distance for glass beads is an inherited trait. For real eggs, no correlation between mothers and daughters was found, possibly because olfaction also plays a role in egg recognition. An increase in reactive distance increases searching efficiency because it increases the area searched per time unit. This means that reactive distance could be a useful quality trait.

In conclusion, searching success in inundative release depends largely on a number of innate capacities of *Trichogramma* females. A high propensity for short flights and jumps increases dispersion within crops, which is extremely important in spotwise release patterns and in young crops where the canopies are not yet closed. However, a high flight propensity may further active or passive long-range dispersal, thus causing a high proportion of females to fly out of the field. High walking velocity and activity (the fraction of total time spent walking) increase the leaf area searched in time and the probability of host encounter. Recognition of semiochemicals of hosts and host plants is important in cases where these cues are associated with host eggs. Although attraction and arrestment by volatile and contact semiochemicals are described for several *Trichogramma* species, their role in host searching in the field is not yet well understood. Intrinsic capacities can translate into searching success only if the females remain active within a large range of extrinsic conditions (environment, crop status, host densities, and so on).

## Quality Control

Inherent attributes that determine performance or quality of a *Trichogramma* species or strain in a specific crop/pest situation are influenced by mass production conditions (Bigler 1989). Behavioral traits responsible for successful host searching, e.g., flight, walking pattern and activity, and recognition of physical structure and chemical cues, may be altered or lost. Salmanova (1991) demonstrated a substantial loss of females' capacity for spontaneous flights in the course of mass production on the factitious host, *Sitotroga cerealella* (Olivier). Temperature adaptation was observed after a few generations reared under constant temperatures, and the normal daily rhythm was lost after fifteen generations (Shchepetilnikova and Kasinskaya 1981). Flight initiation of *T. brassicae* reared in the laboratory on *E. kuehniella* eggs for thirty-nine to forty-two generations was reduced by 20 percent to 75 percent compared to a strain that was continuously produced on its natural host, *O. nubilalis* (Dutton and Bigler 1995).

Bigler et al. (1988) showed that velocity and time spent walking decreased if strains were reared continuously under artificial conditions on the factitious host, *E. kuehniella* eggs. Pizzol and Wajnberg (1994) measured a significant difference in walking speed and time spent walking in two populations of *T. brassicae.* A high variation of walking velocity was demonstrated by Cerutti and Bigler (1995) when they compared fifty-two populations of *T. brassicae* reared under different conditions. They conclude that walking speed and pattern are among other traits useful attributes in assessing the quality of *T. brassicae* strains. Higher parasitism was measured in the field by Dutton et al. (1996) in *T. brassicae* populations with a high walking activity.

The ability to react to host cues can be changed or lost under laboratory-rearing conditions (Salmanova 1991). However, van Bergeijk et al. (1989) showed that *T. maidis* (= *brassicae*) reared for 300 generations in the laboratory did not differ significantly from a wild strain in finding egg masses of *O. nubilalis.* This may indicate that changes in host cues recognition depend on species and rearing conditions.

Quick dispersal after release by flight and jumps within the crop are important features in inundative applications of *Trichogramma.* Because host searching on the plant is predominantly performed by walking, velocity and walking activity have a high impact on success. Measurements of locomotor activity under optimal physical conditions in the laboratory may not necessarily indicate differences between species or populations. However, the reaction to suboptimal field conditions may vary between rearing strains, and specific quality attributes may be altered differently in the field. In quality control, it is therefore crucial to perform measurements of single quality traits under varying physical conditions and thus assess the activity range of locomotion.

The role of chemical and physical host cue recognition in successful searching is not yet elucidated to the point that quality control methods can be recommended. Further research is needed to understand the practical value of host cues in searching success in order to develop simple, quick, and reliable methods and to implement them in quality control programs. Details of quality control in mass production and use of *Trichogramma* are described in Bigler (1989, 1994), Laing and Bigler (1991), Cerutti and Bigler (1991, 1995) and Dutton and Bigler (1995).

## Release Techniques

The objective of a release strategy is to bring the wasps as close as possible to the host eggs at the right time. This, however, is not always feasible. Synchronization is a matter of pest monitoring and wasp availability from the producers. These aspects will not be dealt with in this chapter (see Andow, Chapter 4 in this volume). The release site in a crop and its distance from host eggs depend on the release method. These factors directly influence the attributes required of the wasps as they explore the habitat. For example, the capacity of dispersal within the crop after release will be of increasing importance when the number of release sites per unit area is low and the distance between release sites and host eggs is long.

Obviously, the importance of traits that contribute to the overall performance of the wasps depends on crops and release method. Consequently, quality requirements and assessments of the wasp will be different from system to system. In Table 15.1 we have selected three crops with different plant morphology and architecture in which *Trichogramma* is used. In all cases, the parasites are released as pupae inside the host eggs. The release techniques include (1) broadcast of parasitized eggs from aircraft or tractor, (2) parasitized eggs glued to the inner wall of capsules and dropped onto the soil, and (3) parasitized eggs glued to or wrapped in egg cards and attached to plants. The technique of releasing the wasps after emergence, used in several Latin American countries, is not considered here.

Broadcasting eggs parasitized by *Trichogramma* is a relatively simple technique. The disadvantage is the high proportion of eggs that fall to the ground when they are applied as dry (granular) material. Inert stickers are sometimes used to ensure that the material remains attached to the crop. The loss of parasitized eggs and the success of sticking them to the plant depend mainly on plant morphology, architecture, and canopy density. The same is true of capsules with eggs glued to the inner wall. Adults emerging from eggs or release containers dropped onto soil must fly or walk to the plant. In contrast, females emerging from egg cards hung on plants need not spend time searching for the plant. However, the distribution of the egg cards is generally less uniform in the field than that of the broadcast and capsule methods. Dispersal ability is an important quality attribute of the released wasps in all three release techniques. Because emerging females of bisexual species must mate, broadcast application may hamper the wasp's mating success and negatively affect flight propensity of females (Forsse et al. 1992).

Perception of host habitat cues is usually not an important trait in inundative releases in target crops except if the cues arrest the females. However, when broadcast release is used in forests it may be important to guide the wasps to host sites. The wasp's perception of volatiles emitted by hosts (e.g., sex pheromones) may be useless or even misleading the females if host mating and oviposition sites are not identical (e.g., in different fields).

Because the last step of host searching is performed by walking, these attributes are important for success and are closely linked to the release technique. Assuming

Table 15.1. Examples of rating quality attributes that contribute to the host-searching success of *Trichogramma* used in different crops and released with different techniques.

| Attributes contributing to host searching success | Crops: Cabbage Release technique | | | Crop: Maize Release technique | | | Crop: Pine-forest Release technique | |
|---|---|---|---|---|---|---|---|---|
| | broad-cast | capsules on soil (200/ha) | egg-cards on plant (50/ha) | broad-cast | capsules on soil (200/ha) | egg-cards on plant (50/ha) | broad-cast | egg-cards on trees |
| Jump, hops | ++ | ++ | ++ | ++ | ++ | ++ | ++ | ++ |
| Flight initiation | + | ++ | ++ | + | ++ | ++ | ++ | ++ |
| Short-range flights (< 2cm) | + | ++ | ++ | + | ++ | ++ | + | ++ |
| Long-range flights (1m) | – | + | ++ | – | + | ++ | ++ | + |
| Perception of host habitab cues | – | – | – | – | – | – | ++ | + |
| Perception of volatile host cues | – | – | – | – | – | – | ++ | + |
| Perception of non-volatile host cues | ++ | ++ | ++ | ++ | ++ | ++ | ++ | ++ |
| Walking velocity | ++ | ++ | ++ | ++ | ++ | ++ | ++ | ++ |

– not important, + important, ++ very important

that host eggs are not evenly distributed over the plant, e.g., located on the upper part of a tree canopy or laid on the lower side of maize leaves, it will be advantageous to use specific techniques for placing the parasitized eggs as close as possible to the host eggs. Short walking distances will increase host finding and thus prevent wasps from wasting their time searching plant parts where no host eggs are available.

## Conclusion

Host-searching requirements of inundatively released *Trichogramma* depend mainly on the crop, the distribution of host eggs, and the release technique. The specific interactions of each tritrophic system must be known in order to determine traits that are important for host-searching success. Quality attributes like flight, initiation, short-range flights, and walking velocity seem to be important in all systems. Dispersal and emigration are governed mainly by temperature, wind, and density of the canopy. Dispersal within the crop is random, and a significant proportion of adult parasitoids leave the release fields. Searching on the plant is partly directed by plant structures, though the efficacy compared to random search is not increased. The impact of semiochemicals on host searching in the field remains unclear. The release technique determines to a large extent how close to their hosts the wasps emerge, hosts and thus locomotor requirements of female wasps depend on release technique. The development of a quality control program must rely on thorough knowledge of the crop system, host factors, and the specific host-searching requirements, because traits involved in host-searching are considered of prime importance for the field performance of *Trichogramma.* Quality control must include quantitative assessments of attributes that determine searching efficacy, such as flight initiation and walking activity. A number of traits, considered important for host searching, are presented for different crops and related to behavior and release techniques.

## References

Ables, J. R., D. W. McCommas Jr., S. L. Jones, and R. K. Morrison. 1980. Effect of cotton plant size, host egg location, and location of parasite release on parasitism by *Trichogramma pretiosum. Southwestern Entomologist* 5:261–64.

van Alebeek, F.A.N., G. A. Pak, S. A. Hassan, and J. C. van Lenteren. 1986. Experimental releases of *Trichogramma* spp. against lepidopteran pests in a cabbage field crop in the Netherlands in 1985. *Mededelingen van de Faculteit der Landbouwwetenschappen, Rijksuniversiteit Gent* 51:1017–28.

Andow, D. A., and D. R. Prokrym. 1991. Release density, efficiency and disappearance of *Trichogramma nubilale* for control of European corn borer. *Entomophaga* 36:105–13.

Beevers, M., W. J. Lewis, H. R. Gross, and D. A. Nordlund. 1981. Kairomones and their use for management of entomophagous insects. X. Laboratory studies on manipulation of host finding behavior of *Trichogramma pretiosum* with a kairomone extracted from *Heliothis zea* (Boddie) moth scales. *Journal of Chemical Ecology* 7:635–48.

van Bergeijk, K. E. F. Bigler, N. K. Kaashoek, and G. A. Pak. 1989. Changes in host acceptance and host suitability as an effect of rearing *Trichogramma maidis* on a factitious host. *Entomologia experimentalis et applicata* 52:229–38.

Bieri, M., F. Bigler, and A. Fritschy. 1990. Abschätzung des Sucherfolges von *Trichogramma evanescens* Westw.-Weibchen in einer einfach strukturierten Umgebung. *Mitteilungen der Schweizerischen entomologischen Gesellschaft* 63:337–45.

Biever, K. D. 1972. Effect of temperatures on the rate of search by *Trichogramma* and its potential application in field releases. *Environmental Entomology* 1:194–97.

Bigler, F. 1989. Quality assessment and control in entomophagous insects used for biological control. *Journal of Applied Entomology* 108:390–400.

______. 1994. Quality control in *Trichogramma* production. In *Biological control with egg parasitoids,* ed. E. Wajnberg and S. A. Hassan, 93–111. Oxon, U.K.: CAB International.

Bigler, F., J. Baldinger, and L. Luisoni. 1982. L'impact de la méthode d'élevage et de l'hôte sûr la qualité intrinsèque de *Trichogramma evanescens* Westw. *Les Colloques de l'INRA* 9: 167–80.

Bigler, F., M. Bieri, A. Fritschy, and K. Seidel. 1988. Variation in locomotion between laboratory strains of *Trichogramma maidis* Pint. et Voeg. and its impact on parasitism of eggs of *Ostrinia nubilalis* Hbn. in the field. *Entomologia experimentalis et applicata* 49:283–90.

Bigler, F., S. Bosshart, M. Waldburger, and M. Ingold. 1990. Einfluss der Dispersion von *Trichogramma evanscens* Westw. auf die Parasitierung der Eier der Maiszünslers, *Ostrinia nubilalis. Mitteilungen der Schweizerischen Entomologischen Gesellschaft* 63:381–88.

Buechi R., J. Baldinger, S. Blaser, and R. Brunetti. 1981. Versuche zur Bekämpfung des Maiszünslers, *Ostrinia nubilalis* Hbn., mit der Verwirrungstechnik. *Mitteilungen der schweizerischen entomologischen Gesellschaft* 54:87–98.

Burbutis, P. P., and C. H. Koepke. 1981. European corn borer control in peppers by *Trichogramma nubilale. Journal of Economic Entomology* 74:246–47.

Burbutis, P. P., G. D. Curl, and C. P. Davis. 1977. Host searching behavior by *Trichogramma nubilale* on corn. *Environmental Entomology* 6:400–2.

Cerutti, F., and F. Bigler. 1991. Methods for the quality evaluation of *Trichogramma evanescens* Westw. used against the European corn borer. In *Proceedings of the 5th workshop of the global IOBC working group "Quality control of mass reared organisms,"* ed. F. Bigler, 119–26. Wageningen, The Netherlands: IOBC.

______. 1995. Quality assessment of *Trichogramma brassicae* in the laboratory. *Entomologia experimentalis et applicata* 75:19–26.

Cordillot, F. 1989. Dispersal, flight and oviposition strategies of the European corn borer, *Ostrinia nubilalis* Hbn. Ph.D. dissertation, Naturwissenschaft Fakultäat, Universität Basel.

Dutton A., and F. Bigler. 1995. Flight activity assessment of the egg parasitoid *Trichogramma brassicae* in laboratory and field conditions. *Entomophaga* 40:223–33.

Dutton A., F. Cerutti, and F. Bigler. 1996. Quality and environmental factors affecting *Trichogramma brassicae* efficiency under field conditions. *Entomologia experimentalis et applicata,* in press.

Ferreira, L., B. Pintureau, and J. Voegelé. 1979. Un nouveau type d'olfactomètre; application à la mesure de la capacité de recherche et à la localisation des substances attractives de l'hôte chez les *Trichogrammes* (Hym., Trichogrammatidae). *Annales de Zoologie et Ecologie animal* 11:271–79.

Forsse, E., S. M. Smith., and R. S. Bourchier. 1992. Flight initiation in the egg parasitoid *Trichogramma minutum:* Effects of ambient temperature, mates, food and host eggs. *Entomologia experimentalis et applicata* 62:147–54.

Franz, J. M., and J. Voegelé. 1974. Les *Trichogrammes* en vergers. In *Les organismes auxiliaires en verger de pommiers,* Brochure 3, 201–10. International Organization for Biological Control: Wageningen.

Fye, R. E., and D. J. Larsen. 1969. Preliminary evaluation of *Trichogramma minutum* as a released regulator of lepidopterous pests of cotton. *Journal of Economic Entomology* 62: 1291–96.

Gardner, S. M., and J. C. van Lenteren. 1986. Characterisation of the arrestment responses of *Trichogramma evanescens. Oecologia* 68:265–70.

Gass, T. 1988. Einfluss der Temperatur auf die Bewegungsaktivität und die Parasitierungsleistung von *Trichogramma maidis* Pint. & Voeg. Diplomarbeit Eidgenössische Technische Hochschule Zürich.

Glas, P.C.G., P. H. Smits, P. Vlaming, and J. C. van Lenteren. 1981. Biological control of lepidopteran pests in cabbage crops by means of inundative releases of *Trichogramma* species: A combination of field and laboratory experiments. *Mededelingen van de Faculteit der Landbouwwetenschappen, Rijksuniversiteit Gent* 46:487–97.

Hawlitzky, N., F. M. Dorville, and J. Vaillant. 1994. Statistical study of *Trichogramma brassicae* efficiency in relation with characteristics of the European corn borer egg masses. *Researches in Population Ecology* 36:79–85.

Heiningen, T. G. van, G. A. Pak, S. A. Hassan, and J. C. van Lenteren. 1985. Four years' results of experimental releases of *Trichogramma* egg parasites against Lepidopteran pests in cabbage. *Mededelingen van de Faculteit der Landbouwwetenschappen, Rijksuniversiteit Gent* 50:379–88.

Hendricks, D. E. 1967. Effect of wind on dispersal of *Trichogramma semifumatum. Journal of Economic Entomology* 60:1367–73.

Jones, R. L., W. J. Lewis, M. Beroza, B. A. Bierl, and A. N. Sparks. 1973. Host seeking stimulants (kairomones) for the egg-parasite *Trichogramma evanescens. Environmental Entomology* 2:593–96.

Kaiser, L., M. H. Pham-Delegue, E. Bakchine, and C. Masson. 1989. Olfactory responses of *Trichogramma maidis* Pint et Voeg.: Effects of chemical cues and behavioral plasticity. *Journal of Insect Behavior* 2:701–12.

Kanour, W. W., and P. P. Burbutis. 1984. *Trichogramma nubilale* (Hymenoptera: Trichogrammatidae) field releases in corn and a hypothetical model for control of European corn borer (Lepidoptera: Pyralidae). *Journal of Economic Entomology* 77:103–7.

Keller, M. A., and W. J. Lewis. 1985. Movements by *Trichogramma pretiosum* (Hymenoptera: Trichogrammatidae) released into cotton. *Southwestern Entomologist Suppl.* 8:99–109.

Kolmakova, V. D., and V. A. Molchanova. 1981. Dispersal of *Trichogramma* analyzed through radioactive marking. In *Insect behavior as a basis for developing control measures against pests of field crops and forests,* ed. V. P. Pristavko, 65–72. New Delhi, India: Oxonian.

Kot, J. 1964. Experiments in the biology and ecology of species of the genus *Trichogramma* Westw. and their use in plant protection. *Ekologia Polska—Series A* 12:243–303.

Laing, J. E., and F. Bigler. 1991. Quality control of mass-produced *Trichogramma* species. In *Proceedings of the 5th workshop of the global IOBC working group "Quality control of mass reared organisms,"* ed. F. Bigler, 111–18. Wageningen, The Netherlands: IOBC.

Lewis, W. J., R. L. Jones, D. A. Nordlund, and H. R. Gross. 1975. Kairomones and their use for management of entomophagous insects. II. Mechanisms causing increase in rate of parasitization by *Trichogramma* spp. *Journal of Chemical Ecology* 1:349–60.

Lewis, W. J., R. L. Jones, and A. N. Sparks. 1972. A host seeking stimulant for the egg-parasite *Trichogramma evanescens*: Its source and a demonstration of its laboratory and field activity. *Annals of the Entomological Society of America* 65:1087–89.

Ma, W. Y., J. W. Peng, and Y. X. Zuo. 1988. [Studies on the biology of *Trichogramma dendrolimi* Matsumura.] *Scientia Silvae Sinicae* 24:488–95.

Maini, S., G. Burgio, and M. Carrieri. 1991. *Trichogramma maidis* host-searching in corn vs. pepper. In *Proceedings 4th European Workshop on Insect Parasitoids, Perugia,* ed. F. Bin, 121–28. Florence, Italy: Redia.

Milani, M., P. Zandigiacomo, and R. Barbattini. 1988. *Ostrinia nubilalis* (Hb.)(Lepidoptera Pyralidae) su mais in Friuli. V. Prove di lotta biologica con *Trichogramma maidis* Pint. et Voeg. (Hymenoptera Trichogrammatidae): attività del parassitoide in campo. *Frustula Entomologica* 11: 57–68.

Need, J. T., and P. P. Burbutis. 1979. Searching efficiency of *Trichogramma nubilale. Environmental Entomology* 8:224–27.

Neuffer, U. 1987. Vergleich von Parasitierungsleistung und Verhalten zweier Ökotypen von *Trichogramma evanescens* Westw. Ph.D. thesis, Universität Hohenheim. Stuttgart, Germany.

Noldus, L.P.J.J. 1988. Response of the egg parasitoid *Trichogramma pretiosum* to the sex pheromone of its host *Heliothis zea. Entomologia experimentalis et applicata* 48:293–300.

Noldus, L.P.J.J., and J. C. van Lenteren. 1985a. Kairomones for the egg parasite *Trichogramma evanescens* Westwood. I. Effect of volatile substances released by two of its hosts, *Pieris brassicae* L. and *Mamestra brassicae* L. *Journal of Chemical Ecology* 11:781–91.

______. 1985b. Kairomones for the egg parasite *Trichogramma evanescens* Westwood. II. Effect of contact chemicals produced by two of its hosts, *Pieris brassicae* L. and *Pieris rapae* L. *Journal of Chemical Ecology* 11:793–800.

Noldus, L.P.J.J., J. C. van Lenteren, and W. J. Lewis. 1991a. How *Trichogramma* parasitoids use moth sex pheromones as kairomones: Orientation behaviour in a wind tunnel. *Physiological Entomology* 16:313–27.

Noldus, L.P.J.J., R. P. J. Potting, and H. E. Barendregt. 1991b. Moth sex pheromone adsorption to leaf surface: Bridge in time for chemical spies. *Physiological Entomology* 16:329–44.

Pak, G. A., H. Berkhout, and J. Klapwijk. 1990a. Do *Trichogramma* look for hosts? In Trichogramma *and other egg parasitoids,* ed. E. Wajnberg and S. B. Vinson, 77–80. Paris: INRA.

Pak, G. A., J.W.M. Kaskens, and E. J. de Jong. 1990b. Behavioral variation among strains of *Trichogramma* spp.: Host-species selection. *Entomologia experimentalis et applicata* 56: 91–102.

Pak, G. A., I. van Halder, R. Lindeboom, and J.J.G. Stroet. 1985. Ovarian egg supply, female age and plant spacing as factors influencing searching activity in the egg parasite *Trichogramma* spp. *Mededelingen van de Faculteit der Landbouwwetenschappen, Rijksuniversiteit Gent* 50:369–78.

Pizzol, J., and E. Wajnberg. 1994. Inter-population genetic variation in the walking behavior of *Trichogramma brassicae* females. In Trichogramma *and other egg parasitoids,* ed. E. Wajnberg, 27–30. Paris: INRA.

Pompanon, F., P. Fouillet, and M. Boulétreau. 1994. Locomotor behaviour in females of two *Trichogramma* species: Description and genetic variability. In *Insect parasitoids: Biology and ecology. Proceedings of the 5th European Workshop on Insect Parasitoids,* ed. E. B. Hagvar and T. Hofsvang, 185–90. Ås, Norway: *Norwegian Journal of Agricultural Sciences.*

Salmanova, L. M. 1991. Changes of *Trichogramma* cultures in permanent rearing on the laboratory host *Sitotroga cerealella* Oliv. M. Sc. Dissertation, Lomonosow University, Moscow.

Schaaf, D. A. van der, J.W.M. Kaskens, M. Kole, L.P.J.J. Noldus, and G. A. Pak. 1984. Experimental releases of two strains of *Trichogramma* spp. against lepidopteran pests in a brussels sprouts field crop in the Netherlands. *Mededelingen van de Faculteit der Landbouwwetenschappen, Rijksuniversiteit Gent* 49:803–13.

Schread, J. C. 1932. Behavior of *Trichogramma* in field liberations. *Journal of Economic Entomology* 25:370–74.

Shchepetilnikova, V. A., and L. V. Kasinskaya. 1981. Changes in the environmental preferences of *Trichogramma* effected through conditions of rearing. In *Insect behavior as a basis for developing control measures against pests of field crops and forests,* ed. V. P. Pristavko, 225–31. New Delhi, India: Oxonion.

Shu, S. Q., and R. L. Jones. 1989. Kinetic effects of a kairomone in moth scales of the European corn borer on *Trichogramma nubilale* Ertle and Davis (Hymenoptera: Trichogrammatidae). *Journal of Insect Behavior* 2:123–31.

Skellam, J. G. 1958. The mathematical foundations underlying line transects in animal ecology. *Biometrics* 14:385–400.

Smith, S. M. 1988. Pattern of attack on spruce budworm egg masses by *Trichogramma minutum* (Hym.:Trichogrammatidae) released in forest stands. *Environmental Entomology* 17:1009–15.

Steenburgh, W. E. van. 1934. *Trichogramma minutum* Riley as a parasite of the oriental fruit moth (*Laspeyresia molesta* Busck.) in Ontario. *Canadian Journal of Research* 10:287–314.

Suverkropp, B. P. 1994. Landing of *Trichogramma brassicae* Bezdenko [Hymenoptera: Trichogrammatidae] on maize plants. In *Insect parasitoids: Biology and ecology. Proceedings of the 5th European Workshop on Insect Parasitoids,* ed. E. B. Hagvar and T. Hofsvang, 243–54. Ås, Norway: *Norwegian Journal of Agricultural Sciences.*

Van den Heuvel, H. 1986. Die biologische Bekämpfung des Maisszünslers (*Ostrinia nubilalis* Hübner) mit *Trichogramma maidis* Pint. et Voeg. Report of the Swiss Federal Research Station for Agriculture: Zürich.

Wajnberg, E. 1994. Intrapopulation genetic variation in *Trichogramma.* In *Biological control with egg parasitoids,* ed. E. Wajnberg and S. A. Hassan, 245–71. Oxon, U.K.: CAB International.

Yu, D.S.K., J. E. Laing, and E.A.C. Hagley. 1984. Dispersal of *Trichogramma* spp. (Hymenoptera: Trichogrammatidae) in an apple orchard after inundative releases. *Environmental Entomology* 13:371–74.

Zaborski, E., P.E.A. Teal, and J. E. Laing. 1987. Kairomone-mediated host finding by spruce budworm egg parasite, *Trichogramma minutum. Journal of Chemical Ecology* 13:113–21.

Zaki, F. N. 1985. Reactions of the egg parasitoid *Trichogramma evanescens* Westw. to certain sex pheromones. *Zeitschrift fur angewandte Entomologie* 99:448–53.

# 16

# *Gliocladium* and Biological Control of Damping-Off Complex

*James F. Walter and Robert D. Lumsden*

The fungus *Gliocladium virens* Miller, Giddens and Foster is an important biological control agent (Papavizas 1985). A formulation of a strain of this fungus (strain GL-21) was recently registered with the U.S. Environmental Protection Agency (EPA) by W. R. Grace and Co. of Connecticut. The formulation and registration were subsequently sold to the Thermo Trilogy Corporation. The formulation was developed in cooperation with the Biocontrol of Plant Diseases Laboratory (BPDL), U.S. Department of Agriculture (USDA) (Lumsden et al. 1991). It is intended for use against damping-off diseases of vegetable and ornamental seedlings caused by the soil-borne plant pathogens *Pythium ultimum* and *Rhizoctonia solani* in glasshouse operations (Lumsden and Locke 1989). This fungus is one of the first to be registered for biocontrol of plant diseases; it is available in the United States for commercial use in glasshouse applications.

Certain criteria were considered important in early stages of development of biocontrol agents (Lumsden and Lewis 1989). In developing a screening method for selecting an appropriate microorganism, the following were considered. The screening would involve (1) the use of a relatively uniform, commercially available soilless medium that is used extensively in commercial glasshouses where the disease problem occurs; (2) targeted pathogens that are important in the confines of a glasshouse where use of a biological control agent would be likely to be most successful because of a relatively uniform environment; (3) microorganisms indigenous to the United States because nonindigenous microorganisms might be conceived as more likely problems for the U.S. environment; (4) a single isolate of a biocontrol agent for control of both pathogens in preference to a mixture of isolates; and (5) a high-value crop important in the ornamental production industry to defray the cost of development and registration. On the basis of these factors, consideration of which would make the process for registration and commercialization of an agent for control of plant diseases easier, a screening program was initiated, and more than a hundred isolates of fungi, bacteria, and actinomycetes were tested (Lumsden and Locke 1989). Most of these isolates had some prior history of biocontrol potential or were isolated from survival structures of plant pathogens and thus were likely to have antagonistic capabilities. From these screenings, *G. virens* consistently controlled the pathogens better than other microorganisms tested, and one isolate (GL-21) was effective against both *P. ultimum* and *R. solani.*

Upon selection of *G. virens* as the microorganism of choice, an appropriate formulation of the fungus for ease of preparation, application, and maximum efficacy was chosen. A suitable formulation, previously developed by the BPDL, was based on alginate–wheat bran granules (prill) (Lewis and Papavizas 1987; Fravel et al. 1985). Three U.S. patents (numbers 4,668,512; 4,724,147; and 4,818,530) were issued for this formulation.

The effective *G. virens* isolates, formulated into alginate prill, were thus protected by patents, and the technology was transferred through an exclusive license to W. R. Grace and Co., whose research facilities are located in Columbia, Maryland, near Beltsville where the USDA laboratory is located. The proximity of the two facilities was a major advantage in the development of the product. The cooperative venture was expedited by the Technology Transfer Act of 1986 passed by the U.S. Congress to encourage and promote cooperative projects between government laboratories and private enterprise.

The cooperative agreement involved improvement of the fermentation process, refinement of the formulated *G. virens,* improvement of the shelf life, and establishment of efficacy on several crop plants at several glasshouse locations (Lumsden et al. 1990). Achievement of these objectives was necessary before approaching the EPA for registration and eventual commercialization.

According to subdivision M of the 1989 EPA Pesticide Testing Guidelines (Federal Register 1984), microbial agents for the control of plant pests are treated in many ways similar to chemical pesticides, and companies applying for registration must provide extensive information for approval to use microbial products commercially. Product testing is set up in a tier system that recognizes the inherent risks and degrees of exposure associated with different uses of pesticides. In addition to production and taxonomic data, long- and short-term effects on a variety of organisms including plants, animals, and other nontarget organisms may be necessary (Table 16.1). According to the regulations, studies may be required for effects that are toxicological, mutagenic (causing gene damage), carcinogenic (causing cancer), fetotoxic (toxic to a fetus), teratogenic (causing birth defects), and oncogenic (causing tumors) depending on the envisioned use pattern (Table 16.2).

These requirements were originally designed for evaluation of synthetic chemicals and not living organisms, but some latitude is allowed for the difference. The testing is done on a multitier system so that microorganisms that pass certain requirements at the tier 1 level need not be progressively tested at a more stringent, long-term testing tier 2 level. In addition, the EPA is aware of a cost-benefit mode of assessment and the need to expedite introduction into agricultural use of safe, alternative

Table 16.1. EPA 40CFR regulations regarding data needs for registration of microbial pesticides

| Type of Data | Terrestrial | | Aquatic | | Greenhouse | | Forestry | | Domestic | |
|---|---|---|---|---|---|---|---|---|---|---|
| | Food | Nonfood | Food | Nonfood | Food | Nonfood | Food | Nonfood | Food | Nonfood |
| Acute Avian | R | R | R | R | CR | CR | R | R | CR | CR |
| Avian Injection | R | R | R | R | CR | CR | R | R | CR | CR |
| Wild Mammal | CR | CR | CR | CR | — | — | CR | CR | — | — |
| Freshwater Fish | R | R | R | R | CR | CR | R | R | CR | CR |
| Freshwater Invertebrate | R | R | R | R | CR | CR | R | CR | CR | CR |
| Nontarget Plant | R | R | R | R | — | — | R | CR | — | — |
| Nontarget Fish | R | R | R | R | CR | CR | R | R | — | — |
| Honey Bee | R | R | R | R | CR | CR | R | R | — | — |

R = Required

CR = Conditionally required

*Note:* Tier I data needs for potential environmental hazards to non-target organisms.

Table 16.2. EPA 40CFR regulations regarding data needs for registration of microbial pesticides

| Type of Data | Terrestrial | | Aquatic | | Greenhouse | | Forestry | | Domestic | |
|---|---|---|---|---|---|---|---|---|---|---|
| | Food | Nonfood | Food | Nonfood | Food | Nonfood | Food | Nonfood | Food | Nonfood |
| Acute Oral | R | R | R | R | R | R | R | R | R | R |
| Avian Dermal | R | R | R | R | R | R | R | R | R | R |
| Acute Inhalation | R | R | R | R | R | R | R | R | R | R |
| I.V.I.C.I.P. Injection | R | R | R | R | R | R | R | R | R | R |
| Primary Dermal | R | R | R | R | R | R | R | R | R | R |
| Primary Eye | R | R | R | R | R | R | R | R | R | R |
| Hypersensitivity | R | R | R | R | R | R | R | R | R | R |
| Immune Response | R | R | R | R | R | R | R | R | R | R |
| Tissue Culture (V) | R | R | R | R | R | R | R | R | R | R |
| X Viruses Only | — | — | — | — | — | — | — | — | — | — |

R = Required

*Note:* Tier I toxicology tests that are required.

pest control strategies to minimize the use of chemical pesticides. The EPA is thus willing and able to waive or minimize the impact of certain requirements of the legislated regulations when appropriate. Each situation is judged individually. Table 16.3 outlines trials required for the registration of GL-21 and provides other toxicological information.

In the case of *G. virens,* the EPA chose to waive several of the tests outlined in Table 16.2 or accepted alternate information. These modifications of the requirements were made after appropriate examination of the environmental and worker safety risks involved. Data were requested by the EPA to cover several areas including toxicological data, identification and quantification of toxins, biological characteristics of GL-21, storage, stability, efficacy, and inert ingredients.

## Toxicological Data

Conclusions regarding the safety of the formulation are supported by the demonstrated lack of toxicity in oral, dermal, and pulmonary studies conducted on rats. The toxicological data submitted for this active ingredient included reports of an acute oral toxicity/pathogenicity study, an acute pulmonary toxicity/pathogenicity study, and an acute intravenous toxicity/pathogenicity study. All studies were classified as acceptable. The review of these studies by the EPA indicated that *G. virens* is not toxic to, infective in, or pathogenic to rats by oral or pulmonary routes of exposure and not infective or pathogenic to rats by intravenous injection. Mycelia of the fungus were used for the injections and were found to be acutely toxic and lethal to some test animals due to mechanical clogging of capillaries. However, these mortalities were not considered to be relevant in this case since injection is not a normal route of exposure and the product consists of chlamydospores rather than mycelia.

Acute dermal toxicity testing was not required for *G. virens* based on the fact that the end-use product consists of large pellets (1 to 2 mm in diameter). The label requires that gloves be worn by those handling the product, and application will be by soil incorporation. A primary eye irritation study was not required since protective eye covering is required.

Other toxicity testing was waived based on submission of data that (1) indicated that the fungus does not grow at or near the body temperature of mammals or birds, (2) demonstrated that toxins or antibiotics, which were produced during the manufacture of the product, were not considered to be of toxicological significance, (3) showed that personnel working with *G. virens* strain GL-21 for several years evidenced no adverse toxicological effects attributable to that work, (4) provided the criteria used to determine the extent to which formulated preparations are free from contaminating microorganisms, and (5) confirmed the exempt status of certain inert ingredients. The acute and primary dermal and primary eye tests were waived due to the insoluble pellet form of the product. Furthermore, worker exposure experience from two to three years of working with the material was accepted as evidence of the lack of hypersensitivity or immune response concerns.

Table 16.3. Results of toxicology tests of GL-21.

| Study | Result |
| --- | --- |
| Acute oral toxicity/pathogenicity | No acute toxicity, pathogenic effects, or infections detected. Microbe expelled in feces. |
| Acute pulmonary toxicity/pathogenicity | No acute toxicity, pathogenic effects, or infections detected. Spore takes 2-3 weeks to clear lungs. |
| Acute intravenous toxicity/pathogenicity | No apparent acute toxicity, pathogenic effects or infections detected. Mycelium is trapped in lungs, liver, and spleen and cleared after 14 days. |
| Acute dermal/primal dermal | Waived due to product form. |
| Primary Eye | Waived due to product form. |

| Other Toxicological Information | |
| --- | --- |
| 1. | Study shows GL-21 does not grow at or near body temperature of mammals or birds. |
| 2. | Quantification of antibiotic products shown not to be of toxicological significance. |
| 3. | Documentation of worker exposure to GL-21 production over two to three years shows no toxicological response. |
| 4. | Documentation of methods to validate freedom from contaminating microorganism. |
| 5. | Confirmation of exempt status of inert ingredients generally regarded as safe materials (Food Grade). |
| 6. | Studies showing extent of persistence of GL-21 in environment and extent of spread in soil. |

## Biological Characteristics

### *Biological Properties of* Gliocladium virens

*Gliocladium virens* is a hyphomycete with no confirmed sexual stage. The possible sexual stage is *Hypocrea gelatinosa* (Domsch et al. 1980). It proliferates as asexual conidia that are held in masses of moist spores, or it survives as vegetative segments of the mycelium termed chlamydospores that are usually embedded in organic matter. The spores are not airborne and are dispersed only as spore suspensions in water or carried in soil or in organic debris.

*Gliocladium virens* is a common soil saprophyte and, like many other soil-borne fungi, produces several antibiotic metabolites (Howell and Stipanovic 1983; Jones and Hancock 1987; Taylor 1986) that are thought to enhance its soil competitiveness. The metabolite most likely associated with control of *Pythium* and *Rhizoctonia* damping-off is gliotoxin, a piperazine antibiotic (Lumsden et al. 1992; Roberts and Lumsden 1990). Gliotoxin has antibacterial, antifungal, antiviral, and antitumor activity.

Fig. 16.1 presents the structure, toxicity, and properties of the antibiotics produced by *G. virens* with a description of these compounds. The manufactured product is monitored for the presence of the components by extracting the clarified fermentation broth with chloroform and analyzing the concentration extract by HPLC, and these compounds are absent from the products. It should be noted that at no time have viridin or viridiol been detected in extracted fermentation broth, the manufactured product, or the end-use product. Gliotoxin has been detected in fermentations of *G. virens* conducted at acid to neutral pH (4 to 6.5), and S, S-dimethylgliotoxin is found in fermentations at operating conditions of pH 6 to 7 at levels of less than10 ppm. Even in the case of low pH fermentations, the detected gliotoxin level does not exceed 10 ppm. Therefore, taking the worst case-scenario and assuming the manufactured product contained 10 ppm gliotoxin, this would correspond to an equivalent $LD_{50}$ of the formulated product of about 30g/kg. Furthermore, gliotoxin is known to be sensitive to oxidation, and it is highly likely that any gliotoxin that may be present in the manufactured product is degraded in the drying of the end-use product. Consequently, the amounts of these compounds in the final product are minute and of no toxicological significance.

Under different environmental conditions, however, GL-21 produces gliotoxin and may produce viridin and viridiol. Experimental evidence shows that when a soilless media (such as Redi-Earth) is inoculated with GL-21, chloroform extracts of the media contain gliotoxin. Viridin and viridiol are usually not present. This indicates that, in use, the microorganism will produce these metabolites to some extent. It is unknown how leachable these compounds are so it is not possible to determine exactly how much of these metabolites are present in the media. However, the amounts are likely to be small and not likely to affect workers or nontarget organisms in the greenhouse environment. Gliotoxin is one of several epipolythiodioxopiperizines found in nature. Although it was initially isolated in 1932 from a strain of *Trichoderma,* its true structure was not proven until 1966.

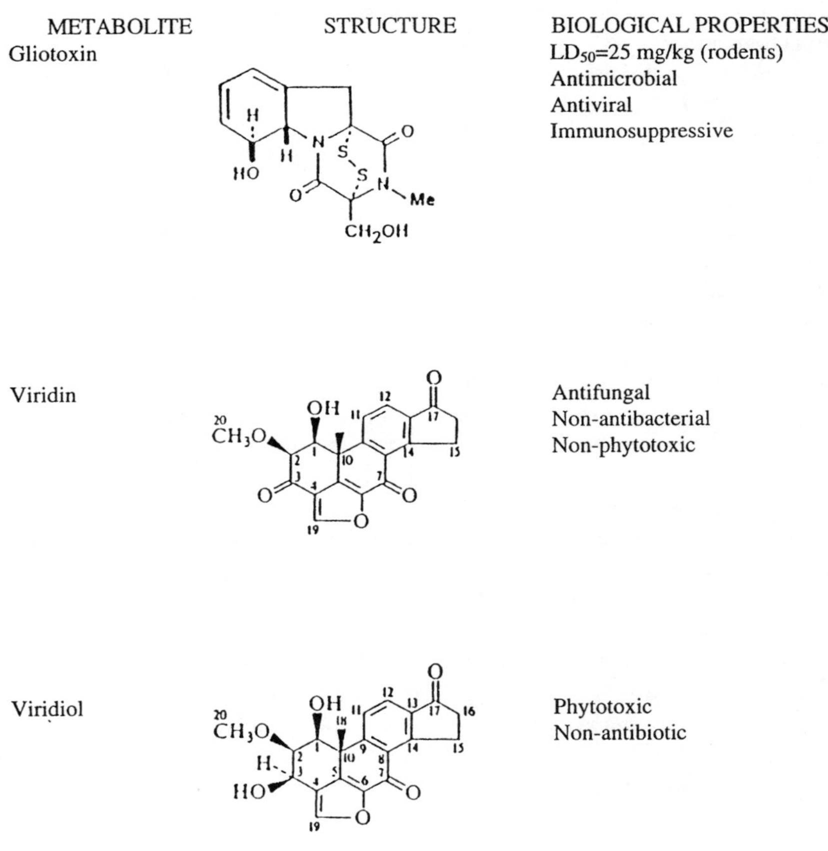


Figure 16.1. Toxicity, structure, and properties of the antibiotics produced by *Gliocladium virens.*

## *Biological Properties of Toxins*

Gliotoxin has been reported to have antibacterial, antifungal, antiviral, and immunosuppressive activities. It has also been shown to inhibit the multiplication of RNA viruses such as poliovirus, herpes, and influenza in cell culture ($LC_{50}$ = 0.002 mg/ml), and various researchers demonstrated that it had significant chemotherapeutic effects on monkeys infected with poliovirus and mice and dogs infected with influenza. However, gliotoxin has mammalian toxicity. Lethal doses of about 45 to 50 mg/kg are reported for rabbits, mice, and rats, and the $LD_{50}$ is about 25 mg/kg. Furthermore, it is toxic to mammalian cells in culture at a concentration of 0.03 to 1 mg/ml. Yet mice have been shown to be able to tolerate a chemotherapeutic program of 5 to 10 mg/kg/day (Waring et al. 1988). The physiological properties of gliotoxin are attributable to its disulfide bridge. Breaking this sulfur linkage inactivates gliotoxin.

Gliotoxin is sensitive to oxidation and heat; it is inactivated by heating for ten minutes at 100°C. A degradation product of gliotoxin is S, S-dimethylgliotoxin, which has no known biological activity.

Viridin is a highly fungistatic metabolic product of certain strains of *Gliocladium virens.* Viridin contains no nitrogen, sulfur, or halogens; its structure is shown in Fig. 16.1. It decomposes without melting at 217 to 223°C and is very soluble in chloroform and insoluble in ether. Most *Trichoderma, Gliocladium,* and *Aspergillus* are not sensitive to viridin. However, viridin is active against *R. solani* to an appreciable extent, less so against *P. ultimum,* and has negligible antibacterial properties (Lumsden et al. 1992). Viridin is also very unstable in neutral to basic aqueous solutions. At pH 7.4 viridin is essentially degraded in five hours, and at pH 8 it is degraded in fifteen minutes. Other substances such as L-tryptophan, L-tyrosine, L-cystine, and glutamic acids are reported to inactivate viridin. In neutral to basic aqueous solutions viridin is reduced to viridiol, and several organisms such as *G. virens* readily convert viridin to viridiol (Jones and Hancock 1987), even at a low pH. Isolation of viridin from fermentation media usually requires a low fermentation pH and extraction with chloroform while the organism is still in the midgrowth phase. Since the GL-21 fermentations are close to neutrality, we have been unable to isolate or detect viridin in our fermentation products.

As opposed to gliotoxin and viridin, viridiol has been reported to be ineffective as an antibiotic; however, it is a potent phytotoxin. Viridiol, and hence viridin, production is very sensitive to media composition. It is reported that the presence of ammonium sulfate and ammonium in the media tends to increase viridiol production, while complex nutrients like asparagine and sucrose or nitrates such as calcium or potassium suppress production. Viridiol concentrations as high as 15 ppm have been reported to be produced in suspension culture at a low pH (Jones and Hancock 1987). In the GL-21 fermentation, however, we have not been able to detect even trace quantities (~0.02 ppm) of viridiol in the media extract. If we grow this organism on solid phase culture (peat vermiculite media), however, chloroform extracts indicate that GL-21 will produce viridiol in detectable quantities (Lumsden et al. 1992).

## Development and Implementation

### *Effect of Storage Conditions on AP-1*

Alginate prill containing GL-21 was stored in commercial containers at 55°C in a dark incubator, at 25°C and 50 percent humidity in the dark, and at 25°C and 50 percent humidity in a 2,500 foot candlelight chamber. Prill was stored in heat-sealed clear polyethylene packages. Several packages were placed in each location. In the case of storage at 25°C in the dark, the polyethylene packets were placed in closed cardboard boxes to simulate the package container. A VWR incubator oven was used as the 55°C storage chamber, while an American Scientific Products

Table 16.4. GlioGard shelf stability at five different temperatures

| Line No. | Time Stored | CFU/gram | | | | |
|---|---|---|---|---|---|---|
| | | 0°C | 10°C | 25°C | 35°C | 55°C |
| (1) | 0 | | | | | |
| (2) | 4 weeks | $2x10^8$ | $2x10^8$ | $2x10^8$ | $2x10^8$ | $2x10^8$ |
| (3) | 8 weeks | $2x10^8$ | $2x10^8$ | $2x10^8$ | $2x10^8$ | $2x10^6$ |
| (4) | 12 weeks | $2x10^8$ | $2x10^8$ | $2x10^8$ | $1.1x10^8$ | $2x10^3$ |
| (5) | 16 weeks | $2x10^8$ | $1.8x10^8$ | $1.8x10^8$ | $7x10^7$ | $1x10^2$ |
| (6) | 20 weeks | $2x10^8$ | $2x10^8$ | $7.1x10^7$ | $4x10^6$ | $1x10^1$ |
| (7) | 24 weeks | $1.8x10^8$ | $1.7x10^8$ | $5.2x10^7$ | $1x10^5$ | $<1x10^1$ |
| (8) | 39 weeks | $1.4x10^8$ | $1.1x10^8$ | $1x10^7$ | $1x10^3$ | $<1x10^1$ |
| (9) | 54 weeks | $1x10^8$ | $8x10^7$ | $3x10^6$ | $1x10^2$ | $<10^1$ |
| (10) | $K_1$ (week $^{-1}$) | | | 0.046 | 0.099 | 1.44 |

growth chamber, with humidity control and 2,500 foot candlelight, was used as the 25°C chamber. The results of the stability test show that exposure to light does not significantly decrease the storage stability of the prill, and the data thus far indicate that this type of package does not impair AP-1 storage stability. Table 16.4 demonstrates the stability of GlioGard at various temperatures.

### *Spectrum of Efficacy*

The soil inoculant *Gliocladium virens* (GL-21) has been shown to be effective against several soil fungal plant pathogens including *Pythium, Rhizoctonia, Sclerotium,* and *Fusarium* (Knauss 1992). It is likely that the soil inoculant may also control other fungal pathogens such as *Phytophthora, Botrytis,* and *Cylindrocladiun,* but these pathogens have not been tested conclusively.

The minimum dose of the inoculant GlioGard, when it is incorporated into a soilless or soil-containing medium, is 2 g/l (1.0–1.25 oz/ft$^3$). If the inoculant is added to the soil surface, it is best added at a rate of 21 to 30 g/m$^2$ (0.7–1.0 oz/ft$^2$) and worked into the surface.

### *Application Conditions*

The soil inoculant works best when it is added to the soil or soilless media zero to three days prior to planting; the best results occur within three days of preincubation. The inoculation time allows the inoculant to germinate, proliferate through the bed and suppress the pathogens already in the media. If the soil or soilless media is presterilized or known to be free of pathogens, however, the inoculant can be added at planting to achieve disease control.

### *Plant Trial Results*

The ability to control both *Pythium ultimum* and *Rhizoctonia solani* damping-off in growth media is demonstrated in Table 16.5. GlioGard was added to moistened Redi-Earth at a rate of 1 g/l three days prior to planting. The media was added to twelve flats seeded with forty zinnia seeds (variety State Fair) per flat. Six flats were then infested with 20 ml of *P. ultimum* spores (Puz-SI) and six flats were infested with a bran culture of *R. solani* (1 g/flat) (RS-SF). Control flats with no GlioGard added prior to planting were also prepared. In six more flats, the seeds were planted, inoculated with *Rhizoctonia,* and drenched with 100 ml of Terrachlor 75WP (3 g/l). Six other flats were planted with seeds, inoculated with *Pythium,* and drenched with 100 ml of metalaxyl, *Pythium,* and *Rhizoctonia* (0.75 g/l).

After three weeks, under greenhouse conditions, the plant stand counts were made; the results are presented in Table 16.5, which shows that GlioGard controlled disease as effectively as the chemical fungicide.

GlioGard was also tested against a series of plant pathogen combinations. In these experiments, either seeds (zinnia, vinca, begonia, geranium, celosia, cotton, sugar beet, soybean) or seedlings (snapdragon, poinsettia, chrysanthemum) were tested in either bench soil (composted loamy soil, peat moss, and perlite in the ratio of 3:3:1) or Redi-Earth media. In all cases the GlioGard was incorporated into the media seventy-two hours before planting or transplant. Healthy checks, pathogen checks with untreated media, and fungicide checks were run with each test plant. The results showed that GlioGard gave excellent control of the pathogens tested in two different types of media.

These studies confirm the previously published work of Lumsden and Locke (1988), who showed that alginate prills of GL-21 controlled damping-off on zinnia, cabbage, and cotton; and of Lewis and Papavizas (1987), who showed that GL-21 controlled *Rhizoctonia* damping-off of cotton, sugar beets, and tomatoes. These studies further demonstrate the wide spectrum of control capability. In addition, Beagle-Ristaino and Papavizas (1985) examined GL-21 control of *Rhizoctonia* on potato; Lewis and Papavizas (1985) showed control of *Rhizoctonia* on cotton, sugar beets, and radishes; and Papavizas and Lewis (1988) used alginate prill of GL-21 to control *Sclerotium rolfsii* on snap bean. All of these cases document the ability of GL-21 to control these soil pathogens.

### *Inert Ingredients*

All the inert ingredients in the GL-21 formulation are on the EPA's exempt inserts list and are generally regarded as safe. The conclusion of all of these trials showed that GL-21 is free of acute and chronic toxicological concerns.

Table 16.5. Control of damping-off caused by *Pythium ultimum* and *Rhizoctonia solani* on zinnia in Redi-Earth soilless mix[1]

| Treatment | Percent Healthy Plant Stand | |
|---|---|---|
| | *Pythium* (PUZ-SI) | *Rhizoctonia* (RS-SF) |
| Healthy Check | 100 A | 100 A |
| Pathogen Check | 24 E | 78 C |
| Fungicide Check | 85 B | 99 A |
| GlioGard | 88 A | 94 AB |

1 Letters designate the degree of disease control, with A being the most effective and E being the least effective. Same letter means the degree of effectiveness is statistically simiar.

### *Marketing Assessment*

Commercialization of *G. virens* is dependent on marketing assessment and determination of market availability and profit margins. Several factors were considered for GL-21: (1) Product would be protected by patents pertaining to the formulation that were issued to the USDA and transferred to industry. (2) The glasshouse bedding plant production industry requires safe, reliable treatments for controlling damping-off diseases, and nonchemical, natural biocontrol systems are favored. (3) A simple, inexpensive fermentation system is available for producing biomass of *G. virens* in large commercial-scale fermentors. (4) Biological pest control agents are cheaper to develop, register, and market than chemical control compounds.

Considering all of these factors, W. R. Grace & Co. concluded that production of a product containing *G. virens* was a sound commercial venture for the horticultural market. Test marketing of GlioGard occurred in 1992 and 1993 with selected growers in Texas and Florida. The product performed as expected under commercial conditions and was readily accepted. However, manufacturing costs were higher than expected. Problems with microbiol contamination plagued the formulation process and caused a higher rejection rate than expected. To solve this problem, the formulation process was modified to include a granulation process. This reduced the product's particle size and its color slightly. Due to the change in product appearance, the product's name was changed to Soilgard and released in 1994. Growers throughout such states as Florida, Texas, California, Wisconsin, Washington and Oregon have accepted this natural alternative to chemical fungicides. In 1996 W. R. Grace sold its biopesticide business to Thermo Trilogy Corporation who continues to manufacture and support Soilgard.

The use of this product will be beneficial in the control of damping-off diseases of seedling plants, which are a major source of economic loss in the glasshouse production of ornamental and food-crop plants. The availability of a biological fungicide will provide a less toxic alternative to the currently registered chemical treatments. In addition, this achievement provides the impetus for future biocontrol technology, development, and application.

## References

Aluko, M. O., and T. F. Hering. 1970. The mechanisms associated with the antagonistic relationship between *Corticium solani* and *Gliocladium virens. Transactions of the British Mycological Society* 55:173–79.

Beagle-Ristiano, J. E., and G. C. Papavizas. 1985. Biological control of *Rhizoctonia* stem canker and black scurf of potato. *Phytopathology* 75:560–64.

Betz, F., A. Rispin, and W. Schneider. 1987. Biotechnology products related to agriculture. Overview of regulatory decisions at the U.S. Environmental Protection Agency. *American Chemical Society Symposium Series* 334:316–27. Washington, D.C.: American Chemical Society.

Coulson, J. R., R. S. Soper, and D. W. Williams. 1991. Biological control quarantine: Needs and procedures. In *Proceedings of a workshop sponsored by USDA-ARS.* Springfield, Virginia: National Technical Information Service.

Domsch, K. H., W. Gams, and T. Anderson. 1980. *Compendium of soil fungi.* London: Academic.

Farr, D. F., G. F. Bills, G. P. Chamuris, and A. Y. Rossman. 1989. *Fungi on plants and plant products in the United States.* St. Paul, Minn.: American Phytopathological Society Press.

Federal Register. 1984. Data requirements for pesticide registration; final rule. 49:42856–42905.

Fravel, D. R., J. J. Marois, R. D. Lumsden, and W. J. Connick Jr. 1985. Encapsulation of potential biocontrol agents in an alginate-clay matrix. *Phytopathology* 75:774–77.

Howell, C. R., and R. D. Stipanovic. 1983. Gliovirin, a new antibiotic from *Gliocladium virens,* and its role in the biological control of *Pythium ultimum. Canadian Journal of Microbiology* 29:321–24.

Jones, R. W., and J. G. Hancock. 1987. Conversion of viridin to viridiol by viridin-producing fungi. *Canadian Journal of Microbiology* 33:963–66.

Knauss, J. F. 1992. *Gliocladium virens,* a new microbial for control of *Pythium* and *Rhizoctonia. Florida Foliage* 18:6–7.

Lewis, L. A., and G. C. Papavizas. 1985. Effect of mycelial preparations of *Trichoderma* and *Gliocladium* on populations of *Rhizoctonia solani* and the incidence of damp-off. *Phytopathology* 75:812–17.

______. 1987. Application of *Trichoderma* and *Gliocladium* in alginate pellets for control of rhizoctonia damping-off. *Plant Pathology* 36:438–46.

Lumsden, R. D., and J. A. Lewis. 1989. Selection, production, formulation and commercial use of plant disease biocontrol fungi, problems and progress. In *Biotechnology of fungi for improving plant growth,* ed. J. M. Whipps and R. D. Lumsden, 171–90. Cambridge: Cambridge University Press.

Lumsden, R. D., and J. C. Locke. 1989. Biological control of damping-off caused by *Pythium ultimum* and *Rhizoctonia solani* with *Gliocladium virens* in soilless mix. *Phytopathology* 79:361–66.

Lumsden, R. D., J. C. Locke, S. T. Adkins, J. F. Walter, and C. J. Ridout. 1992. Isolation and localization of the antibiotic gliotoxin produced by *Gliocladium virens* from alginate prill in soil and soilless media. *Phytopathology* 82:230–35.

Lumsden, R. D., J. C. Locke, J. A. Lewis, S. A. Johnston, J. L. Peterson, and J. B. Ristaino. 1990. Evaluation of *Gliocladium virens* for biocontrol of *Pythium* and *Rhizoctonia* damping-off of bedding plants at four greenhouse locations. *Biological Culture Control Tests* 5:90.

Lumsden, R. D., J. C. Locke, and J. F. Walter. 1991. Approval of *Gliocladium virens* by the U.S. Environmental Protection Agency for biological control of *Pythium* and *Rhizoctonia* damping-off. *Petria* 1:138.

Papavizas, G. C. 1985. *Trichoderma* and *Gliocladium*: Biology, ecology and potential for biocontrol. *Annual Review of Phytopathology* 23:23–54.

Roberts, D. P., and R. D. Lumsden. 1990. Effect of extracellular metabolites from *Gliocladium virens* on germination of sporangia and mycelial growth of *Pythium ultimum. Phytopathology* 80:461–65.

Taylor, A. 1986. Some aspects of the chemistry and biology of the genus *Hypocrea* and its anamorphs, *Trichoderma* and *Gliocladium. Proceedings of the Nova Scotian Institute of Science* 36:27–58.

Waring, P., R. D. Eichner, and A. Mulbacher. 1988. The chemistry and biology of the immunomodulating agent gliotoxin and related epipolythythiodioxopiperizines. *Medical Research Review* 8:499–524.

PART THREE

# Management in Situ

# 17

# Parasitoid Foraging from a Multitrophic Perspective: Significance for Biological Control

*W. Joe Lewis and William Sheehan*

## Introduction

The occurrence of a pest within a crop system is an ecological phenomenon that should be treated in an ecological manner. Managed cropping systems, like any other ecological community, are living ecosystems within and among which physical and biological processes are dynamic and interactive. Despite artificialities created by management procedures, many plants, herbivores, predators, pathogens, and weeds that exist within a particular crop have been molded and tightly woven together by natural selection over evolutionary time. This web of relationships is dynamic: each member of a relationship has a repertoire of offensive and defensive maneuvers and countermoves that it can make in response to other members.

An ecologically based approach to pest management recognizes that any external intervention within this system of relationships causes ripples through the system, the net effect of which must be considered in achieving effective control. Further, it recognizes the importance of natural defenses inherent within the cropping system and seeks to harness them as the foundation upon which to add, in a complementary way, other technologies for managing pest problems. Natural enemies of pests and inherent defenses of plants are two of the key components upon which we should build such strategies.

In this chapter, we examine some of what is known about foraging behavior within one group on natural enemies (insect parasitoids) from a multitrophic perspective. We emphasize interactions between the first trophic level (plants) and the third trophic level (parasitoids) because plants provide the context for other interactions and because we believe these interactions have the greatest potential for manipulation in biological control. Interactions within the third trophic level (e.g., with hyperparasitoids and predators and microbes attacking parasitoids) unfortunately have received relatively little attention to date.

We attempt to identify sources of variability in parasitoid responses to plants and to show how understanding of this variation might be used in improving biological control. We draw heavily on two model systems that have been the focus of research in our laboratories: *Microplitis croceipes* and *Cotesia marginiventris.* Both are endoparasitic, koinobiont hymenopteran parasitoids of noctuid larvae. *Microplitis croceipes,* a host specialist, attacks early and middle instars of species of *Heliothis* and *Helicoverpa. Cotesia marginiventris* is a host generalist that attacks early instar noctuid larvae in at least a dozen genera. Both are plant generalists, at least at the population level, attacking hosts on a wide variety of monocot and dicot plants.

## Sources of Variability in Parasitoid Foraging

One of the themes to emerge over the past decade is that parasitoids exhibit considerable behavioral flexibility during foraging. Once thought of as little automatons, parasitoids are increasingly understood to be highly responsive to both intrinsic and extrinsic sources of variation. For parasitoids of feeding stages of herbivorous insects, plants have long been assumed to be important in foraging (e.g., Vinson 1984). However, there were few quantitative data until the development of assays, particularly flight chambers, for directly studying parasitoid responses to plants. Now it is apparent that plants play a central role in host finding for at least some parasitoids and that most parasitoids learn plant cues associated with host presence. We divide sources of variability into those originating within the parasitoid (intrinsic) and those external to the parasitoid (extrinsic), although in practice both intrinsic and extrinsic sources of variability operate simultaneously.

### *Intrinsic Variability in Parasitoid Responses to Plants*

Three sources of intrinsic variation in responsiveness of parasitoids to various foraging cues have been identified: genotypic variation, phenotypic plasticity (i.e., learning), and physiological state (Lewis et al. 1990). All three sources of variation have been shown to affect parasitoid responses to plants.

***Genotypic Variation.*** Genetic components of foraging behavior have been demonstrated for *M. croceipes.* Flight responses to feeding-damaged plants by females from four isofemale lines were tested for three generations in a flight tunnel and found to be strongly dependent on familial origin (Prevost and Lewis 1990). These results suggest that *M. croceipes* females may inherit at least some components of learning ability of plant-derived as well as of host-derived chemicals.

***Learning.*** Associative learning during foraging, i.e., the ability to associate cues with profitable sites, was once thought to be restricted to "higher" animals. Yet associative learning has been demonstrated in virtually every insect that has been investigated (Papaj and Prokopy 1989). Parasitoids are no exception. Lewis and Tumlinson (1988) were first to demonstrate that *M. croceipes* learn to associate and subsequently fly to volatile plant odors when presented with a host-specific, water-extractable, nonvolatile chemical in host frass. Similarly, *Cotesia marginiventris* (Cresson) was shown to associatively learn plant odors after brief contact experiences (Turlings et al. 1989). Even learning of plant cues during emergence of adult parasitoids from cocoons appears to influence subsequent foraging behavior (Herard et al. 1988).

***Physiological State.*** Learning was also found to interact with parasitoid physiological state in the use of airborne plant-derived odors to find food sources. Female *M. croceipes* associatively learned volatile odors presented while they were feeding on sugar water and subsequently flew to those odors (Lewis and Takasu 1990). Furthermore, wasps learned novel plant odors (vanilla or chocolate) associated with separate host and food resources and then flew to the odor associated with the resource that satisfied their current physiological state (hungry or host-deprived) (Lewis and Takasu 1990). These results demonstrate that wasps can perceive and link multiple cues simultaneously, and that the rank order of responses to these learned cues can be varied in accordance with relative need for the different resources.

A general conceptual model was developed to assess collectively three sources of intrinsic, intraspecific variation in parasitoid foraging (Fig. 17.1; Lewis et al. 1990). Included were considerations of genotypic diversity, influence of different physiological state on the responses by individuals, and plasticity of individual parasitoids caused by preadult and adult experiences. Successful use of parasitoids in biological control was hypothesized to depend on properly matching parasitoids' genotypic and phenotypic behavioral traits with the type of environment in which they must forage (Lewis et al. 1990; Lewis and Martin 1990). A related conceptual model focused on parasitoid learning (Vet et al. 1990). This model specified how potential for learning depends on intrinsic rank-order preference: the less preferred a plant the greater the potential for learning to alter responses to cues from that plant.

### *Extrinsic Variability Affecting Parasitoid Responses to Plants*

The plant component of a parasitoid's environment can be the source of several kinds of variation, each potentially affecting parasitoid foraging: (1) inherent variation in plant attractiveness and accessibility (variation between and within plants or plant parts); (2) herbivore-mediated variation in attractive plant cues (the production of systemic semiochemicals used by parasitoids in locating hosts as a result of host feeding); and (3) associational plant variation (variation resulting from associations with other plants).

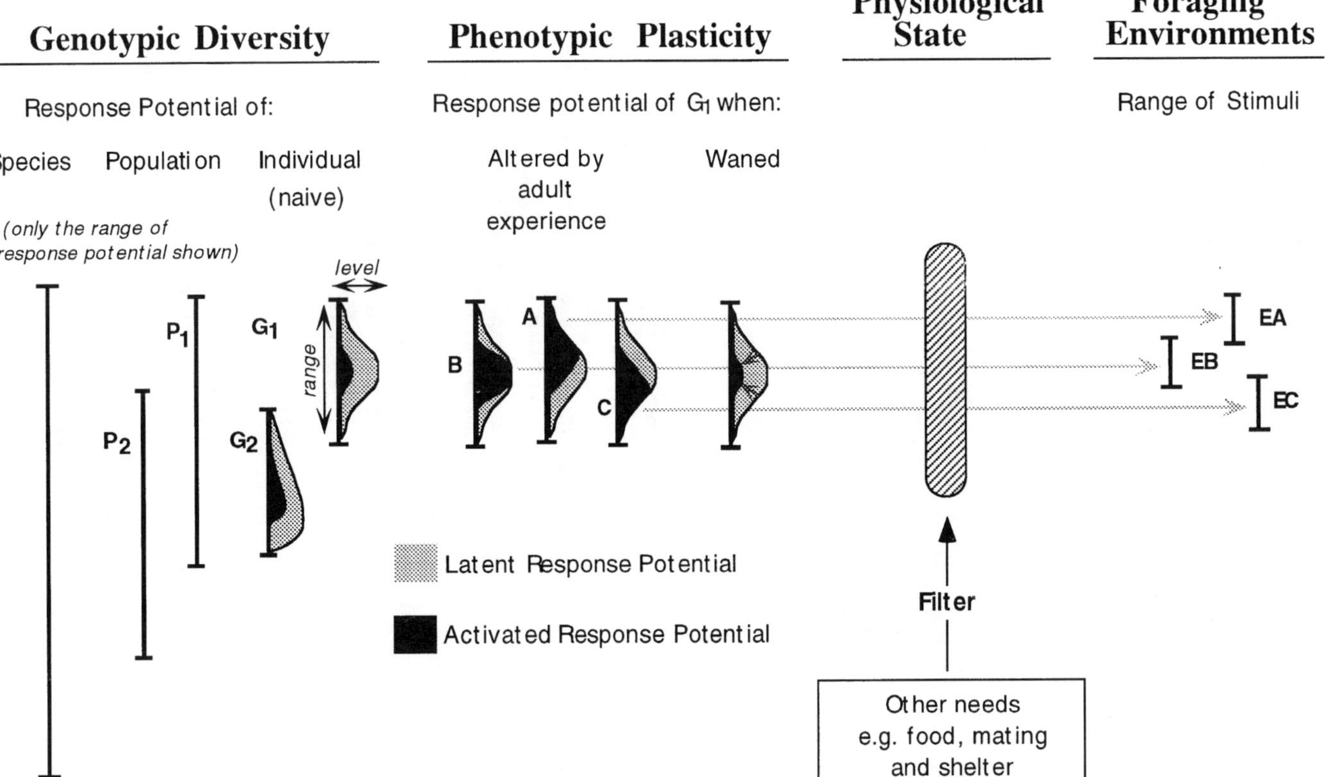


Figure 17.1. A general conceptual model of three sources of intrinsic, intraspecific variation in parasitoid foraging.

***Inherent Variation in Plant Attractiveness and Accessibility.*** Plants and plant parts vary in attractiveness and accessibility to parasitoids and thereby directly affect parasitoid foraging. That plant species differ in attractiveness to parasitoids can be inferred from observations and rearing records (e.g., Lawton 1986; Mueller 1983). Confirmation of differential attractiveness has been provided in laboratory experiments (e.g., Drost et al. 1988; Mueller 1983). Volatiles from cotton leaves and flower buds show different chemical profiles (Turlings et al. 1993), and both *C. marginiventris* and *M. croceipes* can be trained to fly to either plant part, based on profitability of recent foraging experience (Wackers and Lewis 1994). Both species of parasitoid are known to attack hosts feeding on leaves and in buds. Genetic and nutritional variation between individuals of a plant species has been shown to affect herbivore mortality by parasitoids (Letourneau and Fox 1989; Fritz and Nobel 1990; Loader and Damman 1991). Although the authors of these studies speculate that differences in mortality may be caused by treatment effects on parasitoid searching behavior, few data bear directly on this point.

Plants and plant structures are thought to provide refuges from parasitism. For example, *H. zea* is parasitized by *M. croceipes* in exposed, developing heads of whorl stage corn, but hardly at all in the mature ears (unpublished data). Galls provide refuges for some herbivores, and susceptibility to parasitism varies with gall wall thickness resulting both from ontogenetic plant development (Weis and Abrahamson 1985) and from abiotic environmental conditions (Price and Clancy 1986). Considerable evidence suggests that many herbivore species may seek out and exploit host plants less preferred by their natural enemies and with fewer alternative prey species of those natural enemies (Jeffries and Lawton 1984; Hawkins and Gross 1992). Evidence for effects of habitat and host plant selection by parasitoids on herbivore selection of "enemy-free space" is reviewed by Lawton (1986).

***Herbivore-Mediated Variation in Attractive Plant Cues.*** The multitrophic nature of parasitoid foraging is shown most clearly by the recent finding that plant attractiveness to parasitoids increases as a result of host feeding. Working with *C. marginiventris,* Turlings et al. (1990) found that feeding damage stimulated more oriented flights than either frass or larvae. Caterpillar feeding resulted in the production, by the plant, of unique volatile synomones that are used by the parasitoid to locate hosts. The active compounds were not present either at artificially damaged sites or in caterpillar oral secretions; however, when oral secretions were applied to artificially damaged sites, the attractive compounds were produced (Turlings et al. 1990). Recently, a similar role of feeding damage has been demonstrated for *M. croceipes.* At close range (0–10 cm), volatile plant or plant-derived chemicals in caterpillar frass are highly attractive to foraging parasitoids, as noted earlier (Lewis and Tumlinson 1988; Eller et al. 1988). The attractive volatiles in frass are primarily those released by feeding-damaged plants (Turlings et al. 1993).

***Associational Plant Variation.*** A third, less documented source of extrinsic variation in parasitoid response to plants concerns the associations of plants and potential insect hosts in real communities (Price et al. 1980). Monteith (1960) suggested that associated plants might mask the odor of specific host plants, thereby interfering with host location. Zwölfer and Kraus (1957) transplanted unparasitized pupae of fir budworm into artificial leaf rolls on oak; a polyphagous ichneumonid parasitoid of oak tortricids attacked the transplants on oak but not on silver fir growing intermixed with the oak. Others have found higher parasitism in weedy than in simple plant systems. For example, Altieri et al. (1981) found that parasitization by *Trichogramma* spp. was significantly higher in weedy soybeans or in soybeans interplanted with corn than in weed-free soybean monocultures, and Nordlund et al. (1984) found that parasitism by *Trichogramma* spp. was higher on tomato than on adjacent corn or beans. Vegetation characteristics may influence not only host finding by parasitoids, but also patch leaving rates (Sheehan 1989).

## Managing Variability for Increased Biological Control

### *Managing Intrinsic Variability*

Through understanding of the intrinsic variables that affect parasitoid foraging behavior, they potentially can be managed for increased biological control effectiveness.

***Genetic Variability.*** A major hurdle in rearing parasitoids for release in biological control programs is loss of field-adapted traits during culture. While a general decline in vigor would be expected to negatively affect foraging behaviors, typically it is traits like survivorship and fecundity that are measured. However, where parasitoid searching behavior has been measured, it is clear that insects in culture can suffer rapid deterioration of searching behavior (Geden et al. 1992). It will be more difficult to manage the genetic component of foraging variation in inundative (as opposed to inoculative) release programs because of the need to keep parasitoids in culture longer. Natural genetic variation in foraging behavior has been demonstrated (e.g., Prevost and Lewis 1990), but our knowledge is too incomplete for any species to design an effective protocol for screening superior lines of foragers.

***Physiological State.*** Exposing parasitoids to searching kairomones (including plant and plant-derived kairomones) prior to field release changes the motivational state of the parasitoid to an excited state (Gross et al. 1975; Lewis and Martin 1990). Such prerelease stimulation is clearly critical to the establishment of many parasitoids, yet is often neglected. For example, Gross et al. (1975) found that fifteen of sixteen stimulated *M. croceipes* searched plants in field tests, while all but one of twenty-one unstimulated wasps dispersed upon release. Preliminary results from Takasu and Lewis (1993) suggest that manipulating parasitoid hunger levels prior to release may also be used to increase retention in field releases.

***Learning.*** Manipulating phenotypic variation in parasitoids—for example, teaching parasitoids to search particular plants—would seem to have considerable potential for improving effectiveness of released organisms in biological control programs. Only recently have biological control specialists begun to explore ways to incorporate this variable into applied programs. One problem is that a parasitoid that learns one set of information may well abandon (i.e., forget) it when she encounters a nontarget host in the field. Consequently, training parasitoids to search for particular cues may be most relevant to the problem of getting released insects to search the release site long enough to find target hosts. Another problem is that training to artificial cues may reduce performance in natural habitats (Wardle and Borden 1990). Theory may suggest, however, how to select parasitoids with the greatest potential for training: Vet and Dicke (1992) proposed that parasitoids with narrow host ranges but broad host plant ranges should be most phenotypically flexible in foraging behavior. The subject of applying learning to pest management is explored extensively by Prokopy and Lewis (1993).

### *Managing Extrinsic Variability*

Extrinsic plant variability affecting parasitoid foraging may be managed in several ways, including increasing actual plant diversity by intercropping or weedy cropping, adding plant synomones, and altering plant quality.

***Increasing Plant Diversity.*** Increasing agroecosystem diversity through intercropping, no-till agriculture, and so on, may benefit parasitoids by increasing microclimate humidity and plant food (nectar and pollen) availability; generalists may also be more likely to find alternative hosts (Pimentel 1961). On the other hand, odor masking by certain plant combinations could interfere with host location (Monteith 1960), and increased plant density could dilute searching effectiveness (O'Neil and Stimac 1988). Increasing agroecosystem diversity typically results in lower herbivore densities compared with monoculture (Andow 1991), but the pattern for natural enemies is less clear. Sheehan (1986) predicted that agroecosystem diversification should enhance generalist natural enemies more than specialists, a prediction that has received some experimental support (Coll 1993).

***Adding Plant Synomones.*** Application of plant sprays to crops may enhance parasitoid foraging. Altieri et al. (1981) applied crude water extracts of *Amaranthus* sp. and corn to a variety of crops and reported increased parasitization by *Trichogramma* spp. of *H. zea.* There are two major hurdles to overcome. First, the added chemical(s) must not distract parasitoids from finding hosts; second, the chemical(s) should be applied in such a way that parasitoids do not become habituated to it or to them. Limited data suggest that these obstacles can be surmounted. Lewis et al. (1979) found that intermittent application of a host kairomone (moth scale lures) resulted in higher parasitism of moth eggs by *Trichogramma* spp. than continuous application. Sheehan and Lewis (unpublished data) found that application of the attractive plant volatile caryophyllene to cotton in small experimental patches increased retention of released *M. croceipes* without interfering with efficiency. The approach of adding foraging stimuli may be more relevant to short-term foraging (i.e., establishment upon release) than to retaining parasitoids in fields for a long time.

***Manipulation of Plant Qualities.*** Variation in plant quality resulting from plant breeding and fertilization may be manipulated to advantage. Many, perhaps most, elements of plant resistance incorporated into crop plants during breeding programs are likely to be detrimental to parasitoids. For example, sticky trichomes entrap adult parasitoids as well as herbivores (Rabb and Bradley 1968); nectariless cotton (developed to foil adult moth feeding) also deprives parasitoids of needed food for flight; and allelochemicals designed to reduce herbivore feeding may be passed on to developing parasitoids (Campbell and Duffey 1979). However, carefully selected traits, such as *Bt* expression in cotton, may enhance parasitoid activity by extending the window of host vulnerability. Behavioral responses of insects to endogenous plant protectants are complex and need more study (Gould 1988).

## Conclusion

This chapter emphasizes intrinsic variations in the effectiveness of natural enemies. Further, it stresses the interconnectedness of natural enemies with other components of the ecosystem, the dependence of natural enemies on other components of the ecosystem, and the dependence of natural enemies on varied extrinsic resources within the cropping system and in the surrounding vegetation. An understanding of these intrinsic and extrinsic variables and their interdisciplinary management is required in order to ensure reliable performance of natural enemies.

## References

Altieri, M. A., W. J. Lewis, D. A. Nordlund, R. C. Gueldner, and J. W. Todd. 1981. Chemical interactions between plants and *Trichogramma* wasps in Georgia soybean fields. *Protection Ecology* 3:259–63.

Andow, D. A. 1991. Vegetational diversity and arthropod population response. *Annual Review of Entomology* 36:561–86.

Campbell, B. C., and S. S. Duffey. 1979. Tomatine and parasitic wasps: Potential incompatibility of plant antibiosis and biological control. *Science* 205:700–2.

Coll, M. 1993. Response of parasitoids to increased plant species diversity through intercropping. In *Enhancing natural control of arthropod pests through habitat management,* ed. C. H. Pickett and R. L. Bugg.

Drost, Y. C., W. J. Lewis, and J. H. Tumlinson. 1988. Beneficial arthropod behavior mediated by airborne semiochemicals. V. Influence of rearing methods, host plant, adult experience on host-searching behavior of *Microplitis croceipes* (Cresson), a parasitoid of *Heliothis. Journal of Chemical Ecology* 14:1607–16.

Eller, F. J., J. H. Tumlinson, and W. J. Lewis. 1988. Beneficial arthropod behavior mediated by airborne semiochemicals: Source of volatiles mediating the host-location flight behavior of *Microplitis croceipes* (Cresson) (Hymenoptera: Braconidae), a parasitoid of *Heliothis zea* (Boddie) (Lepidoptera: Noctuidae). *Environmental Entomology* 17:745–53.

Fritz, R. S., and J. Nobel. 1990. Host plant variation and mortality of the leaf-folding sawfly on the arroyo willow. *Ecological Entomology* 15:25–35.

Geden, C. J., L. Smith, S. J. Long, and D. A. Rutz. 1992. Rapid deterioration of searching behavior, host destruction, and fecundity of the parasitoid *Muscidifurax raptor* (Hymenoptera: Pteromalidae) in culture. *Annals of the Entomological Society of America* 85: 179–87.

Gould, F. 1988. Evolutionary biology and genetically engineered crops. *BioScience* 38:26–33.

Gross, H. R., W. J. Lewis, R. L. Jones, and D. A. Nordlund. 1975. Kairomones and their use for management of entomophagous insects. III. Stimulation of *Trichogramma achaeae, T. pretiosum* and *Microplitis croceipes* with host-seeking stimuli at time of release to improve their efficiency. *Journal of Chemical Ecology* 1:431–38.

Hawkins, B. A., and P. Gross. 1992. Species richness and population limitation in insect parasitoid-host systems. *American Naturalist* 139:417–23.

Herard, F., M. A. Keller, W. J. Lewis, and J. H. Tumlinson. 1988. Beneficial arthropod behavior mediated by airborne semiochemicals. IV. Influence of host diet on host-oriented flight chamber responses of *Microplitis demolitor* Wilkinson. *Journal of Chemical Ecology* 14:1597–606.

Jeffries, M. J., and J. H. Lawton. 1984. Enemy free space and the structure of ecological communities. *Biological Journal of the Linnean Society* 23:269–86.

Lawton, J. H. 1986. The effect of parasitoids on phytophagous insect communities. In *Insect parasitoids,* ed. J. Waage and D. Greathead, 265–87. New York: Academic.

Letourneau, D. K., and L. R. Fox. 1989. Effects of experimental design and nitrogen on cabbage butterfly oviposition. *Oecologia* 80:211–14.

Lewis, W. J., and W. R. Martin Jr. 1990. Semiochemicals for use with parasitoids: status and future. *Journal of Chemical Ecology* 16:3067–89.

Lewis, W. J., and K. Takasu. 1990. Use of learned odours by a parasitic wasp in accordance with host and food needs. *Nature* 348:635–36.

Lewis, W. J., and J. H. Tumlinson. 1988. Host detection by chemically mediated associative learning in a parasitic wasp. *Nature* 331:257–59.

Lewis, W. J., M. Beevers, D. A. Nordlund, H. R. Gross, and K. S. Hagen. 1979. Kairomones and their use for management of entomophagous insects. IX. Investigations of various kairomone treatment patterns for *Trichogramma* spp. *Journal of Chemical Ecology* 5: 349–60.

Lewis, W. J., L.E.M. Vet, J. H. Tumlinson, J. C. van Lenteren, and D. R. Papaj. 1990. Variations in parasitoid foraging behavior: Essential elements of a sound biological control theory. *Environmental Entomology* 19:1183–93.

Loader, C., and H. Damman. 1991. Nitrogen content of food plants and variability of *Pieris rapae* to natural enemies. *Ecology* 72:1586–90.

Monteith, L. G. 1960. Influence of plants other than food plants of their host on host-finding by tachinid parasites. *Canadian Entomologist* 42:641–52.

Mueller, T. F. 1983. The effect of plants on the host relations of a specialist parasitoid of *Heliothis* larvae. *Entomologia experimentalis et applicata* 34:78–84.

Nordlund, D. A., R. B. Chalfant, and W. J. Lewis. 1984. Arthropod populations, yield and damage in monocultures and polycultures of corn, beans and tomatoes. *Agriculture, Ecosystems and Environment* 11:353–67.

O'Neil, R. J., and J. L. Stimac. 1988. Model of arthropod predation on velvetbean caterpillar (Lepidoptera: Noctuidae) larvae in soybeans. *Environmental Entomology* 17:983–87.

Papaj, D. R., and R. J. Prokopy. 1989. Ecological and evolutionary aspects of learning in phytophagous insects. *Annual Review of Entomology* 34:315–50.

Pimentel, D. 1961. Species diversity and insect population outbreaks. *Annals of the Entomological Society of America* 54:76– 86.

Prevost, G., and W. F. Lewis. 1990. Heritable differences in the response of the braconid wasp *Microplitis croceipes* to volatile allelochemicals. *Journal of Insect Behavior* 3:277–87.

Price, P. W., and K. M. Clancy. 1986. Interactions among three trophic levels: Gall size and parasitoid attack. *Ecology* 67:1593–601.

Price, P. W., C. E. Bouton, P. Gross, B. A. McPheron, J. N. Thompson, and A. E. Weis. 1980. Interactions among three trophic levels: Influence of plants on interactions between insect herbivores and natural enemies. *Annual Review of Ecology and Systematics* 11:41–65.

Prokopy, R. J., and W. J. Lewis. 1993. Application of learning to pest management. In *Insect learning: Ecological and evolutionary perspectives,* ed. D. R. Papaj and A. C. Lewis, 308–42. New York: Chapman and Hall.

Rabb, R. L., and J. L. Bradley. 1968. The influence of host plant on parasitism of eggs of the tobacco hornworm. *Journal of Economic Entomology* 61:1249–52.

Sheehan, W. 1986. Response by specialist and generalist natural enemies to agroecosystem diversification: A selective review. *Environmental Entomology* 15:456–61.

Sheehan, W., and A. Shelton. 1989. Parasitoid response to concentration of herbivore food plants: Finding and leaving plants by *Diaeretiella rapae* (Hymenoptera: Aphidiidae). *Ecology* 70:993–98.

Sheehan, W., F. L. Wackers, and W. J. Lewis. 1993. Discrimination of previously searched host-free sites by *Microplitis croceipes* (Hymenoptera: Braconidae). *Journal of Insect Behavior* 6:323–31.

Takasu, K., and W. J. Lewis. 1993. Host- and food-foraging by the parasitoid *Microplitis croceipes*: Learning and physiological state effects. *Biological Control* 3:70–74.

Turlings, T.C.J., J. H. Tumlinson, and W. J. Lewis. 1990. Exploitation of herbivore-induced plant odors by host-seeking parasitic wasps. *Science* 250:1251–53.

Turlings, T.C.J., J. H. Tumlinson, W. J. Lewis, and L.E.M. Vet. 1989. Beneficial arthropod behavior mediated by airborne semiochemicals. VIII. Learning of host-related odors induced by a brief contact experience with host by-products in *Cotesia marginiventris* (Cresson), a generalist larval parasitoid. *Journal of Insect Behavior* 2:217–25.

Turlings, T.C.J., F. Wackers, L.E.M. Vet, W. J. Lewis, and J. H. Tumlinson. 1993. Learning of host location cues by insect parasitoids. In *Insect learning: Ecological and evolutionary perspectives,* ed. D. R. Papaj and A. C. Lewis, 51–78. New York: Chapman and Hall.

Vet, L.E.M., and M. Dicke. 1992. Ecology of infochemical use by natural enemies in a natural context. *Annual Review of Entomology* 37:141–72.

Vet, L.E.M., W. J. Lewis, D. R. Papaj, and J. C. van Lenteren. 1990. A variable-response model for parasitoid foraging behavior. *Journal of Insect Behavior* 3:471–90.

Vinson, S. B. 1984. How parasitoids locate their hosts: a case of chemical espionage. In *Insect communication,* ed. T. Lewis, 325–48. London: Academic Press.

Wackers, F. L., and W. J. Lewis. 1994. Olfactory and visual learning and this combined influence on host site location by parasitoid *Microplitis croceipes. Biological Control* 4:105–12.

Wardle, A. R., and J. H. Borden. 1990. Learning of host microhabitat form by *Exeristes roborator* (F.) (Hymenoptera: Ichneumonidae). *Journal of Insect Behavior* 3:251–63.

Weis, A. E., and W. G. Abrahamson. 1985. Potential selective pressures by parasitoids on a plant-herbivore interaction. *Ecology* 66:1261–69.

Zwölfer, H., and M. Kraus. 1957. Biocoenotic studies on the parasites of two fir- and oak-tortricids. *Entomophaga* 2:173–96.

# 18

# Altering Community Balance: Organic Amendments, Selection Pressures, and Biocontrol

*Carol E. Windels*

Cultural practices, including addition of organic amendments to soil, augment naturally occurring biological control of plant pathogens (Cook and Baker 1983). Numerous reviews have summarized the beneficial effects of organic amendments on plant growth and disease control (Baker 1991; Baker and Snyder 1965; Bruehl 1975; Cook and Baker 1983; Hoitink and Fahy 1986; Huber and Graham 1992; Jarvis 1992; Linderman 1989; Lumsden et al. 1983b).

In the 1950s and 1960s, considerable research was devoted to establishing principles to explain how crop residues decreased or increased incidence of plant diseases (Lumsden et al. 1983b). Attempts to evaluate amendments in the field that showed potential for reducing disease incidence under controlled conditions in the laboratory or greenhouse, however, often were disappointing. In recent years, concerns about a narrowing arsenal of pesticides, safety of pesticide use, and greater public awareness in conserving the environment have provided new impetus for exploring biological control of plant pests.

Addition of organic matter to soil is one of the most effective ways to change the soil environment and to favor an increase in populations of beneficial soil organisms. Organic materials improve physical properties of soil—aggregate stability, bulk density, porosity, organic matter content, and moisture-holding capacity—that enhance root respiration and growth (Allison 1973). Amendments also are a source of inorganic nutrients to plants and microbes, which, in turn, contribute to soil nutrient availability and soil aggregation. Thus, alterations in physical and chemical properties of soil, as well as in composition of the biological community, directly and indirectly affect the plant and its health.

Soil-borne fungal pathogens are associated with the top few decimeters of soil and thus are potentially amenable to being affected by soil amendments. Recent evidence also indicates that foliar pathogens are affected by soil-incorporated organic amendments or by application to foliage. Weltzien (1990) found that soil incorporation of a compost consisting of horse manure and straw reduced *Erysiphe graminis* on barley and *Sphaerotheca fuliginea* on cucumber. Extracts of the compost also suppressed *Phytophthora infestans* when applied to potato foliage in the field. Organic amendments affect diseases caused by bacteria, viruses, and nematodes (Jarvis 1992), but a discussion of these pathogens is beyond the scope of this chapter.

The use of organic amendments to control soil-borne diseases, while successful in some instances, is poorly understood. Their use appears to be related to suppression of the pathogen in soil or in the rhizosphere through complex interactions that are dependent upon the amendment, pathogen, and environmental factors. This chapter will explore the effects of organic amendments on soil-borne fungal diseases; discuss the direct and indirect effects of organic amendments on the pathogen; provide select examples of successful commercial use of amendments; and briefly highlight some factors that affect disease control.

## Types of Organic Amendments

There is a range of organic amendments of plant and animal origin. Plant residues include remnants of the previous crop, overwintering crops (planted between cultivation of commercial crops, frequently to prevent loss of soil moisture and erosion during the off season), green manure crops (incorporated into soil before crops reach maturity), and amendments (added to soil). Some studies have used simple constituents of plant or animal residues (e.g., sugars and amino acids) or complex materials (e.g., lignin, tannins, and chitin). Animal manures have been applied as raw, treated, or composted materials. Other composts cover an array of materials including municipal solid wastes (leaves, grass clippings, wastepaper, food wastes, land-clearing debris), sewage sludge (several treatments are used to remove human pathogens), and wood-industry wastes. Formulated products are composed of mixtures of these materials, in combination with inorganic materials.

## Effect of Organic Amendments on Plant Pathogens and Disease

### *Increase in Pathogen Number and Disease Incidence*

Sometimes organic amendments result in undesirable effects such as increases in disease incidence (Table 18.1). Some diseases are more severe when the organic material is a suitable substrate for the pathogen. For example, Typhula snowmold occurred when winter wheat was sown after barley was harvested because *Typhula idahoensis* increased its inoculum potential as it saprophytically colonized the barley straw (Huber and Anderson 1976). Incidence of root rot on wheat caused by *Pythium graminicola* increased because the fungus grew on oilcakes of margosa, castor, or groundnut that were added to soil (Kauraw and Singh 1982).

Table 18.1. Examples of organic amendments that increase incidence of soil-borne diseases

| Pathogen | Disease | Crop | Amendment | Citation |
|---|---|---|---|---|
| | | | Crop Residue | |
| *Fusarium nivale*<br>*Typhula idahoensis* | pink snowmold<br>speckled snowmold | winter wheat | *barley, wheat | Huber and Anderson 1976 |
| *Thielaviopsis basicola* | black root rot | cotton | barley | Linderman and Toussoun 1968a, 1968b |
| *Thielaviposis basicola* | black root rot | tobacco | rye, timothy | Patrick and Koch 1963 |
| *Synchytrium endobioticum* | potato wart | potato | *barley, oat | Hampson 1980 |
| | | | Green Manure Crops | |
| *Fusarium oxysporum*, *Pythium* spp., *Rhizoctonia solani* | damping-off | jack pine, black spruce | oat, rye, red clover | Wall 1984 |
| *Aphanomyces cochlioides* | damping-off, root rot | sugar beet | *legumes | Coons *et al.* 1946 |
| | | | Oilcakes | |
| *Pythium graminicola* | root rot | wheat | margosa, castor, groundnut | Kauraw and Singh 1982 |
| | | | Composts | |
| *Fusarium solani* f. sp. *pisi*<br>*Pythium ultimum*<br>*Thielaviopsis basicola* | root rot<br>damping-off<br>black root rot | pea<br>pea<br>bean | municipal sludge | Lumsden *et al.* 1983a |

* Evaluated in the field

In other examples, crop residue decomposition releases soluble and volatile phytotoxins that predispose plants to disease (Patrick and Toussoun 1965). Linderman and Toussoun (1968a, 1986b) found that hydrocinnamic acid, released during decomposition of barley, increased susceptibility of cotton cultivars resistant to *Thielaviopsis basicola.* Also, exposure of sixteen tobacco cultivars to decomposing rye and timothy residues increased susceptibility to Thielaviopsis black root (Patrick and Koch 1963).

### *Decrease in Pathogen Number and Disease Incidence*

Incidence and severity of various diseases (damping-off, root rot, sheath blight, yellows, and wilt) on agricultural, horticultural, floricultural, ornamental, and silvicultural crops have been decreased by crop residues and green manure crops (Table 18.2), constituents of plant or animal residues (Table 18.3), animal manures or composts (Table 18.4), and with formulated mixes (Table 18.5). The soil-borne pathogens and diseases studied are among the most widespread and of major economic importance in the world. Examples include species of *Aphanomyces, Pythium,* and *Phytophthora;* formae speciales of *Fusarium oxysporum; F. solani* f. sp. *phaseoli; Macrophomina phaseoli; Plasmodiophora brassicae; Rhizoctonia solani; Sclerotium rolfsii; Thielaviopsis basicola; Verticillium dahliae;* and *Sclerotina minor.*

Effects of amendments on pathogens and diseases are rarely predictable or consistent. For instance, composted municipal sludge has been shown to increase incidence of Fusarium root rot of pea, Pythium damping-off of pea, and Thielaviopsis black root of bean but decrease incidence of Aphanomyces damping-off of pea, Fusarium wilt of melon, Phytophthora root rot of pepper, Rhizoctonia root rot of bean, and Sclerotinia lettuce drop (Lumsden et al. 1983a). Consequently, the effect of amendments on specific diseases and plant species needs to be evaluated on an individual basis where the crop is grown.

## Organic Amendments and Biological Control

Most reported work on organic amendments has been done in the laboratory and greenhouse, rather than in the field. While these studies have involved unrealistically high amounts of organic matter under artificial conditions, they are of value in that they have enabled researchers to study mechanisms of disease suppression. Organic amendments have been shown to affect pathogens directly by releasing detrimental compounds and decomposition products and indirectly by producing volatile and soluble compounds that alter physical and chemical properties of the soil, affect plant resistance, and provide a substrate for a succession of indigenous microorganisms (and those introduced on the amendment). In turn, antagonistic microbes affect the pathogen via antibiosis, lysis (autolysis and heterolysis), competition, predation, and parasitism. Residues also influence soil fungistasis, which keeps propagules in an inactive state even when they are exposed to a nutrient stimulus (Lockwood 1988).

Table 18.2. Examples of soilborne pathogens where disease incidence was reduced when soil was amended with plant organic matter.

| Pathogen | Disease | Crop | Plant Amendment | Citation |
|---|---|---|---|---|
| *Aphanomyces cochlioides* | damping-off & root rot | sugar beet | *corn, soybean, cereals | Coons *et al.* 1946 |
| | | | oat | Windels and Nielsen 1992 |
| | | | crucifers, soybean | Lewis and Papavizas 1971 |
| *Aphanomyces euteiches* | damping-off & root rot | pea | *kale, cabbage | Papavizas and Ayers 1974 |
| | | | *white mustard | Muehlchen *et al.* 1990 |
| | | | *oat | Tu and Findlay 1986 |
| *Fusarium oxysporum* f.sp. *conglutinans* | yellows | cabbage | *dry cabbage residues (and soil solarization) | Ramirez-Villapudua and Munnecke 1987 |
| *Fusarium oxysporum* f.sp. *radicis lycopersici* | crown & root rot | tomato | **lettuce | Jarvis and Thorpe 1981 |
| *Gaeumannomyces graminis* | take-all | wheat | *oat | Huber and McCay-Buis 1993 |
| *Macrophomina phaseoli* | root rot | cotton | alfalfa meal, barley straw | Ghaffar *et al.* 1969 |
| *Phytophthora cinnamoni* | root rot | avocado | alfalfa meal | Gilpatrick 1969 |
| *Rhizoctonia solani* | root rot | snap bean | cereals, legumes | Davey and Papavizas 1959 |
| | hypocotyl rot | snap bean | *corn | Manning and Crossan 1969 |
| *Streptomyces scabies* | scab | potato | *barley | Weinhold *et al.* 1964 |
| *Thielaviopsis basicola* | root rot | bean, tobacco | alfalfa, cabbage, corn stover | Papavizas and Lewis 1971 |
| | red root | sesame | alfalfa hay, cabbage, corn stover | Adams 1971 |
| *Verticillium dahliae* | wilt | potato | *sudangrass | Davis *et al.* 1994 |

* Evaluated in the field
** Evaluated in commercial greenhouse

Table 18.3. Examples of soilborne pathogens where disease incidence was reduced when soil was amended with constituents of plant or animal residue.

| Pathogen | Disease | Crop | Plant or Animal Constituent | Citation |
|---|---|---|---|---|
| *Aphanomyces cochlioides* | root rot | pea | DL-ß-aminobutryic acid & DL-ß-methylaspartic acid | Papavizas and Ayers 1974 |
| *Fusarium oxysporum* f.sp. *corianderii* | wilt | coriander | oilcakes (linseed, groundnut, sesame) | Srivastava and Sinha 1971 |
| *Fusarium oxysporum* f.sp. *lycopersici* | wilt | tomato | *crab shell powder | Homma *et al.* 1979a |
| *Fusarium oxysporum* f.sp. *raphani* | yellows | radish | *rice straw, fish and shellfish dust | Kato *et al.* 1982 |
| *Fusarium solani* f.sp. *phaseoli* | root rot | snap bean | maltose, dextran, starch cellulose | Papavizas *et al.* 1968 |
| | | | tannins | Lewis and Papavizas 1968 |
| *Rhizoctonia solani* | sheath blight | rice | neem cake, mustard cake, rice chaff, sawdust | Kannaiyan and Prasad 1981 |
| | | | *oilcakes (neem, marotti, rubber seed, punna), coconut pith | Rajan 1980 |
| *Thielaviopsis basicola* | root rot | bean | chestnut, tannin, lecithin | Papavizas *et al.* 1970 |
| *Verticillium dahliae* | wilt | strawberry | chitin, laminarin | Jordan *et al.* 1972 |

* Evaluated in the field

Table 18.4. Examples of soil-borne pathogens where disease incidence was reduced when soil was amended with animal manures or composts.

| Pathogen | Disease | Crop | Amendment | Citation |
|---|---|---|---|---|
| | | | Manure | |
| *Fusarium oxysporum* f.sp. *corianderii* | wilt | coriander | farmyard | Srivastava and Sinha 1971 |
| *Fusarium oxysporum* f.sp. *cucumerium* | wilt | cucumber | fowl + soil fumigation | Wicks *et al.* 1978 |
| *Fusarium oxysporum* f.sp. *lycopersici* | wilt | tomato | *chicken | Homma *et al.* 1979a |
| *Pythium ultimum* | damping-off | lettuce | *chicken (composted) and solarization | Gamliel and Stapleton 1992 |
| *Rhizoctonia solani* | sheath blight | rice | farmyard | Kannaiyan and Prasad 1981 |
| | | | Compost | |
| *Aphanomyces euteiches* | root rot | pea | sewage sludge | Lumsden *et al.* 1983a |
| *Fusarium* species | wilt | floricultural, ornamental crops | **bark | Hoitink *et al.* 1991 |
| *Fusarium oxysporum* f.sp. *corianderii* | wilt | coriander | municipal waste | Srivastava and Sinha 1971 |
| *Fusarium oxysporum* f.sp. *melonis* | wilt | melon | sewage sludge | Lumsden *et al.* 1983a |
| *Pythium* species | damping-off<br>root rot | floricultural & ornamental crops | **bark | Hoitink *et al.* 1991 |
| *Pythium ultimum* | root rot | bean, pea, sugar beet<br>bean, pea | organic household waste<br>domestic waste | Schüler *et al.* 1989<br>van Assche and Uyttebroeck 1981 |

| | | | | |
|---|---|---|---|---|
| *Phytophthora* species | root rot | floricultural & ornamental crops | **bark | Hoitink *et al.* 1991 |
| *Phytophthora capsici* | crown rot | pepper | sewage sludge | Lumsden *et al.* 1983a |
| *Phytophthora palmivora* | root rot | papaya | yard waste | Barkdoll *et al.* 1992 |
| *Rhizoctonia solani* | damping-off | bean, pea | domestic waste | van Assche and Uyttebroeck 1981 |
| *Rhizoctonia solani* | root rot | bean, cotton, radish | sewage sludge | Lumsden *et al.* 1983a |
| *Rhizoctonia solani* | root rot | floricultural, ornamental crops | **bark | Hoitink *et al.* 1991 |
| *Sclerotinia minor* | lettuce drop | lettuce | *sewage sludge | Lumsden *et al.* 1986 |

* Evaluated in the field

** Used commercially in greenhouse industry

Table 18.5. Soil-borne pathogens where disease incidence was reduced when soil was amended with formulated mixes in the field

| Pathogen | Disease | Crop |
|---|---|---|
| S-H Mix[a] | | |
| *Fusarium oxysporum* f.sp. *conglutinans* | yellows | cabbage |
| *Fusarium oxysporum* f.sp. *niveum* | wilt | watermelon |
| *Fusarium oxysporum* f.sp. *raphani* | yellows | radish |
| *Plasmodiophora brassicae* | clubroot | Chinese cabbage |
| *Rhizoctonia solani* | blight | bean |
| *Rhizoctonia solani* | sheath blight | rice |
| *Sclerotium rolfsii* | southern blight | pepper |
| SF-21 Mix[b] | | |
| *Fusarium moniliforme* var. *subglutinans* | damping-off | slash pine seedlings |
| *Pythium aphanidermatum* | damping-off | slash pine seedlings |
| *Rhizoctonia solani* | damping-off | slash pine seedlings |

[a] S-H Mix composition: 4.4% bagasse, 8.4% rice husks, 4.25% oyster shell powder, 8.25% urea, 1.04% potassium nitrate, 13.16% calcium superphosphate, and 60.5% mineral ash; incorporated into soil, 900-1,200 kg/ha (Sun and Huang 1985)

[b] SF-21 Mix composition: 750 g milled pine bark, 150 g $Al_2(SO_4)_3$, 25g KCl, 30 g $CaCl_2$, 10 g triple superphosphate, 35 g $(NH_4)_2SO_4$, and 750 ml 10% glycerine; incorporated into soil, 1% w/w (Huang and Kuhlman 1991a)

Certain segments of the soil microbe population that affect plant pathogens can be enhanced (Linderman 1989; Lumsden et al. 1983b; Owens et al. 1969), but there is a dearth of information on the relative specific effects. The overall effect of organic amendments in the biological control of soil-borne fungal pathogens include (1) reducing the inoculum density, (2) suppressing disease although the pathogen population is not reduced, and (3) preventing buildup of inoculum.

### *Reduce Inoculum of Pathogen*

An example of reducing inoculum of a pathogen by addition of a soil amendment was demonstrated by Papavizas (1968). He amended soil infested with *Thielaviopsis basicola* with 1 percent ground alfalfa hay (w/w) and planted snap bean after soil had incubated for one, five, and twelve weeks. At planting time, populations of *T. basicola* in the amended soil (regardless of incubation period) averaged less than 200 propagules/g of soil compared to the control, which averaged 17,000, 8,000, and 7,000 propagules/g of soil incubated for one, five, and twelve weeks respectively. The alfalfa amendment stimulated maximum germination of endoconida and chlamydospores of *T. basicola* within eighteen hours, which was followed by microbial lysis of germ tubes (Adams and Papavizas 1969). After forty-eight hours, nearly all germ tubes of both spore types had lysed. Alfalfa also suppressed disease by soil fungistasis, in that addition of more alfalfa prevented spores of the pathogen from germinating.

### *Suppress Disease Without Reducing the Pathogen Population*

Lewis and Papavizas (1968) added various tannins (0.5 percent) to soil (w/w) infested with *Fusarium solani* f. sp. *phaseoli.* Three weeks after soils were amended with canaigre tannin, gallotannin, or wattle tannin, pathogen populations increased and averaged 744,000, 296,000, and 332,000 propagules/g of soil, respectively, compared to 80,000 propagules/g of soil in the unamended control. However, root rot indices (0–4 scale) of snap bean planted in soils amended with canaigre tannin, gallotannin, and wattle tannin were 0.4, 0.5, and 0.6, respectively, compared to 2.3 in the unamended control. The authors concluded that tannins probably prevented chlamydospore germination by increasing soil fungistasis.

Homma et al. (1979a) found that addition of chicken manure to field soil suppressed Fusarium wilt of tomato without affecting the pathogen population. Wilt suppression was attributed to increased populations of antagonistic microbes that inhibited conidial germination in the soil, rhizosphere, and rhizoplane and by toxins in the amendment (Homma et al. 1979b).

Similarly, Huisman and Davis (1992) and Davis et al. (1993, 1994) showed a significant reduction of both Verticillium wilt and *V. dahliae* root infections of potato following several green manure treatments. These reductions occurred although inoculum densities of soil-borne *V. dahliae* were not reduced. In fact, *V. dahliae* colonized roots of all green manure crops tested (Huisman, unpublished data). Infection of potato roots by *V. dahliae* was negatively correlated with both nonspecific microbial activities and major changes in populations of nonpathogenic *Fusarium equiseti* and *F. oxysporum.*

In other instances, disease has been suppressed by increasing availability of essential nutrients to plants. In Indiana, growing oat as a precrop to winter wheat in the field reduced take-all caused by *Gaeumannomyces graminis* var. *tritici* by increasing availability of manganese (Huber and McCay-Buis 1993). Rhizosphere soil from take-all infected plants had a large population of manganese-oxidizing organisms compared to rhizosphere soil from healthy wheat plants collected where oat had previously been cropped.

The role of other soil nutrients (micro- and macronutrients) derived from organic amendments may also significantly affect disease (Huber and Graham 1992) but have not been explored. However, it has been shown that nitrogen has a marked effect on disease. Organic amendments stimulate microbial activity, which depletes the concentration of soil nitrogen, or form of nitrogen, so that the infection process of the pathogen is impaired.

### *Prevent Buildup of Pathogen Population*

An example of a plant toxin directly preventing buildup of a pathogen is the saponin avenacin, which is released from undamaged oat roots and is implicated in resistance to several pathogens. Deacon and Mitchell (1985) exposed zoospores of several *Pythium* spp. and an *Aphanomyces* sp. to oat roots. Zoospores accumulated around roots, but did not encyst. Within one to two minutes of exposure, zoospores were immobilized and spherical in shape within 200 µm of undamaged roots and within 5 to 10 mm of damaged oat roots. Zoospores then developed dark internal granules that aggregated into one or two groups while vacuoles developed in another portion of the cytoplasm. By five to six minutes after immobilization, zoospores disintegrated and extruded their contents from the vacuolated region or formed a balloon-like appendage that lysed. Nearly all zoospores lysed within ten to fifteen minutes of immobilization. A commercial preparation of ß-aescin (a saponin) immobilized and lysed zoospores in the same way.

Soil amendments also support microbes that produce antibiotics inhibitory to pathogens and thereby prevent buildup of inoculum. Weinhold et al. (1964) found that growing soybean as a green manure crop prevented buildup of scab (*Streptomyces scabies*) of potato, but that it was ineffective in reducing disease in soils where a high population of the pathogen was established. Populations of *Bacillus subtilis* antagonistic to *S. scabies* predominated in plots where soybean had grown and where disease was suppressed (Weinhold and Bowman 1968). *Streptomyces scabies* was more sensitive to the antibiotic produced by *B. subtilis* (similiar to bacitracin) than were nonpathogenic *Streptomyces* species.

Organic amendments also can indirectly prevent buildup of pathogens by affecting their competitive saprophytic ability. Papavizas and Davey (1960) found that *Rhizoctonia solani* colonized 8 percent and 5 percent of buckwheat stems buried in soil amended with corn and oat, respectively, compared to 60 percent of buckwheat stems in nonamended soil. Snap bean had previously grown in these soils, and rhizosphere populations of streptomycetes antagonistic to *R. solani* were eightfold greater in both the corn- and oat-amended soils compared to the nonamended soil.

## Successful Use of Organic Amendments

There are several examples where diseases are controlled in commercial fields and in the greenhouse industry by amending soils with organic amendments. One successful approach has been to reconstitute a natural, suppressive environment in the field. This strategy is based on minimal inputs to stimulate a high degree of biological diversity and complex interactions in the soil. Another avenue has been to create an artificial, suppressive environment. This has been particularly effective in the greenhouse, where high inputs and low biological diversity occur. In the following examples, the pathogens are controlled through the mechanisms outlined previously but occur in concert with other complex or unique factors in the environment.

***Reconstitute a Natural, Suppressive Environment***

A well-documented example is the control of Phytophthora root rot on avocado (Cook and Baker 1983). Since the late 1960s, avocado growers in Queensland and New South Wales, Australia, have been controlling this disease by adding chicken manure (0.73 t/ha twice annually), by continuously growing legumes and maize as cover crops, and by adding dolomitic limestone if the soil pH is less than 6. These measures simulate rain forest conditions (avocado is a rain forest tree in Central America) and are based on the initial observation by G. Ashburner that Phytophthora root rot occurred in cultivated groves with low organic matter, but not in the rain forest. Suppressiveness of these soils is mediated by several mechanisms: stimulation of a large, heterogenous population of spore-forming, lytic bacteria and actinomycetes; a combination of high organic matter and calcium, which improves soil structure and drainage; high calcium availability, which increases plant resistance to the pathogen; and a high nitrate-nitrogen concentration that is inhibitory to *P. cinnamoni.* However, no single factor accounts for suppression of this disease.

Another example of amendments stimulating complex interactions in soil is the use in Taiwan of Sun-Huang (S-H) mix, which includes wastes from the agriculture, aquaculture, and steel industries as main sources of organic and inorganic matter (Sun and Huang 1985). Ingredients of the patented formulation include 4.4 percent bagasse, 8.4 percent rice husks, 4.25 percent oyster shell powder, 8.25 percent urea, 1.04 percent potassium nitrate, 13.16 percent calcium superphosphate, and 60.5 percent mineral ash (31 percent silicon dioxide, 44 percent calcium oxide, 1.7 percent magnesium oxide, 18 percent aluminum oxide, and 1 percent ferrous oxide). When applied at 900 to 1,200 kg/ha in the field, the S-H mix reduces Fusarium wilts of cabbage, watermelon, and radish, and also is effective in controlling a wide range of other pathogens on vegetable crops (Table 18.5). Mechanisms in suppression of Fusarium wilts by the S-H formulation are complex and incompletely understood. The inorganic components suppress *Fusarium* directly by inhibiting spore germination and enhancing lysis of germ tubes. The inorganic and organic components result in a twenty-five- and a twofold increase in fungus and actinomycetes populations, respectively, but with no effect on bacterial populations. The formulation increases soil pH, which decreases incidence of Fusarium wilt; calcium enhances host resistance; and there is an overall stimulation of root growth.

Formulations that suppress soil-borne diseases may affect pathogens differently. Huang and Kuhlman (1991a, 1991b) developed a formulation that inhibited *Rhizoctonia solani* damping-off of slash pine seedlings by stimulating the proliferation of antagonistic microbes, particularly *Trichoderma harzianum* and *Penicillium ox-*

*alicum.* However, decrease in incidence of damping-off of slash pine seedlings caused by *Pythium aphanidermatum* was directly influenced by inorganic and organic components of the formulation and indirectly influenced by a lower soil pH and stimulation of the microbial population.

### *Create an Artifical, Suppressive Environment*

During the 1960s the nursery industry in the United States began testing wood industry waste as a substitute for peat in container media. Today, composted pine bark is the most widely used peat substitute potting material in commercial greenhouses. Pine bark also has the additional benefit of suppressing several common soil-borne fungal diseases (Hoitink and Fahy 1986; Hoitink et al. 1991).

A brief review of the three-phase, aerobic composting process is outlined to understand the significance of this amendment (Hoitink et al. 1991). In the first twenty-four to forty-eight hours, the temperature rises to 40° to 50°C and destroys sugars and easily biodegradable compounds. In the second phase, temperatures are maintained at 40° to 65°C, which destroys cellulose, less biodegradable compounds, and organisms (plant pathogens, weed seeds, biocontrol agents, except *Bacillus*). During this time, compost piles are frequently turned to expose all parts to high temperatures and ensure a homogeneous product. The third phase is the curing process. Concentrations of readily biodegradable wastes decline and, therefore, heat output and temperatures decline. Mesophilic microorganisms, including biocontrol agents (e.g., species of *Bacillus, Enterobacter, Flavobacterium, Pseudomonas, Streptomyces, Trichoderma* and *Gliocladium*), recolonize the compost. Mature composts consist of lignins, humic substances, and biomass, which are unsuitable substrates for pathogens.

Bark and wood composts are formulated into soil potting mixes, usually 20 percent weight/volume, but pine bark can be as high as 80 percent (Hoitink et al. 1991). Composted pine bark and sphagnum moss are the primary organic components used in preparing potting mixes naturally suppressive to soil-borne plant pathogens. There is a "general suppression" of damping-off and root rot of ornamental and floricultural crops caused by *Pythium* and *Phytophthora,* which are nutrient-dependent pathogens. Composts have a "carrying capacity" where the high microbial activity and biomass suppress germination of sporangia of both pathogens, presumably through microbiostasis. Microbial activity of potting mixes is measured on the basis of rate of hydrolysis of fluorescein diacetate to predict suppressiveness to diseases caused by *Pythium.* The use of composts in "specific suppression" of diseases caused by *Rhizoctonia solani* is more difficult. *Rhizoctonia solani* is a nutrient-independent pathogen that colonizes composts while they mature and is suppressed by a select group of microbes. Addition of specific organisms, such as *Trichoderma* and *Gliocladium,* is beneficial in rendering the compost suppressive to this pathogen.

Growers in the eastern United States use soil mixes supplemented with pine bark to control Fusarium wilt of cylamen (Hoitink et al. 1991). Although the mechanisms of disease suppression are not understood, this biocontrol option is important to the industry because there are no effective fungicides or pathogen-free production techniques.

Mechanisms of disease suppression in composted bark involve several interacting, complex factors (Hoitink et al. 1991). Composts improve the physical structure of the soil, thereby improving aeration in the root system. They support high populations of antagonists based on a microbial "carrying capacity" and phagous organisms (microarthropods) that readily colonize organic matter (multiple species are better than single species). Composts generally are an inadequate food base for pathogens and also release fungicidal-fungistatic compounds.

The odds of improving use of organic amendments to suppress disease in the field may be more easily increased by creating temporary conditions that favor reduction of pathogen populations. The addition of organic amendments to field soils that have been pretreated or cotreated with solarization or sublethal concentrations of soil fumigants looks promising. For example, populations of *Fusarium oxysporum* f. sp. *conglutinans* and symptoms of cabbage yellows were undetected in field plots that previously had been amended with cabbage (1 percent w/w) and solarized for five weeks (Ramirez-Villapudua and Munnecke 1987). The crucifer amendment with solar heating was more effective than either treatment alone. Pathogen populations were nearly eliminated; total fungi decreased by about 20 percent; actinomycetes were unaffected; and bacterial populations increased sixteenfold (Ramirez-Villapudua and Munnecke 1988). The authors proposed that the gases from decomposing cabbage stimulated chlamydospore germination of *F. oxysporum* f. sp. *conglutinans,* which then were attacked by antagonistic soil microbes (stimulated by, or resistant to, the volatiles). Gamliel and Stapleton (1992) also found that solarization of soil supplemented with composted chicken manure improved control of *Pythium ultimum* and *Meloidogyne incognita* and gave lettuce yields greater than those for either treatment alone.

## Factors Affecting Effectiveness of Organic Amendments

Among the factors that affect populations of antagonists and disease control is the amendment itself. Organic materials vary in their chemical nature, in age and maturity, in carbon to nitrogen ratio, and in production of phytotoxins. The amount of the material added to soil, its placement and distribution in the soil profile, and time between incorporation and planting are important. Ideally, application of an amendment should be timed to reduce the pathogen population or to allow maximum protection from disease (peak antagonist activity) when the plant is most vulnerable to attack.

Pathogens are differentially affected by organic amendments depending on the type of pathogen and its survival during saprophytic and overwintering stages and during infection. Pathogens are least tolerant to antagonists just after germination and most tolerant in the resting stage (Patrick and Toussoun 1965). Decomposing amendments can reduce the competitive or saprophytic ability of a pathogen (Papavizas and Davey 1960). An excessively high population of pathogen, or unusually favorable conditions for disease development, can negate beneficial effects of amendments. Also, there are trade-offs: Papavizas (1966) found that cruciferous amendments reduced Aphanomyces root rot but increased Rhizoctonia root rot on pea.

Soil and environmental factors profoundly affect decomposition and effectiveness of organic amendments in disease control (Lumsden et al. 1983b). Soil types differ in chemical and physical properties and in biological composition. Among a multitude of factors, temperature, moisture, and aeration critically affect the decomposition process, microbial activity, and the by-products that are formed.

## Conclusions

Organic amendments consist of plant residues, green manure crops, constituents of plant or animal residues, animal manures, composts, and formulations. Organic amendments affect pathogens as a direct food source, stimulate antagonistic organisms, and improve the physical properties of soil and nutrient status of the crop, which can enhance disease resistance. The succession of organisms that utilize a substrate during decomposition affect the activity, survival, and population of pathogens and disease suppression through antibiosis and lysis, competition, parasitism, or predation. Populations of soil bacteria (including actinomycetes) and fungi are affected by the type and maturity of the amendment, kinds and amounts of decomposition products, proportion of available nutrients in relation to resistant components, carbon to nitrogen ratio of soil, and the physical and chemical environment.

Considerable evidence has accumulated to show that application of organic amendments has potential, and is a feasible approach, to control plant diseases. However, the effects are usually indirect. That is, amendments affect plants and soil microorganisms that in turn affect pathogens and disease. Moreover, certain combinations of host and pathogen behave differently in response to organic amendments and must be studied individually to arrive at practical disease control measures. Part of the difficulty in making recommendations is that more information is needed to elucidate factors that affect disease development in specific host-pathogen combinations in relation to organic amendments, as well as factors that affect the ecology of soil microorganisms. Current information suggests that complete control of any disease depends on a combination of environmental and biological inputs, of which organic amendments is but one. It is imperative to understand how organic amendments alter the physical, chemical, and biological composition of soil so that optimal plant productivity and consistent, effective control of soil-borne plant pathogens can be attained.

## References

Adams, P. B. 1971. Effect of soil temperature and soil amendments on Thielaviopsis root rot of sesame. *Phytopathology* 61:93–97.

Adams, P. B., and G. C. Papavizas. 1969. Survival of root-infecting fungi in soil. X. Sensitivity of propagules of *Thielaviopsis basicola* to soil fungistasis in natural and alfalfa-amended soil. *Phytopathology* 59:135–38.

Allison, F. E. 1973. *Soil organic matter and its role in crop production.* Amsterdam: Elsevier.

van Assche, C., and P. Uyttebroeck. 1981. The influence of domestic waste compost on plant diseases. *Acta Horticulturae* 126:169–78.

Baker, K. F., and W. C. Snyder. 1965. *Ecology of soil-borne plant pathogens.* Berkeley: University of California Press.

Baker, R. 1991. Biological control: Eradication of plant pathogens by adding organic amendments to soil. In *CRC handbook of pest management in agriculture,* ed. D. Pimentel, 299–309. Boca Raton, Fla.: CRC Press.

Barkdoll, A. W., D. J. Mitchell, P. A. Rayside, and R. A. Nordstedt. 1992. Effect of yard waste compost on Phytophthora root rot of papaya. *Phytopathology* 82:1087–88.

Bruehl, G. W. 1975. *Biology and control of soil-borne plant pathogens.* St. Paul, Minn.: American Phytopathological Society Press.

Cook, R. J., and K. F. Baker. 1983. *The nature and practice of biological control of plant pathogens.* St. Paul, Minn.: American Phytopathological Society Press.

Coons, G. H., J. E. Kotila, and H. W. Bockstahler. 1946. Black root of sugar beets and possibilities for its control. *American Society of Sugar Beet Technology Proceedings* 4:364–80.

Davey, C. B., and G. C. Papavizas. 1959. Effect of organic soil amendments on the Rhizoctonia disease of snap beans. *Agronomy Journal* 51:493–96.

Davis, J. R., O. C. Huisman, L. H. Sorensen, and A. T. Schneider. 1993. Increases of disease incidence on potato by *Verticillium dahliae* in fallow field soils. Abstracts for 6th International Congress of Plant Pathology, 265. Palais des Congrès de Montreal.

Davis, J. R., O. C. Huisman, D. T. Westermann, L. H. Sorensen, A. T. Schneider, and J. C. Stark. 1994. The influence of cover crops on the suppression of Verticillium wilt of potato. In *Advances in potato pest biology and management,* ed. G. W. Zehnder, M. L. Powelson, R. K. Jansson, and K. V. Ramen, 332–41. St. Paul, Minn.: American Phytopathological Society.

Deacon, J. W., and R. T. Mitchell. 1985. Toxicity of oat roots, oat root extracts, and saponins to zoospores of *Pythium* spp. and other fungi. *Transactions of the British Mycological Society* 84:479–87.

Gamliel, A., and J. J. Stapleton. 1992. Effect of organic amendments and solarization on pathogen control, rhizosphere microbiology, plant growth, and volatiles in soil. *Phytopathology* 82:1115.

Ghaffar, A., G. A. Zentmeyer, and D. C. Erwin. 1969. Effect of organic amendments on severity of Macrophomina root rot of cotton. *Phytopathology* 59:1267–69.

Gilpatrick, J. D. 1969. Effect of soil amendments upon inoculum survival and function in Phytophthora root rot of avocado. *Phytopathology* 59:979–85.

Hampson, M. C. 1980. Pathogenesis of *Synchytrium endobioticum.* Effect of soil amendments and fertilization. *Canada Journal of Plant Pathology* 2:148–51.

Hoitink, H.A.J., and P. C. Fahy. 1986. Basis for the control of soilborne plant pathogens with composts. *Annual Review of Phytopathology* 24:93–114.

Hoitink, H.A.J., Y. Inbar, and M. J. Boehm. 1991. Status of compost-amended potting mixes naturally suppressive to soilborne diseases of floricultural crops. *Plant Disease* 75:869–73.

Homma, Y., C. Kubo, M. Ishii, and K. Ohata. 1979a. Effect of organic amendments to soil on suppression of disease severity of tomato *Fusarium*-wilt. *Bulletin of the Shikoku Agriculture Experiment Station* 34:89–101. (In Japanese, with English summary)

______. 1979b. Mechanism of suppression of tomato *Fusarium*-wilt by amendment of chicken manure to soil. *Bulletin of the Shikoku Agriculture Experiment Station* 34:103–21 (In Japanese, with English summary)

Huang, J. W., and E. G. Kuhlman. 1991a. Formulation of a soil amendment to control damping-off of slash pine seedlings. *Phytopathology* 81:163–70.

______. 1991b. Mechanisms inhibiting damping-off pathogens of slash pine seedlings with a formulated soil amendment. *Phytopathology* 81:171–77.

Huber, D. M., and G. R. Anderson. 1976. Effect of organic residues on snowmold of winter wheat. *Phytopathology* 66:1028–32.

Huber, D. M., and R. D. Graham. 1992. Techniques for studying nutrient-disease interactions. In *Methods for research on soilborne phytopathogenic fungi,* ed. L. J. Singleton, J. D. Mihail, and C. M. Rush, 204–14. St. Paul, Minn.: American Phytopathological Society Press.

Huber, D. M., and T. S. McCay-Buis. 1993. A multiple component analysis of the take-all disease of cereals. *Plant Disease* 77:437–47.

Huisman, O. C., and J. R. Davis. 1992. Effect of cropping sequences on the efficiency of colonization of potato roots by *Verticillium dahliae* and *Fusarium* species. *Phytopathology* 82:1089.

Jarvis, W. R. 1992. *Managing diseases in greenhouse crops.* St. Paul, Minn.: American Phytopathological Society Press.

Jarvis, W. R., and H. J. Thorpe. 1981. Control of fusarium foot and root rot of tomato by soil amendment with lettuce residues. *Canada Journal of Plant Pathology* 3:159–62.

Jordan, V.W.L., B. Sneh, and B. P. Eddy. 1972. Influence of organic soil amendments on *Verticillium dahliae* and on the microbial composition of the strawberry rhizosphere. *Annals of Applied Biology* 70:139–48.

Kannaiyan, S., and N. N. Prasad. 1981. Effect of organic amendments on seedling infection of rice caused by *Rhizoctonia solani. Plant and Soil* 62:131–33.

Kato, K., M. Fukaya, and I. Tomita. 1982. Effect on control of radish yellows by successive use of some soil amendment input. *Research Bulletin of the Aichi Agriculture Research Center* 14:162–69 (In Japanese, with English summary)

Kauraw, L. P., and R. S. Singh. 1982. Effect of organic amendment of soil on the incidence of root rots of wheat. *Indian Journal of Mycology and Plant Pathology* 12:271–77.

Lewis, J. A., and G. C. Papavizas. 1968. Survival of root-infecting fungi in soil. VII. Decomposition of tannins and lignins in soils and their effects on Fusarium root rot of bean. *Phytopathologische Zeitschrift* 63:124–34.

______. 1971. Damping-off of sugarbeets caused by *Aphanomyces cochlioides* as affected by soil amendments and chemicals in the greenhouse. *Plant Disease Reporter* 55:440–44.

Linderman, R. G. 1989. Organic amendments and soil-borne diseases. *Canadian Journal of Plant Pathology* 11:180–83.

Linderman, R. G., and T. A. Toussoun. 1968a. Breakdown in *Thielaviopsis basicola* root rot resistance in cotton by hydrocinnamic (3-phenylpropionic) acid. *Phytopathology* 58:1431–32.

______. 1968b. Predisposition to Thielaviopsis root rot of cotton by phytotoxins from decomposing barley residues. *Phytopathology* 58:1571–74.

Lockwood, J. L. 1988. Evolution of concepts associated with soilborne plant pathogens. *Annual Review of Phytopathology* 26:93–121.

Lumsden, R. D., J. A. Lewis, and P. D. Millner. 1983a. Effect of composted sewage sludge on several soilborne pathogens and diseases. *Phytopathology* 73:1543–48.

Lumsden, R. D., J. A. Lewis, and G. C. Papavizas. 1983b. Effect of organic amendments on soilborne plant diseases and pathogen antagonists. In *Environmentally sound agriculture,* ed. W. Lockeretz, 51–70. New York: Praeger.

Lumsden, R. D., P. D. Millner, and J. A. Lewis. 1986. Suppression of lettuce drop caused by *Sclerotinia minor* with composted sewage sludge. *Plant Disease* 70:197–201.

Manning, W. J., and D. F. Crossan. 1969. Field and greenhouse studies on the effects of plant amendments on Rhizoctonia hypocotyl rot of snapbean. *Plant Disease Reporter* 53: 227–31.

Muehlchen, A. M., R. E. Rand, and J. L. Parke. 1990. Evaluation of crucifer green manures for controlling Aphanomyces root rot of peas. *Plant Disease* 74:651–54.

Owens, L. D., R. G. Gilbert, G. E. Griebel, and J. D. Menzies. 1969. Identification of plant volatiles that stimulate microbial respiration and growth in soil. *Phytopathology* 59: 1468–72.

Papavizas, G. C. 1966. Suppression of Aphanomyces root rot of peas by cruciferous soil amendments. *Phytopathology* 56:1071–75.

______. 1968. Survival of root-infecting fungi in soil. VI. Effect of amendments on bean root rot caused by *Thielaviopsis basicola* and on inoculum density of the causal organism. *Phytopathology* 58:421–28.

Papavizas, G. C., and W. A. Ayers. 1974. *Aphanomyces species and their root diseases in pea and sugarbeet.* Agricultural Research Service, U.S. Department of Agriculture Technical Bulletin 1485.

Papavizas, G. C., and C. B. Davey. 1960. Rhizoctonia disease of bean as affected by decomposing green plant materials and associated microfloras. *Phytopathology* 50:516–22.

Papavizas, G. C., and J. A. Lewis. 1971. Black root rot of bean and tobacco caused by *Thielaviopsis basicola* as affected by soil amendments and fungicides in the greenhouse. *Plant Disease Reporter* 55:352–56.

Papavizas, G. C, J. A. Lewis, and P. B. Adams. 1968. Survival of root-infecting fungi in soil. II. Influence of amendment and soil carbon-to-nitrogen balance on Fusarium root rot of beans. *Phytopathology* 58:365–72.

______. 1970. Survival of root-infecting fungi in soil. XIV. Effect of amendments and fungicides on bean root rot caused by *Thielaviopsis basicola. Plant Disease Reporter* 54:114–18.

Patrick, Z. A., and L. W. Koch. 1963. The adverse influence of phytotoxic substances from decomposing plant residues on resistance of tobacco to black root rot. *Canadian Journal of Botany* 41:747–58.

Patrick, Z. A., and T. A. Toussoun. 1965. Plant residues and organic amendments in relation to biological control. In *Ecology of soil-borne plant pathogens,* ed. K. F. Baker and W. C. Snyder, 440–59. Berkeley: University of California Press.

Rajan, K. M. 1980. Soil amendments in plant disease control. *IRRN* 5:4.

Ramirez-Villapudua, J., and D. E. Munnecke. 1987. Control of cabbage yellows (*Fusarium oxysporum* f. sp. *conglutinans*) by solar heating of field soils amended with dry cabbage residues. *Plant Disease* 71:217–21.

______. 1988. Effect of solar heating and soil amendments of cruciferous residues on *Fusarium oxysporum* f. sp. *conglutinans* and other organisms. *Phytopathology* 78:289–95.

Schüler, C., J. Biala, C. Bruns, R. Gottschall, S. Ahlers, and H. Vogtmann. 1989. Suppression of root rot on peas, beans and beet roots caused by *Pythium ultimum* and *Rhizoctonia solani* through the amendment of growing media with composted organic household waste. *Journal of Phytopathology* 127:227–38.

Srivastava, U. S., and S. Sinha. 1971. Effect of various soil amendments on the wilt of coriander (*Coriandrum sativum* L.). *Indian Journal of Agricultural Science* 41:779–82.

Sun, S., and J. Huang. 1985. Formulated soil amendment for controlling Fusarium wilt and other soilborne diseases. *Plant Disease* 69:917–20.

Tu, J. C., and W. I. Findlay. 1986. The effects of different green manure crops and tillage practices on pea root rots. *British Crop Protection Conference—Pests and Diseases* 3C–1: 229–36.

Wall, R. E. 1984. Effects of recently incorporated organic amendments on damping-off of conifer seedlings. *Plant Disease* 68:59–60.

Weinhold, A. R., and T. Bowman. 1968. Selective inhibition of the potato scab pathogen by antagonistic bacteria and substrate influence on antibiotic production. *Plant and Soil* 28:12–24.

Weinhold, A. R., J. W. Oswald, T. Bowman, J. Bishop, and D. Wright. 1964. Influence of green manures and crop rotation on common scab of potato. *American Potato Journal* 41:265–73.

Weltzien, H. C. 1990. The use of composted materials for leaf disease suppression in field crops. *BCPC Monograph on Organic and Low Input Agriculture* 45:115–20.

Wicks, T. J., D. Volle, and B. T. Baker. 1978. The effect of soil fumigation and fowl manure on populations of *Fusarium oxysporum* f. sp. *cucumerinum* in glasshouse soil and on the incidence of cucumber wilt. *Agricultural Record* 5:4–8.

Windels, C. E., and J. Nielsen. 1992. Incubation time and soil temperature effects on sugar beet and Aphanomyces damping-off after soil-incorporation of a green oat crop. *Phytopathology* 82:1088.

# 19

# Interference of Fungicides with Entomopathogens: Effects on Entomophthoran Pathogens of Green Peach Aphid

*Abdelaziz Lagnaoui and Edward B. Radcliffe*

## Introduction

Potato is sprayed routinely with fungicide for control of foliar pathogens, especially early blight, *Alternaria solani* (Ell. and G. Martin) Sor., and late blight, *Phytophthora infestans* (Mont.) de Bary. In Minnesota, Nanne and Radcliffe (1971) showed that three fungicides then commonly used on potato—captafol, mancozeb, and Bordeaux mixture—all favored green peach aphid, *Myzus persicae* (Sulzer), by suppressing entomophthoran (Order: Entomophthorales) fungi. In that experiment, fungicide-treated potatoes had late-season green peach aphid populations 1.7 to 2.6 times greater than did nonfungicidal controls. Aphids in nonfungicidal controls were 22.4 percent infected by entomophthoran pathogens, whereas those in fungicidal treatments were only 4.0 to 5.2 percent infected. Enhancement of green peach aphid on potato is of concern because this aphid species is the most efficient vector of both potato leafroll virus and potato virus Y (Radcliffe et al. 1991). Annually, these two viruses cost the U.S. potato industry tens of millions of dollars in losses and control costs.

The contribution of naturally occurring entomopathogenic fungi to regulation of aphid pests on agricultural crops has excited little enthusiasm on the part of crop protection specialists. Most who have written on the subject appear to have concluded that entomopathogenic fungi are usually of minor importance in the population dynamics of aphids. Reasons suggested for this presumed general ineffectiveness are inadequate inoculum levels, infection being too dependent upon specific environmental conditions, and dissemination too dependent upon uniformly distributed and abundant hosts (Hall and Bell 1960, 1961; Shands et al. 1958, 1962,

1963, 1972; MacLeod et al. 1966; Yendol 1968; Wilding 1969; Missonnier et al. 1970; Dean and Wilding 1971; Remaudière and Michel 1971; Robert et al. 1973; Rabasse and Robert 1975; Wilding 1975; Dedryver 1978; Papierok and Wilding 1981; Remaudière et al. 1981; Soper et al. 1981; Soper and MacLeod 1981; Latgé et al. 1983; Milner and Bourne 1983; Holdom 1984).

However, outbreaks of many agriculturally important aphid species often end in sudden drastic collapse caused by epizootics of entomophthoran fungi. At low aphid densities, infection incidence is also characteristically low; even so, the regulating effect of entomophthoran fungi on populations can be important. Crops in which aphid species are commonly regulated by entomophthoran fungi include alfalfa, e.g., pea aphid, *Acyrthosiphon pisum* (Harris), spotted alfalfa aphid, *Therioaphis maculata* (Buckton), and *Acrythosiphon kondoi* Shinji (Hall and Dunn 1957; van den Bosch et al. 1959; Evlakhova and Voronina 1965; Voronina 1971; Wilding 1975; Milner et al. 1980, 1982; Hutchison and Hogg 1985); fava (field) bean, e.g., bean aphid, *Aphis fabae* Scopoli (Berthelem et al. 1969; Dedryver 1976, 1978; Wilding et al. 1978, 1979; Wilding and Perry 1980); sugar beet, e.g., bean aphid (Gustafsson 1969); and small grain cereals (Dean and Wilding 1971, 1973; Latteur 1973; Papierok et al. 1984).

There is a small but increasing body of literature reporting that agricultural use of pesticides can adversely affect the beneficial contribution of entomophthoran pathogens (Byrde 1966; Cadatal and Gabriel 1970; Bailiss et al. 1978; Wilding et al. 1978, 1979; Elkassabany et al. 1992; Lagnaoui 1990). These field observations are supported by numerous laboratory studies showing that most fungicides and some insecticides inhibit germination of conidia and growth of mycelia of entomophthoran fungi (Hall and Dunn 1959; Yendol 1968; Soper et al. 1974; Fritz 1976, 1977; Delorme and Fritz 1978; Öncüer and Latteur 1979; Wilding and Brobyn 1980; Lagnaoui 1990).

We conducted field and laboratory studies to examine the effects of several commonly used potato fungicides and insecticides on three species of fungi: *Pandora* (= *Ernyia*) *neoaphidis* (Remaudière et Hennebert), *Entomophthora planchoniana* Cornu (Entomophthtoraceae), and *Conidiobolus obscurus* (Hall and Dunn) Remaudière and Keller (Ancylistaceae). It is these studies (Lagnaoui 1990) that we review in this chapter.

## Effects of Potato Fungicides and Insecticides on Entomophthoran Pathogens and the Population Dynamics of Green Peach Aphid

Effects of several commonly used potato fungicides and two insecticides on the incidence of entomophthoran mycoses in green peach aphid populations on Russet Burbank potatoes were assessed in field experiments conducted at the University of Minnesota Agricultural Experiment Station in Rosemount, Minnesota, in 1985 and 1986. Seasonal changes in green peach aphid densities on the crop in response to the interacting effects of the entomophthoran pathogens and pesticides used were monitored each year.

Two experiments were conducted in 1985. In one experiment, potatoes were sprayed with azinphosmethyl (Guthion 2S) insecticide at 0.84 kg AI/ha to enhance green peach aphid numbers (Lowery and Sears 1986). In the other experiment, no insecticidal sprays were applied. Each experiment had six fungicidal treatments: benomyl (Benomyl 50WP), captafol (Difolatan 80SP), chlorothalonil (Bravo 500F), copper hydroxide (Kocide 50WP), mancozeb (Dithane 80WP), and triphenyltin hydroxide (Du-Ter 30F). All pesticides were applied weekly using a high-pressure tractor-drawn boom sprayer with drop nozzles delivering 935 liters of water per hectare.

In 1986, a single experiment was conducted with insecticides applied as split-plot treatments to produce two densities of green peach aphid. Azinphosmethyl at 0.84 kg AI/ha was used to induce high green peach aphid densities; methoxychlor (Methoxychlor 2E) at 0.34 kg AI/ha was used to induce moderate green peach aphid densities (Hanafi et al. 1989). Fungicidal treatments and application were identical to those used in 1985 except that metalaxyl (Ridomil 2WP) was substituted for benomyl.

All fungicides used in these experiments except benomyl were registered for use on potato when the research was done. Each fungicide was applied at the upper rate labeled for use on potato. Control treatments not sprayed with fungicide were included in each experiment.

Green peach aphid apterae were counted weekly on midplant leaves in each plot from July 10 to September 3 in 1985, and from July 10 to September 12 in 1986. Incidence of entomophthoran mycoses were estimated twice in 1985, on August 13 and September 3, and weekly in 1986, from July 10 to September 12. Samples of green peach aphid apterae were collected from each plot. These were held individually in glass vials on excised potato foliage that was replaced every twenty-four hours. Parasitism was estimated two to five days later using the live sample method of Dean and Wilding (1971). Cadavers were placed on water agar in plastic petri dishes and incubated twenty-four hours at 100 percent relative humidity. Plates of *P. neoaphidis* and *C. obscurus* were incubated at 20°C, plates of *E. planchoniana* at 18°C. Presence or absence of rhizoids and cystoids provided tentative identification of the pathogens, but identifications were confirmed by examination of primary and secondary conidia.

Summer 1985 was cooler and drier than average for Rosemount (Seeley and Spoden 1986). Temperatures June through August ranged from 2.2° to 37.2°C with daily maxima averaging 26.4°C. Monthly day-degree totals (base 4.4°C) for this period departed from thirty year norms by –20, +24, and –40 CDD, respectively. Precipitation in June was 80 mm, in July 51 mm, and in August 124 mm: departures from norms of –41, –40, and +15 mm, respectively. Supplemental irrigation was not available, so soil moisture was often suboptimal. The potato canopy never closed in 1985; conditions thus were not ideal for the development of entomophthoran mycoses or aphid reproduction.

Summer 1986 was again cooler than usual for Rosemount, but precipitation was above average (Seeley and Spoden 1987). Temperatures June through September ranged from 4.1° to 33.1°C with daily maxima averaging 24.9°C. Monthly day-degree totals for this period departed from norms by +54, +29, –53, and +13 CDD, respectively. Precipitation in June was 167 mm, in July 95 mm, in August 101 mm, and in September 222 mm: departures from norms of +45, +4, –8, and +141 mm, respectively. The potato canopy closed early in 1986, providing a favorable microclimate for development of entomophthoran mycoses and aphid reproduction.

### *Aphid Populations*

*Aphid Populations 1985.* In the experiment not sprayed with insecticide (low aphid densities), mean densities of green peach aphid apterae across treatments ranged from 0 apterae per leaf on July 10 to 4.7 per leaf on September 3. However, treatment differences in aphid numbers were not significantly different on any sampling date.

In the experiment sprayed with azinphosmethyl (high aphid densities), mean densities of green peach aphid apterae across treatments ranged from 0 apterae per leaf on July 18 to 41 apterae per leaf on September 3. The nonfungicidal control had fewer aphids than any fungicidal treatment except benomyl on all sampling dates. The only date that treatments differed significantly ($p = 0.05$) was September 12, when captafol and mancozeb both had more aphids than did the nonfungicidal control. Aphid numbers in the copper hydroxide, triphenyltin hydroxide, and chlorothalonil treatments were greater than in the nonfungicidal control, but these differences were not statistically significant.

*Aphid Populations 1986.* In plots sprayed with methoxychlor (moderate aphid densities), mean densities of green peach aphid apterae across treatments ranged from 0 apterae per leaf on July 10 to 92 per leaf on September 12. On September 12, all fungicidal treatments had significantly ($p = 0.05$) more aphids than did the nonfungicidal control. From August 29 to September 12, green peach aphid numbers in metalaxyl, captafol, and mancozeb treatments increased 179 to 277 percent, whereas in the chlorothalonil, copper hydroxide, and the nonfungicidal control, green peach aphid numbers decreased 37 to 69 percent.

In plots sprayed with azinphosmethyl (high aphid densities), mean densities of green peach aphid apterae across treatments ranged from 0 apterae per leaf on July 10 to 149 apterae per leaf on September 12. On September 12, the metalaxyl, captafol, and mancozeb treatments had significantly ($p = 0.05$) more aphids than did the nonfungicidal control. From August 29 to September 12, green peach aphid numbers in the metalaxyl, captafol, and mancozeb treatments increased 128 to 146 percent, whereas in the chlorothalonil, copper hydroxide, and the nonfungicidal control, green peach aphid numbers decreased 69 to 95 percent.

### *Entomophthoran Mycoses*

*Incidence of Entomophthoran Mycoses 1985.* In the experiment not sprayed with insecticide, incidence of entomophthoran mycoses never exceeded 5 percent in any treatment. In the experiment sprayed with azinphosmethyl, infection incidence on September 3 ranged from 2.0 percent (benomyl) to 12.7 percent (copper hydroxide) in fungicidal treatments and was 32.4 percent in the nonfungicidal control (Table 19.1).

***Incidence of Entomophthoran Mycoses 1986.*** On August 29, incidence of entomophthoran mycoses in plots sprayed with methoxychlor ranged from 2 percent (metalaxyl) to 8.7 percent (chlorothalonil) in fungicidal treatments and was 8.2 percent in the nonfungicidal control (Table 19.1). Over the next two weeks, aphid densities and infection incidence increased greatly. On September 12, infection incidence ranged from 5.6 percent (metalaxyl) to 22.1 percent (chlorothalonil) in fungicidal treatments and was 26.3 percent in the nonfungicidal control.

On August 29, infection incidence in plots sprayed with azinphosmethyl ranged from 2.5 percent (metalaxyl) to 9.6 percent (chlorothalonil) in fungicidal treatments and was 14.4 percent in the nonfungicidal control (Table 19.1). On September 12, infection incidence ranged from 11.9 percent (metalaxyl) to 33.8 percent (chlorothalonil) in fungicidal treatments and was 53.9 percent in the nonfungicidal control.

The entomophthoran fungi causing mycoses in green peach aphid in 1985 were not identified to species. In 1986, *P. neoaphidis* was the most prevalent entomophthoran pathogen of green peach aphid. Across all treatments and sampling dates, *P. neoaphidis* accounted for 67 percent of total green peach aphid mortality attributable to entomophthoran fungi, *E. planchoniana* caused 22 percent, *C. obscurus* 8 percent, and undetermined species 2 percent.

***Association of Entomophthoran Mycoses with Environmental Conditions.*** Relationships among incidence of entomophthoran mycoses, predisposing environmental conditions, and aphid densities were tested with data from the 1986 nonfungicidal control plots using simple and multiple linear regression models. Parameters considered included average temperature, cumulative rainfall, hours of relative humidity above 90 percent, hours of leaf wetness, and aphid density.

In plots sprayed with methoxychlor, peak incidence of mycoses was correlated (all regressions had 8 d.f.) with cumulative precipitation ($r^2 = 0.67$) and duration of leaf wetness ($r^2 = 0.49$) over the preceding seven days. Incidence of entomophthoran mycoses showed only weak correlation with aphid density ($r^2 = 0.18$). However, addition of aphid density improved correlation in multiple regression models more than did the addition of any other variable. Use of aphid density, relative humidity, duration of leaf wetness, and cumulative precipitation over the preceding seven days yielded the best correlation ($r^2 = 0.91$).

In plots sprayed with azinphosmethyl, peak incidence of mycoses was correlated (all regressions had 8 d.f.) with cumulative precipitation ($r^2 = 0.64$) and duration of leaf wetness ($r^2 = 0.47$) over the preceding seven days. Incidence of entomophthoran mycoses showed only weak correlation with aphid density ($r^2 = 0.21$). Again, addition of aphid density improved correlation in multiple regression models more than

Table 19.1. Infection of green peach aphid apterae by entomophthoran pathogens

| | Insecticide, year, percent infection, date[1] | | | | | |
|---|---|---|---|---|---|---|
| | Azinphosmethyl, 1985 | | Methoxychlor, 1986 | | Azinphosmethyl, 1986 | |
| Fungicide | August 13 | September 3 | August 29 | September 12 | August 29 | September 12 |
| Benomyl | 1.5a | 2.4a | — | — | — | — |
| Metalaxyl | — | — | 2.0c | 5.6a | 2.5de | 11.9e |
| Captafol | 5.4d | 11.2bc | 2.2c | 8.1d | 5.0d | 20.8d |
| Triphenyltin hydroxide | 2.0ab | 4.2a | — | — | — | — |
| Mancozeb | 3.5bc | 8.4d | 5.1ab | 11.3cd | 6.0cd | 23.6cd |
| Copper hydroxide | 5.8bc | 12.7c | 8.0a | 16.1ab | 7.9bc | 31.8b |
| Chlorothalonil | 5.0bc | 9.5bc | 8.7a | 22.1a | 9.6bc | 33.8b |
| Control | 6.9d | 32.4d | 8.2a | 26.6a | 14.4a | 53.9a |

[1] Insecticides applied weekly to enhance green peach aphid populations; methoxychlor to produce moderate population densities, azinphosmethyl to produce high population densities. Percent infection determined from field collection and laboratory culture of ca. 50 aphids per replicate per treatment per sampling date. Means within column followed by the same letter are not significantly different ($P<0.05$) by Duncan's Multiple Range Test.

*Note:* Rosemount, Minnesota, 1985–1986.

did the addition of any other variable. Use of aphid density, relative humidity, duration of leaf wetness, and cumulative precipitation over the preceding seven days yielded the best correlation ($r^2 = 0.94$).

## Effects of Fungicides and Insecticides on the Germination of Conidia and Growth of Mycelia

Effects of several common potato fungicides and insecticides on the germination of conidia and growth of mycelia of *P. neophidis, E. planchoniana,* and *C. obscurus* were measured in laboratory bioassays. For this, we used both in vivo and in vitro cultures of each pathogen. In vivo cultures were isolated from field-collected green peach aphids. In vitro isolates, *P. neoaphidis* (ARSEF 1600), *E. planchoniana* (RS 1927), and *C. obscurus* (ARSEF 129), were provided from the USDA collection by R. A. Humber of the Boyce Thompson Institute in Ithaca, New York. Pathogen cultures were maintained on Sabouraud egg milk agar. The green peach aphids used were from a Minnesota laboratory colony established from field-collected aphids. This colony was maintained on Chinese cabbage and sprayed periodically with azinphosmethyl to maintain its parasitoid-free status. Pesticides were used in these experiments at concentrations equivalent to 0.1X, 1X, and 2X, the upper rates labeled for use on potato, assuming application in 935 liters of water per hectare.

### *Effects on Conidia*

Pesticides tested for effects on conidia were the fungicides captafol, chlorothalonil, copper hydroxide, mancozeb, and metalaxyl, and the insecticides azinphosmethyl and methoxychlor. Pesticides were tested at 0.1X, 1X, and 2X rates against *P. neoaphidis* and *C. obscurus,* but only at the 1X rate against *E. planchoniana.* Pesticide residues were deposited on the washed, autoclaved microscope slides by spraying to run-off with a hand-pumped atomizer. After application of the pesticides, the slides were air dried for two hours, then transferred to a controlled environment chamber where they were exposed to showers of conidial spores of the desired pathogen.

Germination of conidia of all three pathogens was reduced significantly ($p = 0.05$) by all fungicides except chlorothalonil (Table 19.2). Captafol, mancozeb, and metalaxyl severely inhibited germination of conidia even at 0.1X rates of application. Azinphosmethyl and methoxychlor did not cause statistically significant reductions in germination of conidia, but effects were negative and appeared to be dose-dependent.

### *Effects on Mycelia*

Pesticides tested for effects on mycelia were the fungicides benomyl, captafol, chlorothalonil, copper hydroxide, mancozeb, metalaxyl, and triphenyltin hydroxide, and the insecticides azinphosmethyl and methoxychlor. All pesticides were tested at

Table 19.2. Rating of effects of 1x rates of pesticides on green peach aphid and associated species of entomophthoran pathogens.

| | *Myzus persicae* | | *Pandora neoaphidis* | | *Entomophthora planchoniama* | | *Conidiobolus obscurus* | |
|---|---|---|---|---|---|---|---|---|
| Pesticide | Population[1] | Individual[2] | Conidia[3] | Myceliae[3] | Conidia[3] | Mycelia[3] | Conidia[3] | Mycelia[3] |
| *Insecticides* | | | | | | | | |
| Azinphosmethyl | +3 | +1 | -1 | -1 | -2 | -1 | -1 | -1 |
| Methoxychlor | +2 | -1 | -2 | -1 | -2 | -1 | -1 | -1 |
| *Fungicides* | | | | | | | | |
| Benomyl | -1 | -2 | — | -3 | — | -3 | — | -3 |
| Captafol | +3 | -1 | -3 | -3 | -3 | -3 | -3 | -3 |
| Chlorothanonil | +1 | -1 | -1 | -2 | -1 | -1 | -2 | -2 |
| Copper hydroxide | +2 | -1 | -2 | -2 | -2 | -3 | -3 | -3 |
| Mancozeb | +3 | -1 | -3 | -3 | -3 | -3 | -3 | -3 |
| Metalaxyl | +3 | -1 | -3 | -3 | -3 | -3 | -3 | -3 |
| Triphenyltin hydroxide | +1 | -3 | — | -3 | — | -3 | — | -3 |

[1] Based on aphid densities on September 12, 1986, in azinphosmethyl-sprayed plots of all treatments except benomyl and triphenyltin hydroxide. Benomyl and triphenyltin hydroxide are rated by comparison to aphid densities on September 3, 1985, in the mancozeb and control plots sprayed with azinphosmethyl. Ratings are aphid density relative to nonfungicide control: +3 = >100X density, +2 = 10-100X density, +1 = 1-10X density, -1 = 0.1-1X density.

[2] Based on our previous field experience with these insecticides or laboratory dip tests of the fungicides: -1 = <33% mortality relative to control, -2 = 33-67% mortality, - 3 = >67% mortality.

[3] Effects on conidia are based on percent inhibition of germination, and effects on mycelia are based on growth in treatment as a percentage of growth in nonfungicidal control: - 1 = <33% reduction, -2 = 33-67% reduction, -3 = >67% reduction.

0.1X, 1X, and 2X concentrations. Sabouraud egg milk agar incorporating the desired concentration of pesticide was poured into Petri dishes and allowed to solidify. A 5 mm core was cut from the center of each agar plate and replaced with a 5 mm plug taken from an agar plate with an actively growing culture of the desired fungal pathogen. *P. neoaphidis* and *C. obscurus* were then incubated at room temperature (20° to 25°C), 90 percent relative humidity, and a photoperiod of sixteen hours. Plates of *E. planchoniana* were treated similarly, except they were incubated in a controlled environment chamber at 18°C. After three weeks, mean diameter of each mycelial mat was estimated. Effects on growth of mycelia were expressed as percentages compared to growth in controls without pesticide.

Growth of mycelia of all three entomophthorans was significantly ($p = 0.05$) inhibited by benomyl, captafol, triphenyltin hydroxide, mancozeb, and metalaxyl (Table 19.2). Copper hydroxide significantly ($p = 0.05$) inhibited growth of *C. obscurus* and *E. planchoniana,* but had little effect on *P. neoaphidis.* Chlorothalonil did not have a significant effect on growth of mycelia of any entomophthoran species even at the 2X rate. Inoculants were not killed by 1X concentrations of any pesticide. When transferred back to media without a pesticide, inoculants inhibited by 1X pesticide concentrations recovered and grew normally. *Conidiobolus obscurus* was the species most vulnerable to pesticide poisoning; it was killed by exposure to 2X concentrations of benomyl, metalaxyl, and triphenyltin hydroxide. Azinphosmethyl and methoxychlor caused no significant ($P = 0.05$) inhibition of growth of mycelia even at 2X concentrations.

## Pathogenicity of Primary Spores to Green Peach Aphid

Pathogenicities of primary spores of *P. neoaphidis, E. planchoniana,* and *C. obscurus* to green peach aphid apterae were measured in laboratory bioassays. Both in vivo and in vitro isolates were used. in vivo isolates were produced on site from cadavers of aphids recently killed in the field. in vitro isolates, *P. neoaphidis* (ARSEF 1600), *E. planchoniana* (RS 1927), and *C. obscurus* (ARSEF 129), were from the USDA collection.

Green peach aphids from the previously described laboratory culture were used in the bioassays. First instar nymphs were transferred to excised Chinese cabbage and exposed to showers of primary spores. Two exposure techniques were used: (1) in glass petri dishes with the spore source placed 2 cm above exposed nymphs, and (2) in lamp chimneys with the spore source placed 8 cm above exposed nymphs. Inoculum dosage was controlled by varying exposure time and was quantified from a glass cover slip exposed next to the aphids. After exposure, the aphids were transferred to nontreated Chinese cabbage seedlings and held in a controlled environment chamber at 75 percent relative humidity and a sixteen hour photoperiod. Dead aphids were collected daily for the next seven days and transferred to 100 percent relative humidity for twenty-four hours to induce outgrowth and sporulation. Virulence ratios were calculated for each pathogen species by dividing in vitro $LC_{50}$ by in vivo $LC_{50}$.

Pathogenicity to green peach aphid was greatest for *E. planchoniana,* intermediate for *P. neoaphidis,* and least for *C. obscurus.* There was close linear relationship ($r^2$ = 0.88 to 0.99) between log-dose and probit mortality for all six fungal isolates.

In petri dish bioassay, the $LC_{50}$ for *Pandora neoaphidis* spores of in vivo origin was 12.1/mm$^2$, slope 1.79; the $LC_{50}$ for spores of in vitro origin was 17.9/mm$^2$, slope 2.51. In lamp chimney bioassay, the $LC_{50}$ for *P. neoaphidis* spores of in vivo origin was 16.1/mm$^2$, slope 1.69; the $LC_{50}$ for spores of in vitro origin was 23.2/mm$^2$, slope 1.32.

In petri dish bioassay, the $LC_{50}$ for *Entomophthora planchoniana* spores of in vivo origin was 8.8/mm$^2$, slope 1.93; the $LC_{50}$ for spores of in vitro origin was 20.6/mm$^2$, slope 2.81. In lamp chimney bioassay, the $LC_{50}$ for *E. planchoniana* spores of in vivo origin was 12.6/mm$^2$, slope 1.79; the $LC_{50}$ for spores of in vitro origin was 23.7/mm$^2$, slope 1.38.

In petri dish bioassay, the $LC_{50}$ for *Conidiobolus obscurus* spores of in vivo origin was 21.0/mm$^2$, slope 3.02; the $LC_{50}$ for spores of in vitro origin was 26.5/mm$^2$, slope 2.20. In lamp chimney bioassay, the $LC_{50}$ for *C. obscurus* spores of in vivo origin was 28.3/mm$^2$, slope 2.18; the $LC_{50}$ for spores of in vitro origin was 36.4/mm$^2$, slope 1.85. Only for *C. obscurus* were significantly ($p$ = 0.05) more spores required to kill an aphid when the distance between the source and the target was increased from 2 cm (petri dish bioassay) to 8 cm (lamp chimney bioassay).

### *Virulence of in vivo and in vitro sources*

In general, spores of in vivo origin were more virulent than were spores of in vitro origin. In petri dish bioassays, virulence ratios for *P. neoaphidis, E. planchoniana,* and *C. obscurus* were 1.48, 1.26, and 2.34, respectively. In lamp chimney bioassays, virulence ratios for *P. neoaphidis, E. planchoniana,* and *C. obscurus* were 1.12, 1.34, and 1.88, respectively. For *P. neoaphidis,* differences in the pathogenicity of in vivo and in vitro isolates were not statistically significant ($p$ = 0.05). For *E. planchoniana* and *C. obscurus,* spores of in vivo origin were significantly ($p$ = 0.05) more pathogenic than spores of in vitro origin.

For *P. neoaphidis* and *E. planchoniana,* the probit regression lines comparing pathogenicity of isolates of in vivo and in vitro origin were not parallel ($p$ = 0.05, one-tailed paired comparison $t$ test) in lamp chimney bioassay. *P. neoaphidis* probit regression lines were not parallel ($p$ = 0.05) in petri dish bioassay. *C. obscurus* probit regression lines were parallel in both types of bioassay. Thus, only for *C. obscurus* can the differences in pathogenicity of spores of in vivo and in vitro origin be considered significant at all concentrations.

Our results are consistent with the observation of Holdom (1983) that fungi maintained on culture media are less infectious than those isolated from cadavers. A possible explanation for poorer sporulation and lower infectivity with isolates from in vitro culture may be that the culture media is deficient in some essential nutrient.

## Toxicity of Potato Pesticides to Green Peach Aphid

There are virtually no data on the direct toxicity of fungicides to green peach aphid on potato, but some fungicides are known to have aphidicidal properties (Delorme 1977; Vickerman 1977; Bailiss et al. 1978). We used dip tests as described by Busvine (1980) to test benomyl, captafol, chlorothalonil, copper hydroxide, mancozeb, metalaxyl, and triphenyltin for toxicity to green peach aphid. Green peach aphids from the previously described laboratory culture were used in these tests.

Concentrations of fungicide equivalent to the upper rates recommended for field use were generally toxic to green peach aphid in our laboratory tests (Table 19.2). Most toxic were benomyl, chlorothalonil, and copper hydroxide, which caused cumulative mortality of 43, 16, and 18 percent at twenty-four hours exposure and cumulative mortality of 65, 21, and 21 percent at forty-eight hours exposure. Less toxic were captafol, mancozeb, and metalaxyl which caused cumulative mortality of 8, 12, and 12 percent at twenty-four hours exposure, and cumulative mortality of 9, 12, and 15 percent at forty-eight hours exposure.

## Implications

When green peach aphid densities are low, entomophthoran pathogens appear not to play a key role in their population dynamics; in this respect these pathogens resemble green peach aphid parasitoids. However, our experiments have demonstrated that entomophthoran fungi can play a much greater role in the population dynamics of green peach aphid than is generally recognized.

Fungicides used to protect potato from foliar pathogens can be highly detrimental to entomophthoran fungi. Thus, the potential exists for upsetting biological control and triggering green peach aphid outbreaks when certain fungicides are used. Captafol, mancozeb, and metalaxyl were particularly prone to enhancing green peach aphid populations, whereas chlorothalonil, copper hydroxide, and triphenyltin hydroxide resulted in only modest increases. The benomyl treatment had fewer aphids than did the nonfungicidal control. Captafol, mancozeb, and metalaxyl severely inhibited germination of the conidia of *P. neoaphidis, E. planchoniana,* and *C. obscurus.* Copper hydroxide was intermediate in effect and chlorothalonil essentially without effect. Benomyl, captafol, mancozeb, metalaxyl, and triphenyltin hydroxide severely inhibited the growth of mycelia of all three fungi. Copper hydroxide inhibited mycelial growth of *E. planchoniana* and *C. obscurus,* but had only intermediate effect on *P. neoaphidis.* Chlorothalonil had little effect on mycelia growth of species. It is evident that careful selection of fungicides is a must if we wish to benefit from the natural control of green peach aphid afforded by entomophthoran fungi. The insecticides azinphosmethyl and methoxychlor had only minor effects on entomophthoran fungi. All fungicides were, to some extent, toxic to green peach aphid. Most toxic were benomyl, chlorothalonil, and copper hydroxide; least toxic were captafol, mancozeb, and metalaxyl.

On potato, green peach aphid is rarely abundant except when populations are enhanced by inappropriate selection of pesticides, either insecticides targeted against other insect pests or fungicides applied for control of foliar pathogens. Therefore, even when green peach aphid is not the target pest, the possible effects on this species should always be considered in choosing potato pesticides. This consideration is especially important in seed potato production (because of the role of green peach aphid as a vector of potato viruses) and in growing cultivars susceptible to "net necrosis" (a tuber condition caused by infection with potato leafroll virus).

## References

Bailiss, K. W., G. A. Partis, C. J. Hodgson, and E. V. Stone. 1978. Some effects of benomyl and carbendazim on *Aphis fabae* and *Acyrthosiphon pisum* on field bean (*Vicia faba*). *Annals of Applied Biology* 89:443–49.

Berthelem, P., J. Missonnier, and Y. Robert. 1969. Le puceron noir de la fève, *Aphis fabae* Scop. (Hom., Aphididae) et la culture de la féverole de printemps (*Vicia faba* L.). *Annales de Zoologie, Écologie Animale* 1:183–96.

Busvine, J. R. 1980. Method for adult aphids—FAO method no. 17. In *Recommended methods for measurement of pest resistance to pesticides,* 103–6. *FAO Plant Production and Protection Paper* 21.

Byrde, R.J.W. 1966. The vulnerability of fungus spores to fungicides. In *The fungus spore* (Proceedings of the 18th Symposium of the Colston Research Society, University of Bristol), ed. M. F. Madelin, 289–97. London: Butterworth.

Cadatal, T. D., and B. P. Gabriel. 1970. Effect of chemical pesticides on the development of fungi pathogenic to some rice insects. *Philippine Entomologist* 1:379–95.

Dean, G.J.W., and N. Wilding. 1971. *Entomophthora* infecting the cereal aphids *Metapolophium dirhodum* and *Sitobion avenae. Journal of Invertebrate Pathology* 18:169–76.

______. 1973. Infection of cereal aphids by the fungus *Entomophthora. Annals of Applied Biology* 74:133–38.

Dedryver, C. A. 1976. Évolution comparée des mycoses à *Entomophthora* sur le puceron noir de la fève (*Aphis fabae* Scop.) en cultures de betteraves et de féveroles. *Sciences Agronomiques Rennes* 1976: 189–99.

______. 1978. Facteurs de limitation des populations d'*Aphis fabae* dans l'Ouest de la France. 3. Répartition et incidence des différentes espèces d'*Entomophthora* dans les populations. *Entomophaga* 23:137–51.

Delorme, R. 1977. Étude de l'activité aphicide de quelques fongicides systématiques. *Phytiatrie-Phytopharmacie* 26:99–106.

Delorme, R., and R. Fritz. 1978. Action de divers fongicides sur le développement d'une mycose à *Entomophthora aphidis. Entomophaga* 23:389–401.

Elkassabany, N. M., D. C. Steinkraus, P. J. McLeod, J. C. Correll, and T. E. Morelock. 1992. *Pandora neoaphidis* (Entomophthorales: Entomophthoraceae): A potential biological control agent against *Myzus persicae* (Homoptera: Aphididae) on spinach. *Journal of the Kansas Entomological Society* 65:196–99.

Evlakhova, A. A., and E. Voronina. 1965. Study on the entomophthorosis of Aphididae for their practical utilization. *Proceedings 12th International Congress of Entomology, London, 1964,* Section 11: Insect Pathology, 751.

Fritz, R. 1976. Action de quelques fongicides sur la croissance mycélienne de trois espèces d'Entomophthorales. *Entomophaga* 21:239–49.

______. 1977. Action de quelques fongicides sur des Entomophthorales pathogènes de pucerons. *Phytiatrie-Phytopharmacie* 26:193–200.

Gustafsson, M. 1969. On species of the genus *Entomophthora* Fres. in Sweden. III. Possibility of usage in biological control. *Lantbrukshogskolans Annaler* 35:235–74.

Hall, I. M., and J. V. Bell. 1960. The effect of temperature on some entomophthoraceous fungi. *Journal of Insect Pathology* 2:247–53.

______. 1961. Further studies on the effect of temperature on the growth of some entomophthoraceous fungi. *Journal of Insect Pathology* 3:289—96.

Hall, I. M., and P. H. Dunn. 1957. Entomophthorous fungi parasitic on the spotted alfalfa aphid. *Hilgardia* 27:159–81.

______. 1959. The effect of certain insecticides and fungicides on fungi pathogenic to the spotted alfalfa aphid. *Journal of Economic Entomology* 52:28–29.

Hanafi, A., E. B. Radcliffe, and D. W. Ragsdale. 1989. Spread and control of potato leafroll virus in Minnesota. *Journal of Economic Entomology* 82:1201–6.

Holdom, D. G. 1983. In vitro culture of the aphid pathogenic fungus *Entomophthora planchoniana* Cornu (Zygomycetes: Entomophthorales). *Journal of the Australian Entomological Society* 22:188.

______. 1984. Moisture requirements and field occurrence of *Entomophthora planchoniana* Cornu. *Proceedings of the 4th Australian Applied Entomological Research Conference* 1984:368–74.

Hutchison, W. D., and D. B. Hogg. 1985. Time-specific life tables for the pea aphid, *Acyrthosiphon pisum* (Harris), on alfalfa. *Researches on Population Ecology* 27:231–53.

Lagnaoui, A. 1990. Effects of potato fungicides on entomophthoraceous fungi and the population dynamics of the green peach aphid, *Myzus persicae* (Sulzer). M.Sc. thesis, University of Minnesota, St. Paul.

Latgé P., P. Silvie, B. Papierok, G. Remaudière, C. A. Dedryver, and J. M. Rabasse. 1983. Advantages and disadvantages of *Conidiobolus obscurus* and of *Erynia neoaphidis* in the biological control of aphids. In *Aphid antagonists,* ed. R. Cavalloro, 20–32. Rotterdam: Balkema.

Latteur, G. 1973. Étude de la dynamique des populations des pucerons des céréales. Premières données relatives aux organismes aphidiphages en trois localitiés différentes. *Parasitica* 29:134–51.

Lowery, D. T., and M. K. Sears. 1986. Stimulation of reproduction of the green peach aphid (Homoptera: Aphididae) by azinphosmethyl applied to potatoes. *Journal of Economic Entomology* 79:1530–33.

MacLeod, D. M., J. W. Cameron, and R. S. Soper. 1966. The influence of environmental conditions on epizootics caused by entomogenous fungi. *Revue Roumaine de Biologie Serie de Botanique* 11:125–34.

Milner, R. J., and J. Bourne. 1983. Influence of temperature and duration of leaf wetness on infection of *Acyrthosiphon kondoi* with *Erynia neoaphidis. Annals of Applied Biology* 102:19–27.

Milner, R. J., R. S. Soper, and G. G. Lutton. 1982. Field release of an Israeli strain of the fungus *Zoophthora radicans* (Brefeld) Batko for biological control of *Therioaphis trifolii* (Monell) f. *maculata. Journal of the Australian Entomological Society* 21:113–18.

Milner, R. J., R. E. Teakle, G. G. Lutton, and F. M. Dare. 1980. Pathogens (Phycomycetes: Entomophthoraceae) of the blue-green aphid *Acrythosiphon kondoi* Shinji and other aphids in Australia. *Australian Journal of Botany* 28:601–19.

Missonnier, J., Y. Robert, and G. Thoizon. 1970. Circonstances épidémiologiques semblant favoriser le développement des mycoses à entomophthorales ches trois aphides, *Aphis fabae* Scop. *Capitophorus horni* B;auorner et *Myzus persicae* (Sulz.). *Entomophaga* 15:169–90.

Nanne, H. W., and E. B. Radcliffe. 1971. Green peach aphid populations on potatoes enhanced by fungicides. *Journal of Economic Entomology* 64:1569–70.

Öncüer, C., and G. Latteur. 1979. The effect of 10 fungicides on the infectivity of conidia of *Entomophthora obscura* Hall & Dunn on the surface of sterile soil. *Parasitica* 35:3–15.

Papierok, B., and N. Wilding. 1981. Étude du comportement de plusieurs souches de *Conidiobolus obscurus* (Zygomycètes Entomophthoraceae) vis-à-vis des pucerons *Acyrthosiphon pisum* et *Sitobion avenae* (Hom: Aphididae). *Entomophaga* 26:241–49.

Papierok, B., P. Silvie, J. P. Latgé, C. A. Dedryver, J. M. Rabasse, and G. Remaudière. 1984. Biological control of cereal aphids with Entomophthorales. In *C.E.C. Programme on Integrated and Biological Control, Final Report 1979–1983,* ed. R. Cavalloro and A. Piavaux, 339–52. Luxembourg: EUR 8689.

Rabasse, J. M., and Y. Robert. 1975. Facteurs de limitation des populations d'*Aphis fabae* dans l'Ouest de la France. 2. Incidence des mycoses à *Entomophthora* sur les populations des hôtes primaires et de la féverole de printemps. *Entomophaga* 20:49–63.

Radcliffe, E. B., K. L. Flanders, D. W. Ragsdale, and D. M. Noetzel. 1991. Pest management systems for potato insects. In *CRC Handbook of Pest Management,* 2d ed. Vol. III, ed. D. Pimentel, 587–621. Boca Raton, Fla.: CRC Press.

Remaudière, G., and M. F. Michel. 1971. Première expérimentation écologique sur les entomophthorales (Phycomcètes) parasites de pucerons en vergers de pêchers. *Entomophaga* 16:75–94.

Remaudière, G., J. P. Latgé, and M. F. Michel. 1981. Ecologie comparée Entomophthoracées pathogènes de pucerons en France littorale et continentale. *Entomophaga* 26:157–78.

Robert, Y., J. M. Rabasse, and P. Scheltes. 1973. Facteurs de limitation des populations d'*Aphis fabae* Scop. dans l'Ouest de la France. 1.Épizootiologie des maladies à Entomophthorales sur féverole de printemps. *Entomophaga* 18:61–75.

Seeley, M. W., and G. J. Spoden. 1986. C.A.W.A.P. 1985 crop season climatic data, University of Minnesota Agricultural Experiment station research locations. St. Paul: University of Minnesota Agricultural Extension Service AG-BU-2201-1986.

______. 1987. C.A.W.A.P. 1986 crop season climatic data, University of Minnesota Agricultural Experiment station research locations. St. Paul: University of Minnesota Agricultural Extension Service AG-BU-2291-1987.

Shands, W. A., I. M. Hall, and G. W. Simpson. 1962. Entomophthoraceous fungi attacking the potato aphid in northeastern Maine in 1960. *Journal of Economic Entomology* 55: 174–79.

Shands, W. A., G. W. Simpson, and I. M. Hall. 1963. Importance of entomogenous fungi in controlling aphids on potatoes in northeastern Maine. *Maine Agriculture Experiment Station Technical Bulletin* 6.

Shands, W. A., G. W. Simpson, I. M. Hall, and C. C. Gordon. 1972. Further evaluation of entomogenous fungi as a biological control agent of aphid control in northeastern Maine. *USDA Technical Bulletin* 58.

Shands, W. A., C. G. Thompson, G. W. Simpson, and H. E. Wave. 1958. Preliminary studies of entomogenous fungi for the control of potato-infesting aphids in Maine. *Journal of Economic Entomology* 51:184–86.

Soper, R. S. 1981. Role of entomophthoran fungi in aphid control for potato integrated pest management. In *Advances in potato pest management,* ed. J. H. Lashomb and R. Casagrande, 153–77. Stroudsburg, Pa.: Hutchison Ross.

Soper, R. S., and D. M. MacLeod. 1981. Descriptive epizootiology of an aphid mycosis. *USDA Technical Bulletin* 1632.

Soper, R. S., F. R. Holbrook, and C. C. Gordon. 1974. Comparative pesticide effects on *Entomophthora* and the phytopathogen *Alternaria solani. Environmental Entomology* 3:560–62.

van den Bosch, R., E. I. Schlinger, E. J. Dietrick, and I. M. Hall. 1959. The role of imported parasites in the biological control of the spotted alfalfa aphid in southern California in 1957. *Journal of Economic Entomology* 52:142–54.

Vickerman, G. P. 1977. The effects of foliar fungicides on some insect pests of cereals. *1977 British Crop Protection Conference—Pests and Diseases,* 121–30.

Voronina, E. G. 1971. Entomophthorosis epizootics of the pea aphid *Acyrthosiphon pisum* (Homoptera: Aphidoidea). *Entomologicheskoe Obozrenie* 50:780–90; *Entomological Review* 50:444–53.

Wilding, N. 1969. Effect of humidity on the sporulation of *Entomophthora aphidis* and *E. thaxteriana. Transactions of the British Mycological Society* 53:126–30.

______. 1975. *Entomophthora* species infecting pea aphis. *Transactions of the Royal Entomological Society* 127:171–83.

Wilding, N., and P. J. Brobyn. 1980. Effects of fungicides on development of *Entomophthora aphidis. Transactions of the British Mycological Society* 75:297–302.

Wilding, N., and J. N. Perry. 1980. Studies on *Entomophthora* in populations of *Aphis fabae* on field beans. *Annals of Applied Biology* 94:367–78.

Wilding, N., P. J. Brobyn, and S. K. Best. 1978. *Entomophthora* species controlling bean aphids: The effect of fungicides on *Entomophthora. Annual Report Rothamstead Experiment Station, 1977 (part 1),* 102–3.

______. 1979. *Entomophthora* species controlling bean aphids: The effect of fungicides on *Entomophthora. Annual Report Rothamstead Experiment Station, 1978 (part 1),* 97.

Yendol, W. G. 1968. Factors affecting germination of *Entomophthora* conidia. *Journal of Invertebrate Pathology* 10:116–21.

# About the Book and Editors

Recent interest in nonchemical methods of pest control has brought renewed attention to the biological control of plant pests in the fields of entomology, plant pathology, and weed science. *Ecological Interactions and Biological Control* addresses issues of theory and practice common to all three fields. Focusing on systems rather than on individual problems, the contributors are able to look at the larger issues of how ecological theory has aided biological control and vice versa. Most important, they suggest ways to integrate theory and practice more closely in order to contribute to the future development of biological control.

**David A. Andow** and **David W. Ragsdale** are associate professors in the Department of Entomology and **Robert F. Nyvall** is professor in the Department of Plant Pathology, all at the University of Minnesota.

# About the Contributors

Neil A. Anderson is professor in the Department of Plant Pathology at the University of Minnesota in St. Paul, where he studies mycology and the genetics of plant pathogens.

Franz Bigler is a research entomologist at the Swiss Federal Research Station for Agronomy in Zurich, Switzerland, where he studies biological control and quality control of *Trichogramma* wasps.

Fabio Cerutti is now at the Bundesamt für Landwirtschaft in Bern, Switzerland.

Robert L. De Haan is assistant professor in the Biology Department at Dordt College in Sioux, Iowa.

Eric C. Eckwall is a Ph.D. student in the Microbiology, Immunology, and Molecular Pathobiology Program at the University of Minnesota in St. Paul.

Nancy J. Ehlke is associate professor in the Department of Agronomy and Plant Genetics at the University of Minnesota in St. Paul, where she studies weed control.

Linda A. Gilkeson is in the Pesticide Management Branch at British Columbia Environment in Victoria, where she promotes IPM programs and alternatives to pesticides.

Robert M. Goodman is professor in the Department of Plant Pathology at the University of Wisconsin in Madison.

Jo Handelsman is professor in the Department of Plant Pathology at the University of Wisconsin in Madison, where she studies the molecular ecology of beneficial bacteria, including potential biocontrol agents and nitrogen fixers.

Peter Harris has studied and implemented classical biocontrol of weeds in Canada and is now at the Agriculture and Agri-Food Canada Research Centre in Lethbridge, Alberta.

Ann C. Kennedy is a soil scientist in the Land Management and Water Conservation Unit of USDA, ARS at Washington State University in Pullman, where she is investigating the role of soil microorganisms in sustainable agriculture.

Linda L. Kinkel is associate professor in the Department of Plant Pathology at the University of Minnesota in St. Paul, where she studies the epidemiology and ecology of plant-associated microbes.

Timothy J. Kurtti is professor in the Department of Entomology at the University of Minnesota in St. Paul, where he studies insect pathology and medical entomology.

Abdelaziz Lagnaoui is an entomologist at the International Potato Center in Lima, Peru.

Kurt J. Leonard is director of the USDA, ARS Cereal Rust Lab at the University of Minnesota in St. Paul, where he conducts research on the genetics and population dynamics of cereal rust diseases.

W. Joe Lewis is research entomologist at the USDA, ARS laboratory in Tifton, Georgia, where he conducts research on insect behavior, learning, semiochemicals, and parasitoid-host relations.

Steven E. Lindow is professor in the Department of Plant and Microbial Biology at the University of California–Berkeley, where he studies microbial interactions on the phylloplane.

Daqun Liu is professor and dean associated with the Department of Plant Protection at the Agriculture University of Hebei in Baoding, Hebei, People's Republic of China.

Robert D. Lumsden is a research scientist in the Biocontrol of Plant Disease Laboratory in the Plant Sciences Institute of USDA, ARS in Beltsville, Maryland, where he conducts research on biocontrol of soil-borne plant pathogens.

Mark S. McClure, who studies competition among Homoptera, is chief scientist at the Valley Laboratory of the Connecticut Agricultural Experiment Station in Windsor, Connecticut.

Roberte M. D. Makowski is former research scientist and section head of the Biocontrol of Weeds Section at the Agriculture Canada Research Station in Regina, Saskatchewan, and is now science director at Wilmington College in Delaware and weed science instructor at the University of Delaware.

Bruce D. Maxwell is associate professor in the Department of Plant and Soil Science at Montana State University in Bozeman, Montana, where he studies weed ecology.

Jocelyn Milner is a researcher in the Department of Plant Pathology at the University of Wisconsin–Madison.

Ulrike G. Munderloh is in the Department of Entomology at the University of Minnesota, St. Paul, where she studies microsporidia and spirochetes, including Lyme disease spirochete.

Danial H. Putnam is now associated with the Department of Agriculture and Sciences at the University of CaliforniaDavis.

Edward B. Radcliffe is professor in the Department of Entomology at the University of Minnesota, St. Paul, where he studies the management of pests of potato.

Janet L. Schottel is professor in the Department of Biochemistry at the University of Minnesota in St. Paul, where she studies the mechanisms of pathogenicity and biocontrol of plant disease.

William Sheehan was a post-doctoral research associate at the USDA, ARS laboratory in Tifton, Georgia. He now resides in Athens, Georgia.

Sallie P. Sheldon is associate professor in the Department of Biology, Middlebury College in Middlebury, Vermont.

Laura Silo-Suh is a researcher at the VA Hospital in Memphis, Tennessee.

Bas P. Suverkropp studies the biology of *Trichogramma* at the Swiss Federal Research Station for Agronomy in Zurich, Switzerland.

James F. Walter, formerly a manager of biochemical engineering with W. R. Grace and Co., is now with Thermo Trilogy Corporation, where he continues work to commercialize *Gliocladium virens.*

Carol E. Windels is associate professor in Plant Pathology at the Northwest Experiment Station of the University of Minnesota, Crookston, where her research emphasis is on diseases of sugar beets.

Donald L. Wyse is professor in the Department of Agronomy and Plant Genetics at the University of Minnesota in St. Paul, Minnesota, where he is the director of the Minnesota Institute for Sustainable Agriculture and studies weed biology.

# Index